Concrete Face Rockfill Dams

Concrete Face Rockfill Dams

Paulo T. Cruz
Dam Consulting Engineer, São Paulo, Brazil

Bayardo Materón
Bayardo Materón & Associates, São Paulo, Brazil

Manoel Freitas
Hydrogeo Engharia S/C Ltda, São Paulo, Brazil

CRC Press
Taylor & Francis Group
Boca Raton London New York Leiden

CRC Press is an imprint of the
Taylor & Francis Group, an **informa** business
A BALKEMA BOOK

Covers Illustrations:
Front, top: Campos Novos Dam
Front, bottom: Deformed mathematical model of the Campos Novos Dam (deplacements after reservoir filling (Xavier et al., 2007).
Back, bottom: Seepage weir in operation at the Campos Novos Dam

Originally published as:
Barragens de Encrocamento com Face de Concreto/Concrete Face Rockfill Dams – Paulo T. Cruz, Bayardo Matéron, Manoel Freitas – Bilingual edition: Portuguese/ English – São Paulo: Oficina de Textos, 2009

This current English edition: *Concrete Face Rockfill Dams*

CRC Press/Balkema is an imprint of the Taylor & Francis Group, an informa business

Typeset by Vikatan Publishing Solutions (P) Ltd, Chennai, India
Printed and bound in Great Britain by TJ International Ltd, Padstow, Cornwall

Published by: CRC Press/Balkema
P.O. Box 447, 2300 AK Leiden, The Netherlands
e-mail: Pub.NL@taylorandfrancis.com
www.crcpress.com – www.taylorandfrancis.co.uk – www.balkema.nl

Library of Congress Cataloging-in-Publication Data

Cruz, Paulo T.
[Barragens de encroçamento com façe de concreto. English]
Concrete face rockfill dams / Paulo T. Cruz, Bayardo Materón, Manoel Freitas.
p. cm.
"A Balkema book."
Includes bibliographical references and index.
ISBN 978-0-415-57869-1 (hardcover : alk. paper) 1. Earth dams–Design and construction. 2. Earth dams–Materials. 3. Rockfills. 4. Concrete. 5. Earthwork. I. Materón, Bayardo. II. Freitas, Manoel. III. Title.

TC543.C7813 2010
627'.83–dc22

2010012534

ISBN: 978-0-415-57869-1 (Hbk)

Contents

List of figures

List of tables

Foreword

The design and construction of concrete face rockfill dams (CFRDs) have improved substantially since the 1970s. The construction of CFRDs over 150 m high is possible mainly because of the new technologies that were developed back in that era.

CFRDs are long-term safe structures with static and dynamic stability. This was recently demonstrated by the Zipingpu CFRD, 156 m high and built in 2006 in the province of Sichuan, China. The dam was hit by an earthquake of magnitude 8.0 on the Richter scale in May 2008, and the epicenter was just 20 km away from where the dam is located. Aside from damage to the slabs and dam crest, the structure's performance remained safe and sound after the severe shake.

CFRDs can be a low-cost and effective alternative to other rockfill structures, such as impervious core (clay, asphalt) and concrete structures (CCR or arch) in both narrow valleys ($A/H^2 < 4$) and wider ones ($A/H^2 \geq 4$).

The economic appeal of CFRDs over other structures is a consequence of the flexibility it allows in construction. The rockfill for the simple zoning can be taken from the excavation site where the dam will be built. Foundation treatments (excavations, superficial treatments, and the grouting curtain placed outside the dam) are also easier to undertake. Another reason for their appeal is that CFRDs are technically feasible in zones where soil for an impervious core is scarce and in places with high precipitation levels, factors that can be an impediment to the impervious core solution.

In places where the riverbed is in thick alluvium (over 20 m), the construction of a CFRD is made feasible by connecting an articulated plinth to a diaphragm wall crossing through the thick alluvium. This turns out to be a great advantage over other alternatives that require the complete excavation of the alluvium.

CFRDs are stable with steep slopes – such as 1.3(H):1.0(V); 1.4(H):1.0(V); 1.5(H):1.0(V) – allowing a narrow offset. The consequence is cost efficiency in the construction of diversion structures and intake tunnels. Rockfill structures as CFRDs may even take low levels of overtopping if the rockfill has been reinforced – leading to further possible cost reductions in the diversion structures. CFRDs also make it possible to place the rockfill in both abutments to build the plinth and to initiate the grouting process before diverting the river.

Technological developments in construction equipment in the past 20 years, such as in hauling trucks and rockfill compactors, allow (providing it is adequately planned) high production outputs to be obtained, reaching peak levels of over 1 million m^3/month.

Slip forms, 12 m–18 m wide, allow the CFRD slab to be built in two or three stages, a significant advantage when it comes to reducing construction time and costs.

In Latin America, especially in Brazil, this type of dam has been widely accepted and often preferred to earth core rockfill dams and concrete structures with RCC. The 11 Brazilian concrete face rockfill dams only account for 3.6% of the worldwide total of over 300 CFRDs (built or under construction) that are over 30 m in height. China has built around 180 CFRDs. However, despite the low numbers, Brazil holds an outstanding position in the field for three main reasons.

1 Foz do Areia (1975–1980), 160 m high, was the highest CFRD in the world at the time it was built and set a milestone in this type of construction. It involved developing new design and construction methods that yielded high production efficiencies. Campos Novos (2001–2006), 202 m high, also held the title of the highest dam of this type prior to the completion of the Shuibuya (China) in 2008, 233 m in height.
2 The Brazilian tradition of disseminating detailed performance evaluations of the construction of its dams in many domestic and international papers has contributed to engineering developments on new projects.
3 Both independent consultants and Brazilian companies are well recognized worldwide, especially independent consultants who hold high positions on international boards of consultants appointed to the most important projects throughout the world.

The history of Brazilian dams starts in the early 1900. It has had its setbacks and has been on and off, going through periods of intense activity as well as troughs as a consequence of a volatile government, which is ultimately responsible for green power and irrigation projects.

The downturns in construction led to an analysis of, and reflection on, the Brazilian dams already built at the time and this resulted in the papers published by the Brazilian Committee on Dams (1982–2000, 2009) and in works of synthesis, such as 100 *Barragens Brasileiras* (Cruz, 1996).

CFRDs are trailblazers, for they have their own path of development (different from earth and earth core rockfill structures), which is based on the experience of engineers and – to a certain extent – on some numerical analysis and modeling.

Figure A shows the progress of these dams after the development of vibratory roller compactors in the 1960s.

Today, China is considering projects that demand CFRDs between 250 m–340 m high.

As a consequence of the experience acquired, gradual changes have been made to construction techniques on new projects. The purpose is to reduce leakage, optimize costs, and simplify construction methods.

In some recent cases (2003, 2005, 2006, 2007) slab disruptions occurred after reservoir filling, as at TSQ1, Barra Grande, Campos Novos and Mohale, taking designers, constructors, and consultants by surprise. These events were rapidly analyzed and corrections promptly applied – incorporating changes in the central compressive joint fillers in order to mitigate the high compression stresses developed between slab lanes – to dams undergoing construction. In addition, heavy compaction and slab

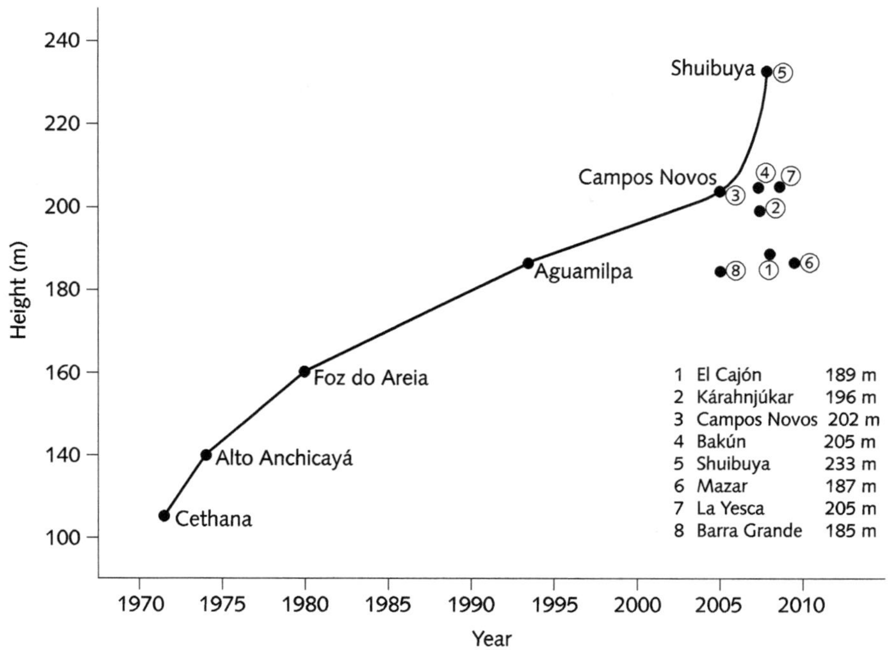

Figure A **The height evolution of dams after the introduction of vibratory rollers in the 1960s.**

design changes were made to reduce the stresses. Kárahnjúkar, Shuibuya, Bakún, La Yesca, El Cajón and Caracoles, to mention a few, were treated in this way and displayed positive performance results.

Some CFRDs, while still at the design stage, have been planned using a mix of numerical analysis and empirical criteria. However, the final design criteria have been drawn up mainly through experience rather than by modeling or testing.

For the Cethana Dam, Australia, Boughton (1970) and Wilkins (1970) developed an elastic analysis of the rockfill in order to predict the behavior of the dam. Sigvaldason et al. (1975) developed finite element analysis for the slab design and the plinth for the Alto Anchicayá Dam (which is set on very steep abutments). Similar methods were applied for the Foz do Areia and Aguamilpa dams.

The empirical methods which are used in the design and construction of CFRDs are made upon are sometimes supported by mathematical modeling in order to predict stresses and deformations passed on by the rockfill to the face slab during the construction, reservoir filling, and operation.

In this book special attention is given to rockfill and to the deformation observed in prototypes as well as to the mathematical approaches which may be used for the design of CFRDs.

Laboratory testing using the large equipment (oedometers and triaxial apparatus) associated with fill measurements has been conducted to obtain stress-strength parameters for the finite element analysis used to predict the performance of these dams. Unfortunately, there is not yet a refined enough mathematical model that realistically

Figure B Artistic view of Shuibuya CFRD (233 m, China) (Gezhouba Group).

simulates the behavior of dams. The actual criteria used in the designs are still based on experience and on the behavior of similar structures.

Chapter 11 was written by Professor Xu Zeping, from the *China Institute of Water Resources and Hydropower Research* (IWHR) in Beijing. He is a world renowned researcher and has visited some CFRD projects in Brazil, and given lectures at the Instituto de Engenharia in São Paulo. His chapter explains the application of numerical methods to the design and performance of CFRDs. It also draws on some Brazilian works on the same subject.

Twenty-eight CFRDs from around the world, mostly projects in which the authors have actively participated, are described in detail in chapter 3. Other chapters contain case studies in which the behavior observed in the dams has no recorded precedents.

Shuibuya is today the highest CFRD structure in the world and it keeps performing magnificently (Figure B), proof of the high technology embodied in the design and construction of these structures.

Acknowledgments

The publication of this book has only been possible thanks to the trust and financial support of our collaborators: The Brazilian Committee on Dams, Engevix Engenharia, Intertechne Consultores Associados, Construções e Comércio Camargo Corrêa and Construtora Norberto Odebrecht.

Engevix and Intertechne are world renowned corporations, having been pioneers in the development of many project designs featured in this book, both in Brazil and worldwide. Camargo Corrêa and Norberto Odebrecht are two construction companies whose stories blend in with the history of concrete face rockfill dams in Brazil; furthermore, they are internationally recognized in CFRD construction.

We are also thankful to two great dam engineers: Dr Edilberto Maurer, for his generous considerations in the introduction and his contributions in chapter 1, which have broadened our knowledge of the history of concrete face rockfill dams and, in particular, the introduction of this type of dam to Brazil; and Professor Xu Zeping for his inputs to this publication on numerical analysis and for the information on the development of CFRDs in China.

Since its foundation in 1961, the Brazilian Committee on Dams (CBDB) in association with the International Commission on Large Dams has promoted conferences on Brazilian dams all over the world. Such conferences have had great impact in Brazil as well as worldwide. The Brazilian Committee on Dams has largely contributed to the propagation of information on Brazilian dams by means of its own publications and by giving support to publications such as this one.

Engevix Engenharia is a 43-year old Brazilian engineering company that was involved in the design of Itaipu and Tucuruí Dams among many other projects. It developed the design of Itá, Itapebi, Quebra-Queixo, Barra Grande, Campos Novos (at 202 m, the highest CFRD in the world in 2006), Monjolinho and Paiquerê CFRDs in Brazil. It also designed the Baines CFRD in Namibia and took part, through Braspower, in the design of the Shuibuya CFRD in China. Engevix organized and had an active participation, along with the CBDB, in two international symposia on CFRD (1999 and 2007) in Florianópolis. It also participated in the preparation of the *J. Barry Cooke Volume – CFRD* in homage to Dr Barry Cooke, handed out to the attendees of the 20th Conference of ICOLD (Beijing, September 2000).

Intertechne is a Brazilian engineering consulting firm, founded in 1987 by a group of Brazilian engineers who had worked together since the 1960s in the design and construction of hydraulic, hydroelectric and transmission systems. The firm has been able to widen its reach by incorporating new talents in the development of new technology analysis and in the design of engineering projects. It has consolidated itself as one of the most cutting-edge corporations in design and consulting, within a large spectrum of enterprises in Brazil and overseas, by carefully selecting which projects to get involved with. In CFRDs, it has participated, along with Milder Kaiser, in the design of Foz do Areia CFRD, a pioneer of this type of dam in Brazil and the highest in the world at the time of its completion. Intertechne has also taken part in the design of Segredo, Itapebi, Pichi Picún Leufú, Bakún and, El Cajón dams and, more recently, two dams in Africa: Lauco and Caculo.

The active presence of Construtora Camargo Corrêa, both in Brazil and overseas, in large-scale construction projects such as hydroelectric schemes, has gained this company many important awards. Moreover, the international recognition for its capacity for innovation, its *au courant* technology, and its inestimable human resources are pillars upon which it prides itself. The continuous improvements – *kaisen*, as the Japanese would say it – and the frantic search for paramount technology, including but not limited to the innovations they have instated in this market, is what has led the company to earn its worldwide faultless reputation. Construtora Camargo Corrêa built the CFRDs of Machadinho, Barra Grande and Campos Novos. The CFRD of Porce III, Colombia, is under construction and the CFRD of Paiquerê has just began construction.

Construtora Norberto Odebrecht is today one of the largest contractors both in Brazil and overseas. It mainly focuses on the construction of dams, and in its résumé there are almost one hundred dams built for power generation as well as for other purposes. It is known internationally in the field of engineering and its capacity to build dams is evidenced by the 14 dams built simultaneously all over the globe in 2008. Construtora Norberto Odebrecht built the Foz do Areia, Xingó, Itapebi and Itá CFRDs. In Argentina, it built Pichi Picún Leufú, it built El Limón in Peru, it has started the construction of Bakún Dam in Malaysia, and Tocoma Dam is under construction in Venezuela. Within the last year it has been recognized by ENR (Engineering News-Record) as one of two main companies dedicated to the construction of hydropower dams operating at a global level.

Our special thanks to Oficina de Textos, a pioneer in this subject when it first released the book *100 Brazilian Dams* back in 1996. It has once again been a groundbreaker by publishing this bilingual, Portuguese-English, book on concrete face rockfill dams.

To all of you, our sincere thank you.

The authors

About the authors

Paulo Teixeira da Cruz obtained a BSc in Civil Engineering from the Mackenzie University School of Engineering (1957), holds a Masters degree from the Massachusetts Institute of Technology, and a PhD degree in Geotechnical Engineering from University of São Paulo – where he has worked for over 40 years.

His first work with dams was on the historical Três Marias Dam. In the past 50 years of his professional life he has worked on countless dams all over Brazil, including the world-renowned Itaipu and Tucuruí dams. Dr Cruz was on the board of consultants for the Campos Novos Dam and since the 1980s he has been providing his expertise as an independent consultant. He is the author of *100 Brazilian Dams – history cases, material, construction, and design* (1996), in which the Brazilian know-how in dam design and construction is consolidated. Today, Dr Cruz is the current Vice-President of the CFRD International Society.

Bayardo Materón is a Civil Engineer graduate from Cauca University, Popayán, Colombia (1960) and holds a MSc degree in Civil Engineering from Purdue University, Indiana, USA (1965). He works as a consulting engineer in the field of rockfill dams and hydropower construction methods. Since the completion of the Alto Anchicayá CFRD dam in 1974, he has been involved with many leading engineering organizations, and has advised on the design and construction of rockfill dams and hydro projects. A member of several boards of consultants for different projects under construction, Mr Materón is the current President of the CFRD International Society. He has participated in the design and construction of some of the world's highest CFRDs such as Alto Anchicayá, Salvajina, Porce III, Ranchería (Colombia); Foz do Areia, Xingó, Segredo, Itá, Itapebi, Machadinho, Campos Novos, Barra Grande (Brazil); Aguamilpa, El Cajón, La Yesca, La Parota (Mexico); Antamina, Torata, Olmos (Peru); Caracoles, Punta Negra (Argentina); Messochora (Greece); Kannaviou (Cyprus); Bakún (Malaysia); Mohale (Lesotho, Africa); Tiangshenqiao 1 (China); Merowe (Sudan); Berg River, Braamhoek (South Africa); Santa Juana, Puclaro, Punilla, Ancoa, Carén (Chile); Kárahnjúkar (Iceland); and Siah Bishe (Iran).

Manoel de Souza Freitas Jr holds a BSc in Civil Engineering (1969) from the São Carlos School of Engineering, University of São Paulo, Brazil. Since 1970 he has participated as a geotechnical engineer in several dam project designs and construction activities in water supply and hydroelectric power generation. He has currently been operating as an independent consultant for several construction companies, and a consultant for the World Bank and the Inter American Bank in hydro projects in Brazil. Mr Freitas has participated in the Tianshengqiao 1 Project (1,200 MW, China) as a chief engineer and manager, and has been working as an independent consultant for several large CFRDs such as Barra Grande, Campos Novos and Mazar (Republic of Ecuador).

Introduction

The initiative of engineers and friends Bayardo Materón, Paulo Cruz, and Manoel de Freitas to publish a book on the state of the art on concrete face rockfill dams is a merit in itself, and it is sure to become an indispensable source of information to anyone who wishes to endeavor in this field.

In matters of technology, references to the conceptual historical evolution is of undeniable worth, by making adequate use of the failures of the past and by having an understanding of its successes as a reference guide, it is possible to minimize the impact of such failures and learn from successful projects in developing new designs.

Particularly when it comes to CFRDs, the dominant empirical character evident in the development of new designs makes it even more relevant, as the essential teachings necessary for the evolution of new technologies are drawn from either the failures or successes of the past.

Therefore, the authors – and we – have high expectations that the compilation herein becomes a resource for all those who are variously involved in the conception, design and execution of CFRDs undertakings.

The release of this book, so rich in content, at an event of major importance to Brazilian dam engineering – the 23rd International Congress on Large Dams in Brasilia – is a great happening. It is important to remember that this is the most distinguished event in dam engineering in the world, and the pre-eminent event of the International Commission on Large Dams (ICOLD/CIGB).

For the Brazilian Dams Committee (CBDB), the local representative of the ICOLD, the opportunity to organize and host such a prestigious event in this field in the country is of pivotal relevance and a great honor.

Our sincere compliments and wishes for success to the authors.

Edilberto Maurer
President of CBDB – Brazilian Committee on Dams
Vice-President of ICOLD – International Commission on Large Dams
March 2009

Chapter 1

An overall introduction to concrete face rockfill dams

1.1 A PANORAMA OF CFRDs IN THE WORLD

The concept of building a rockfill dam as a stable structure with an external impervious face has proved to be a safe and economically efficient alternative to other dam structures. Examples of these solutions can been found in technical papers dating as far back as the beginning of the twentieth century.

La Granjilla Dam, 13 m high and 460 m long, was built in Spain as early as 1660 with an impervious face of mortar and lime. The dam has an upstream slope of 0.16(H):1(V) and a downstream slope of 1(H):1(V), and its embankment is made of soil and rockfill.

Saturnino de Brito Dam, in Poços de Caldas, Minas Gerais, Brazil, was also built with an impervious face. This dam was built in the beginning of the twentieth century.

It is also worth mentioning Nissastrom Dam, 15 m high and completed in 1950 in Sweden, and Quioch Dam, 38 m high and 320 m long (built near Inverness, Scotland in 1954). These were the first two compacted CFRDs.

In the United States, the construction of rockfill dams began early in the modern era, between 1850 and 1870. These were built to store water for those exploring for gold in the Sierra Nevada mountains in California. The upstream face sealing of those dams was initially comprised wooden boards, although this was later replaced by concrete. Two of the old rockfill dams with wood sealing are the English Dam in California, 24 m high and built in 1856, and the Meadow Lake Dam, 23 m high and built in the same state in 1903. Given the risk of fire outbreaks in the dry season and the dam's vulnerability to deterioration in the long run, the upstream sealing was eventually replaced with concrete.

In the USA, there are many old rockfill dams: Morena (54 m and built in 1895); Strawberry (50 m, 1916); Salt Springs (100 m, 1931); Cogswell (85 m, 1934); Lower Bear Nr. 1 (71 m, 1952); Lower Bear Nr. 2 (50 m, 1952); Courtright (98 m, 1958); Wishon (82 m, 1958); New Exchequer (150 m, 1966); Balsam Meadows (40 m, 1988); and Spicer Meadow (82 m, 1988). Many of these are in California. In Canada it is worth mentioning Outardes 2 CFRD (55 m, 1978) and Toulnustouc CFRD (77 m, 2006).

In Australia, CFRD construction began in the late 1960s with the construction of Kangaroo Creek Dam (59 m), followed by the dams of Pindari (45 m, 1969) and Cethana (110 m, 1971). Cethana Dam is a landmark in modern CFRD history for

Figure 1.1 Saturnino de Brito Dam, 1933 (Brazil). Courtesy of Eng. Cícero M. Moraes.

its use of rockfill compaction and for the criteria used for dimensioning the face slab. The experience in Australia with concrete face dams extends to the dams of Winneke (85 m, 1979), Mackintosh (75 m, 1981), Mangrove Creek (80 m, 1981), Murchison (94 m, 1982), Reece (122 m, 1986), Crotty (82 m, 1991), and Pindari (83 m, 1994).

Countries such as France, Germany, United Kingdom, Portugal, Spain, Romania, Albania, Greece, Turkey, Bulgaria and Iceland, just to mention a few, have also successfully constructed CFRDs.

In Portugal it is worth mentioning the CFRDs of Salazar (70 m, 1949) and the historical dam of Paradela (112 m, 1955), which were followed by Vilar (55 m, 1965) and Odeleite (61 m, 1988).

Spain has developed about 22 CFRD projects since the 1960s: Pias (47 m, 1961); Piedras (40 m, 1967); Amalahuigue (60 m, 1983), and San Anton (68 m) and Bejar (71 m) in the same year; Guadalcacin (78 m, 1988); and the Alfilorios (67 m, 1990).

France built several concrete face dams between 1960 and 1980. These include the dams of Fades (70 m, 1967), Gandes (44 m) in the same year, and Rouchain (60 m) in 1976.

Two dams over 100 m in height were completed in 2001 in Turkey: Kürtün (133 m) and Dim (135 m). In Greece it's worth mentioning the Messochora CFRD (150 m) completed in 2006.

In 2007, Kárahnjúkar CFRD (196 m) was completed in Iceland. The extreme climate conditions, with low temperatures during most of the year, required the engineers to adopt innovative construction techniques. Kárahnjúkar Dam has performed excellently when assessed against behavior deformations and leakage.

In Mexico, the historical highlights are the dams of Pinzanes (67 m, 1956) and San Ildefonso (62 m, 1959). Since the 1990s, Mexico has focused on the construction of high CFRDs, with structures such as Aguamilpa (187 m, 1992), El Cajón (188 m, 2006) and La Yesca (220 m), which started being constructed in 2007. In Central America, there is Fortuna CFRD (105 m and built in two stages in 1984 and in 1994) and Guaigui CFRD in the Dominican Republic.

In South America, CFRD design and construction techniques have contributed – in the last three decades – to the development of technology that has been used in several countries.

Chile was the second country in South America to adopt CFRDs as an alternative to other dam structures with the Caritaya Dam (40 m, 1937). Subsequently, Cogotí CFRD (85 m) was concluded in 1939. Other important Chilean projects were the CFRDs of Santa Juana (110 m, 1995), Corrales (70 m, 2000), Puclaro (85 m, 2000), and more recently, and still under construction, El Bato (55 m), Ancoa (115 m) and Puntilla del Viento (105 m).

In Colombia, the construction of CFRDs started with Alto Anchicayá (140 m) in 1974, an unprecendented height for a dam in that country. Golillas (125 m, 1978) and the Salvajina (148 m, 1984) are two other large Colombian projects that should be referenced along with Sogamoso Dam (190 m, 2009). At present, the dams of Porce III (155 m) and El Cercado (120 m) are under construction.

In Argentina the construction of the Pichi Picún Leufú (50 m), Los Molles (46 m) and Potrerillos CFRDs, which were built over thick alluvium deposits, was remarkable, as was Los Caracoles Dam (131 m), still under construction. Punta Negra Dam (86 m) was scheduled to start construction in 2009.

In Venezuela attention ought to be on the CFRDs of Neveri Turimiquire (115 m, 1981), Caruachi (80 m), Macágua (20 m) and Yacambu (162 m) all built in 1996, and Tocoma CFRD (40 m) currently under construction.

Also in Latin America, it is worth mentioning the Peruvian experience in the construction of Aguada Blanca Dam (45 m, 1970), which was treated with a steel face, Torata CFRD (130 m and built in two stages, 2000 and 2001), Antamina (110 m, 2002), and Limón (43 m), with a second stage 83 m high under construction as this book is being written.

In Ecuador, the Mazar CFRD (166 m) has been under construction since 2005 in a narrow and very assymmetrical canyon, with a valley shape ratio A/H^2 of 1.7.

The history of CFRDs in Brazil starts when Companhia Paranaense de Energia COPEL, a company owned by the state of Paraná, in southern Brasil, was granted the concession to explore a stretch of the high Iguaçu River in the same state.

For a full development of the project, it became necessary to build a 160 m high dam. After analyzing the possible alternatives, it was concluded that a concrete face rockfill dam was the most economical and the more suitable solution, considering the local topography, geology and climate.

To support the decision of building not only the first compacted dam of this type in Brazil, but one that would be the highest in the world at the time (160 m, with a face area of 139,000 m^2 and volume of 14,000,000 m^3), Copel decided to have a series of guaranties.

It called for a board of consultants, with a well-known specialist in the construction of large projects, our dear late J. Barry Cooke, supported by James Libby (who has since deceased), an expert in geotechniques, Prof. Victor F. B. de Mello, who has recently passed away, and an expert in hydraulic works, the well-known local engineer Nelson Sousa Pinto.

It also called on an expert on this type of dam who had worked on the construction of the Alto Anchicayá, the highest in the world at that time, Eng. Bayardo Materón, one of the authors of this book.

Following the successful performance of Foz do Areia, concluded in 1980, other CFRD projects started to be developed in Brazil.

In 1992 the Segredo Dam (145 m) was completed on the same Iguaçu River; in 1993 Xingó Dam (150 m) was completed in northeast Brazil on the São Francisco River; in 1999 Itá Dam (125 m) on the border of Santa Catarina and Rio Grande do Sul; in 2002, Machadinho Dam (125 m) near Itá; and recently in 2006 the CFRDs of Barra Grande and Campos Novos were finished, both dams also in Santa Catarina and, respectively, 185m and 202 m high. In planning is the Paiquerê Dam, 150 m high, which was scheduled to start in 2009.

In Africa it is relevant to mention Nakhla Dam, 46 m high (1961), and Dchar El Oued, 101 m high (1999), in Morocco; Shiroro Dam, 130 m high in Nigeria (1984); Mohale, 145 m high in Lesotho (2006); Mukorsi, 89 m high in Zimbabwe (2002); Berg River, 60 m high in South Africa (2006); and Merowe, 53 m high in Sudan (2008). These dams are all remarkable examples of CFRD construction over the last decades.

In the Middle East, of interest are the Siah Bishe projects (one 100 m high and the other 76 m high) and Narmashir, 115 m high, both located in Iran and under construction.

In Asia, memorable examples of CFRD construction can be found in Korea, Philippines, Indonesia (Cirata, 125 m high and built in 1987), Thailand, Laos, Malaysia, Sri Lanka (Kotmale, 97 m high and concluded in 1985), Japan, Pakistan and India. An important example is Bakun CFRD, 205 m high in Malaysia, which is currently undergoing construction.

China has been leading the world in the last 20 years. The country is important for CFRD construction, both in terms of quantity of projects and the dimensions of the dams.

The first dumped rockfill dam with an upstream face slab in China was the Baihua Dam, 49 m high and built in 1966. Subsequently the Nanshan and Sanduxi dams, both about 50 m high, were constructed using a similar technology. In 1982, the CFRD of Kekeya, 41 m high, was built by using compacted rockfill founded on sediments consolidated by the construction of a diaphragm wall.

The use of modern construction techniques, such as slipping forms for the concrete slab, began in 1985 with the building of the Xibeikou CFRD, 95 m high. In the late 1990s, over 40 CFRD projects had been completed and approximately 30 CFRDs were under construction in China – 12 of them over 100 m high. Important CFRDs that ought to be mentioned are: Tianshengqiao 1, 178 m high, finished in 2000; Hongjiadu, 184 m high and built in 2004; Baiyun, 120 m high; Gaotang 111 m high; Panshitou, 101 m high; Chaishitan, 103 m high; Gudongkou, 120 m high; Qinshan, 122 m high; Wuluwati, 135 m high; Heiquan, 124 m high; Yutiao, 110 m high; Panshitou, 101 m high; Baixi, 124 m high; and, more recently, the Zipingpu, 156 m high and built in 2007, Sanbanxi, 186 m high and finished in 2008, and the highest CFRD ever built, Shuibuya Dam, 233 m high, completed in 2007.

An updated list of CFRDs is presented in the Yearbook 2009 published by International Water Power & Dam Construction, England.

1.2 IMPORTANT EVENTS RELATED TO CFRD

The history and evolution of CFRDs can also be found in the records of symposia, conferences and congresses. These are some of the significant events.

- 1985 – Symposium on Concrete Face Rockfill Dams – Detroit – USA.
- 1988 – 16th International Congress on Large Dams – San Francisco – USA.
- 1993 – Symposium on CFRD – Beijing – China.
- 1999 – 2nd Symposium on CFRD – Florianópolis – Brazil.
- 2000 – International Symposium on CFRD – in homage to J. Barry Cooke – Beijing – China.
- 2004 – International Conference of Hydropower – Yichang – China.
- 2005 – Symposium on 20 Years for Chinese CFRD Construction – Yichang – China
- 2007 – Workshop on High Dams Know How – Yichang – China.
- 2007 – 5th Dam Engineering Conference – Lisbon – Portugal.
- 2007 – 3rd Symposium on CFRDs – Florianópolis – Brazil.

China and Brazil are the leading countries in the field of CFRD construction.

The empirical approach to the design and construction of CFRD is supported by the overall performance of its end results, and this approach is emphasized in numerous published presented at the symposia, conferences and congresses listed above. Examples of this approach are made clear in extracts from two papers presented below. In the 2000 symposium on CFRDs held in China, Barry Cooke summed up his experience working on dozens of the 197 CFRDs that had been completed at the time:

1 *The safety of the dam is assured.*
2 *Face cracks may occur and, therefore, there can be leakage.*
3 *The semi-pervious rockfill face will limit the leakage.*
4 *Leakage can be sealed substantially by the underwater placement of impervious silty fine sand.*
5 *Little change in the present design practice is foreseen.*

Some quite surprising accidents occurred on the concrete faces of CFRDs TSQ-1 in China (2003 – 2004), Campos Novos in Brazil (2005), Barra Grande in Brazil (2006), and Mohale in Lesotho (2006). These were discussed in the 2007 symposium in Florianópolis:

Pinto's analysis (Pinto, 2007) of these mishaps is reproduced here:

> "CFRDs remain as inherently safe despite the reported accidents as they always have been." This statement by Barry Cooke has become a widely quoted cliché on this issue. With the plinth on an unerodible foundation and the rockfill zoning providing a positive gradient of permeability in a downstream direction, CFRDs

can safely discharge flow-throughs many times the registered values of 1.0 – 1.5 m^3/s. The ongoing operational ability of the Barra Grande and Mohale dams was not affected during the mishaps, and there was never any concern about their safety either. The Campos Novos reservoir emptied out for reasons other than the ones we've been discussing. It was caused by problems found in a diversion tunnel, and the dam has been fully operational since it was refilled following repair works.

CFRD design has evolved empirically, that is it has been guided by experience and not by theory alone. One of the consequences of the compression problems commented on above has been that the empirical procedure has been questioned and more emphasis has been placed on FEM modeling as an alternative method for the design of a dam. However, the complexity of the physical problem sets a natural limit to what can be accomplished analytically by a mathematical model. Significant experience in the field is still the best and essentially the only foundation for the design of CFRDs – including the very high dams.

An analysis of the experience of recent construction projects has resulted in much greater emphasis on rockfill compaction, stronger central face slabs, and the use of soft joints, and these are practical and effective provisions to extend the domain of CFRDs. This has led to design improvements to the very high CFRDs presently under design and/or construction. The behavior of the dams will provide a practical way to ratify these design decisions and will serve as a basic reference for future dams.

Although numerical modeling is a promising design tool, it is still under development. It is of great help the design of dams because it allows for easier parametric analyses and for comparisons with alternative solutions. However, it has not yet reached a satisfactory stage where it can become the main instrument for CFRD design. When the nature of CFRDs is taken into account, it is doubtful that it ever will unless, of course, the interpretation of the physical phenomena is much improved. The main design decisions are bound to continue to be dependent on the empirical approach for a long time still."

1.3 CFRD IN SEISMIC AREAS – A HISTORICAL EVENT

Another aspect of CFRD behavior foreseen by Cooke and Sherard in their 1987 papers and discussed in Bulletin 70 of ICOLD, published in 1989, is their high resistance to earthquakes. Extracts from item 3.4 of the Bulletin are reproduced below:

- Dynamic analyses that have been performed as part of the design activity of CFRDs built in recent years have been primarily used to demonstrate that by using state-of-art static design procedures the dams are stable. The dynamic analyses have not had significant influence on the main design decisions.
- For areas of very high seismicity (peak base acceleration larger than 0.5 g and earthquake magnitude 8½), flatter slopes are required in order to limit the movements to acceptable values in the upper third of the dam. In these situations, the provision of a conservative extra freeboard will be advisable.

Twenty-one years elapsed between 1987 and May 12, 2008 when the Sichuan earthquake of magnitude 8 hit the Zipingpu CFRD, standing 156 m high, in China.

Three papers were presented on the subject, two by Xu Zeping (Strong earthquake on May 12 in China and its impact on dam safety, and Performance of Zipingpu CFRD during the strong earthquake). A third paper was presented by the Hydro China Technical Division to the CFRD International Society Seminar, held in December 2008 in Hong Kong, China.

This is an extract from this third paper, *Effects of the 12 May 2008 Sichuan earthquake on dams*:

> "The earthquake of May 12, 2008 had a magnitude of 8.0 on the Richter scale and because of the earth's shallowness (about 10 km deep), the ground shake in the epicenter must have been very severe. The Zipingpu Dam site was located 17.17 km from the epicenter.
>
> The shaking was very severe during the earthquake to the point where people weren't able to stand on the crest of the dam. A peak acceleration of 2.06 g (dam axis or vertical wise) and 1.65 g (stream wise) was recorded on the top of the downstream slope at the central section of the dam. The strongest shock period lasted for almost a full minute. A peak acceleration of greater than 0.5 g was calculated in the dam bedrock. The earthquake intensity and the dynamic peak acceleration that actually took place on the dam were much greater than had been foreseen by the engineers working on the original design criteria. This was the first case in the world in which a macro-seism occurred so close to such a high CFRD.

Figure 1.2 Displacement of the access way and the crest of the dam – this measured up to 630 mm. (See colour plate section).

The intensity at the epicenter was 11. The seismic acceleration peak value in the base rock of the dam body exceeded 500 gal. The intensity on the crest of the dam was over 9 based on calculations of peak acceleration taken by seismic acceleration measuring tools. This significantly exceeded the seismic intensity allowed for in the design.

The dam suffered compressive stresses on its face slab as can be seen in Figures 1.2, 1.3 and 1.4.

The maximum settlement measured at the crest during the earthquake was of 810.3 mm; although, a more careful analysis concluded that the settlement could have reached 100 cm. This value is about 0.64% of the height of the dam.

The horizontal displacement at El. 854.0 was 270.8 mm.

Figure 1.5 shows the horizontal displacements of the dam after the earthquake.

On the downstream slope the rock blocks were loosened off (Fig. 1.6)

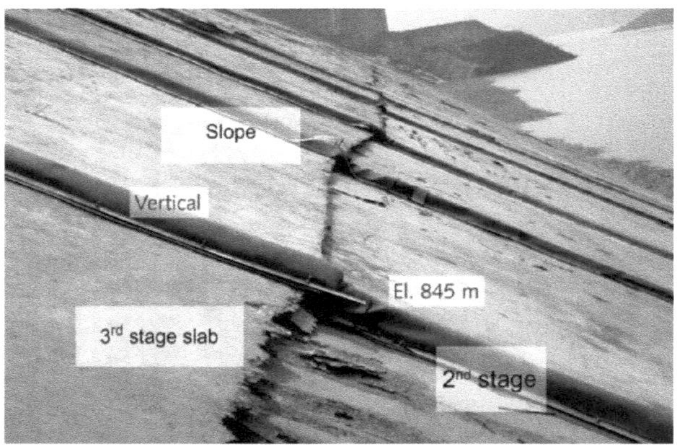

Figure 1.3 Horizontal joint offset at the El. 845 m crossing through 26 slabs. (See colour plate section).

Figure 1.4 Compressive failure on slabs #23 and #24 at the central part of the dam. (See colour plate section).

Leakage went from 10.38 ℓ/s on May 10, 2008, before the earthquake, to 18.82 ℓ/s on June 1, 2008 until it finally stabilized at 19 ℓ/s.

On the day of the earthquake, the reservoir level was low (El. 830.0 m) accounting for 30% of its full capacity.

The main point to note is that this CFRD resisted the intensive shaking, and repairs are already finished.

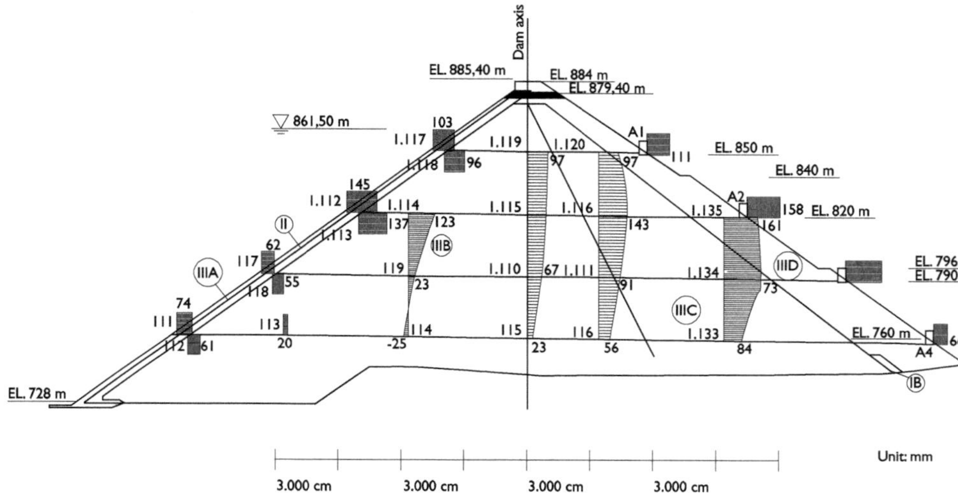

Figure 1.5 Horizontal displacements during the earthquake at Zipingpu Dam (China).

Figure 1.6 Rockfills loosened off on the downstream slope of Zipingpu Dam. (See colour plate section).

1.4 HIGH DAMS IN THE NEAR FUTURE

By the end of 2008, 294 CFRDs had been completed, 26 were under construction and other 58 were being designed according to International Water Power & Dam Construction's Yearbook 2008.

Qian (2008), in a paper on "Immediate Development and Future of 300 m High CFRD", has a table showing that pre-feasibility studies are being made into seven extremely high dams in China: Cihaxia, 253 m high and 700 long; Maji, 300 m high and 800 m long; Linghekou, 305 m high; Songta, 307 m high; Gushui, 310 m high and 540 m long; Shuangjiangkou, 314 m high; and Rumei, 340 m high and 800 m long.

Only the first dam of those seven has been confirmed as a CFRD. The design of the others is still being discussed.

Qian (2008) also reports that Banduo CFRD, 250 m high, is currently under construction in China.

1.5 THOUGHTS ON VERY HIGH CFRDs

When considering very high CFRD, several factors should be taken into account.

Compacted rockfills have been proved to be stable structures under both static and dynamic conditions, even when having relatively steep slopes and heights over 200 m. The main concern with these rockfill structures has been their performance in respect of settlement and face deflections during construction, reservoir filling and operation. Face leakage and rainfall have induced settlement and face deflections, and the progressive accommodation of the rockfill rock blocks, mostly in the downstream zones. Creep effects have to be taken into consideration.

The shape of the valley, expressed by the ratio A/H^2, has been proved to be an important factor affecting the internal distribution of the stresses within the rockfill embankment, and thus, the transmission of forces from the embankment to the concrete face slab.

A 300 m high dam should be designed with a better compaction intensity and improved dam zoning in order to reduce its compressibility. Construction techniques have improved significantly in the last two decades. Heavier equipment and thinner layers of rockfill should be applied in critical areas in order to obtain a higher compressibility modulus. Rather than relying purely on empirical knowledge, other approaches to design should be considered and different alternatives compared.

The design of the joints should be reviewed in order to provide appropriate openings in the joints at the center area slab to absorb compressive deformations.

The use of downstream weir systems for leakage control is important for monitoring CFRD performance during and after reservoir impounding.

In seismic areas, special attention must be given to the downstream slope. Enlarging the upper quarter of the dam with flatter slopes to improve the embankment mass in order to support seismic accelerations should be carefully analyzed.

A rational analysis of the valley, materials, face slab details, and construction techniques will confirm whether it is feasible to build very high dams in any given site.

The role of instrumentation as a tool to analyze dam performance deserves attention. Internal cells in the rockfill and near the upstream face which measure the displacement towards the center of the valley may help to predict the stresses that eventually will develop on the concrete face.

Chapter 2

Design criteria for CFRDs

2.1 INTRODUCTION

Two papers written by J. Barry Cooke and James L. Sherard, *Concrete Face Rockfill Dams I: Assessment*, and *Concrete Face Rockfill Dams II: Design*, both published in Journal of Geotechnical Engineering (vol. 113, n. 10, October 1987) of the American Society of Civil Engineers consolidated the basis for the design and construction of CFRDs. The papers are based on the design, the experience, and the performance of the dams in service at that time (1985–1987).

Foz do Areia Dam (Brazil, 1980), the highest CFRD dam in the world at that time at 160 m high, is repeatedly mentioned in these papers.

In the past 21 years (1987–2008), the number of dams designed and built that are over 50 m high has skyrocketed to 390 worldwide. Today, the highest is the Shuibuya (233 m high, China). However, ruptures on the concrete faces of some CFRDs, particularly in the central compressive joints of four large side dams between 2003 and 2006, surprised world specialists because these had not been foreseen either by the consultants and designers or by the constructors.

Figure 2.1 shows Campos Novos Dam (202 m high, Brazil), one of the dams in which the concrete face suffered ruptures. Details of these incidents are described in chapters 3, 8, 10 and 11.

These incidents led to a review of Barry Cooke and James Sherard's guides for the design and construction of CFRDs. The review, as discussed here, made adjustments to the master work of these two great engineers who had the courage to introduce and establish an alternative to the more traditional designs, aiming at economy, safety and speed in construction schedules.

The four dams that suffered ruptures on their concrete faces were observed and repaired, and they are currently in operation with no risk of failure.

The authors emphasize the importance of monitoring the slab and the rockfill by an effective instrumentation system. Unfortunately, in some current projects the owners have reduced CFRD instrumentation in order to save money. This decision has jeopardized further analysis of slab failures, as occurred in the cases mentioned above, and will not help to improve mathematical analysis models and FEM for future CFRD projects.

The review of the design criteria proposed by Cooke and Sherard (1987) takes into account the performance of CFRDs, new approaches to foundation treatments

Figure 2.1 Campos Novos Dam. (See colour plate section).

based on rock classification, the use of lower quality construction materials, and new construction techniques for the plinth, the concrete of the face slab, and the new extruded concrete curb.

2.2 ROCKFILL EMBANKMENT

2.2.1 Foundation excavation and treatment criteria

Considering that *"all the rockfill is downstream from the water thrust, the base width is more than 2.6 times the height (of the dam) and essentially all the water load is taken into the rock foundation upstream from the dam axis"*, Cooke and Sherard (1987) conclude that *"the criteria for foundation excavation and treatment under the downstream half of the dam are therefore lower than for CCRDs"* (clay core rockfill dams).

The specifications are as follows:

"Typical currently accepted good practice requires trimming overhangs and vertical faces higher than about 2 m for a horizontal distance downstream of the toe slab equal to about 30% of the dam height, or about 10 m minimum. In the rest of the foundation area, cliffs and overhangs are left in place. In this main foundation area, abutment voids or zones of lower density under overhangs are arched over by the highly frictional rockfill embankment, and have no influence on the face slab movements or CFRD performance.

Over most of the foundation area, the excavation can be made with earth-moving equipment, removing soil-like surface deposits and exposing the points of in situ hard rock. Dozer removal is adequate; ripping is seldom required. Under an upstream portion of high-dam foundations, most of the soil and soft weathered rock between hard rock points should be excavated with a backhoe or similar equipment, but final cleanup by hand labor is not required. Under the downstream portion of the foundation, and most of the upstream portion for dams of low to moderate height, soil and surface material between the hard rock points is left in place. Alluvial gravel deposits in the riverbed (but not sand) are commonly left in place except for a short distance downstream from the toe slab. Such gravel deposits usually have a high modulus of compressibility, far exceeding that of well-compacted rockfill. Settlement is not a problem, but alluvium that is judged to be possibly subject to liquefaction is removed."

There are many dams where the alluvium has been left in place. Itapebi Dam (120 m high, Brazil, see Figure 2.2) is a case in which a 15 m thick sandy alluvium layer was protected by filters and left in place, with the exception of 40 m at the upstream area and 30 m near the downstream toe, which were removed for stability reasons. The end of the construction settlements measured at Itapebi dam are shown in Figure 2.2.

A similar concept was applied at the Aguamilpa Dam (187 m, Mexico, Figure 2.3). The natural alluvium was left in the foundation, except for an area downstream from the plinth.

Table 2.1 sets out the materials and compaction specifications for each zone of the Aguamilpa Dam (Mexico). In higher dams, however, layer thickness, the number of passes and the roller weight should be increased. In chapter 3, the zones of all dams are named according to the terms used by Barry Cooke and James Sherard.

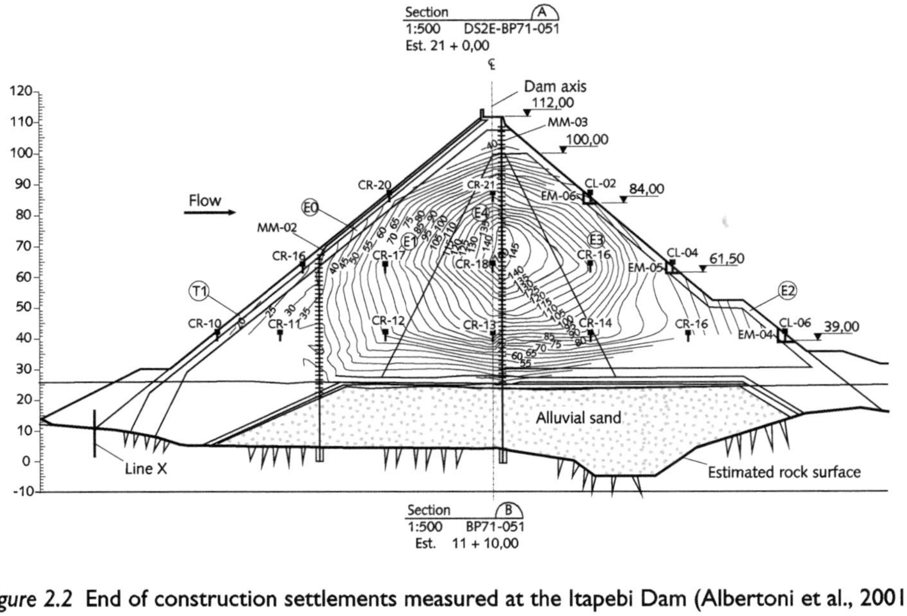

Figure 2.2 End of construction settlements measured at the Itapebi Dam (Albertoni et al., 2001).

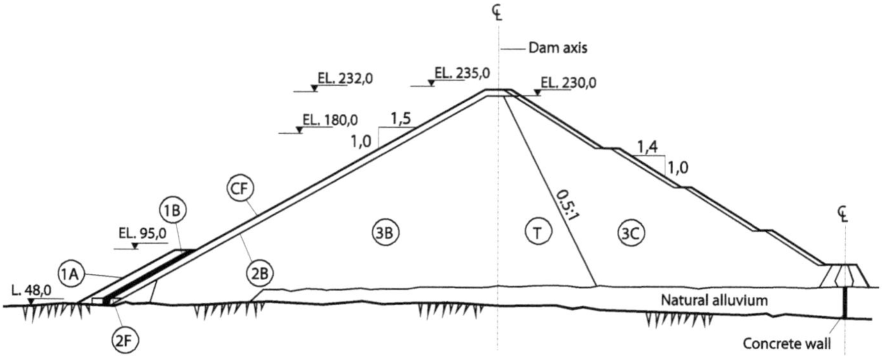

Figure 2.3 Aguamilpa Dam (Gómez, 1999).

Table 2.1 Zoning, specifications and materials at Aguamilpa Dam (Mexico).

Material	*Specification*	*Zone*	*Compaction procedure*	*Compaction specification*
Soil	Random	1B	Not compacted, just placed 80 cm	Construction equipment
	Fine silty sand $d_{max.}$ 0,2 cm	1A	Not compacted, just placed 30 cm	Construction equipment
Filter	Alluvial gravel and silty sand mixture, $d_{max.}$ 3,8 cm	2F or 2A	Compacted, layer 30 cm	4 passes 100 kN SDVR
Rockfill or gravels	Crushed alluvial gravel and sand mixture, $d_{max.}$ 7,6 cm	2B	Compacted, layer 30 cm	4 passes 100 kN SDVR 6 passes of 40 kN or 130 kN PC
	Aluvial, $d_{max.}$ 40 cm	3B	Compacted, layer 60 cm	4 passes 100 kN SDVR
Rockfill	Rockfill 3C with reduced $d_{max.}$ 50 cm	T	Compacted, layer 60 cm	4 passes 100 kN SDVR
Rockfill downstream	Rockfill $d_{max.}$ 100 cm	3C	Compacted, layer 120 cm	4 passes 100 kN SDVR
Rockfill	Placed	4	Placed	Oversizes
Slab	Concrete face	CF	–	–
River bed	Natural alluvium	NA	–	–

Pichi Picún Leufú (50 m high, Argentina) is another interesting case where natural alluvium to a depth of 30 m was left in the foundation, but removed from an area within 9.0 m of the "X" plinth line. Alto Anchicayá (140 m high, Colombia), Golillas, Salvajina, Potrerillos, Caracoles and Limon are also examples of dams where alluvium has been left as foundation material.

A few other requirements for foundation treatment have been modified in the last 20 years because the relatively simple requirements suggested by Cooke and Sherard in 1987 led to excessive displacements of the concrete face. The requirements for the compaction of the downstream shell have been made more stringent. This is discussed in chapters 3, 4, 7 and 8.

2.2.2 Zoning designations

The zoning designations are one of the most important sections of Cooke and Sherard's 1987 papers. They clearly define the functions and requirements of each embankment zone and provide an easy terminology for the exchange of information between CFRD designers and constructors.

In spite of the fact that many dams zones have been named with different numbers and letters, the basic concepts of zoning are still present in a large majority of dams.

Figure 2.4, reproduced from Cooke and Sherard (1987), shows the zone designations.

Zone 1 – impervious.

Zone 2 – the filter or transition zone directly under the concrete slab.

Zone 3 – the main rockfill.

In chapter 3, the tables that are presented with information on the cross sections of individual dams adopt this numbering system.

The required materials and compaction specifications for each zone were described by Cooke and Sherard in the 1987 papers:

> "**Zone 1** – A blanket of compacted impervious soil was placed on the lower part of the concrete face at Alto Anchicayá Dam, since the dam height was breaking precedent. This detail has since then been repeated on the Areia, Khao Laem, and Golillas Dams and on several other high dams. The purpose is to cover the perimeter joint and slab in the lower elevations with impervious soil, preferably silt, which would seal any cracks or joint openings. The minimum practical construction thickness of impervious soil can be used directly adjacent to the concrete slab and rock foundation, and then covered with less costly waste material for stability

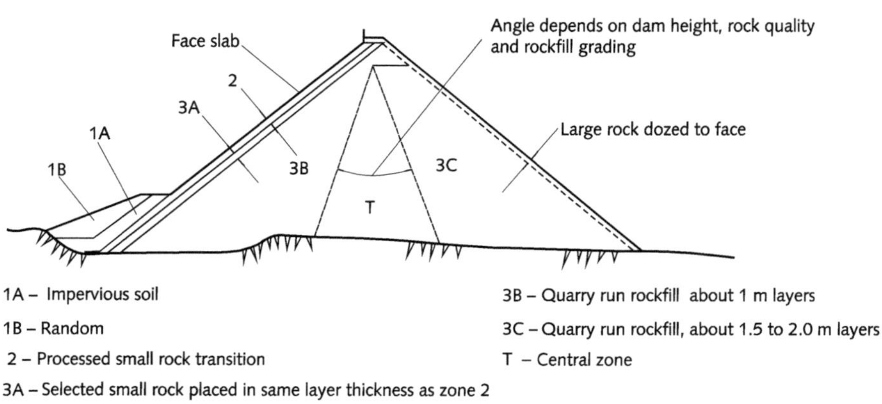

Figure 2.4 Zone designations for CFRD of sound rockfill (Cooke & Sherard, 1987).

(zone 1B, Fig. 2.4). As suggested by Nelson Pinto, a strip of silty fine sand could be located over the perimeter joint.

Dams without this upstream zone 1 have been completely successful, indicating that this zone 1 is not necessarily useful. It is useful only if a problem develops. Hence, there is a question in any given case about whether the cost is justified, except perhaps for the highest dams. The writers believe that it is generally desirable to consider placing zone 1 to a level several meters above the original riverbed.

Zone 2 – The early and primary purpose of the thin zone of finer rock directly under the slab was to provide uniform and firm support for the concrete slab. Crusher-run minus 15 to 7.5 cm rockfill has been used. Recently there has been a trend toward making zone 2 grading have a sufficient quantity of sand-sized particles and fines to improve workability, reduce excess concrete and have reliably low permeability, and have an approximate filter grading. Such select grading of zone 2 provides a semi-impervious barrier, preventing any large leakage, even if a leakage path develops through a crack in the concrete slab or a defective water-stop. The semi-pervious property is of value near the perimeter joint, and to an elevation where flood retention during diversion could rise, before the placement of the concrete face.

The select type of zone 2 material, one with 40% sand-sized particles and fines, should be placed and compacted in layers 0.4 m or 0.5 m thick with a smooth steel-drum vibratory roller. For this kind of material, the compaction is not strongly influenced by the water content. Generally the material is delivered to the site in a "damp" condition, commonly with water content (of the minus N°. 4 fraction) in range of 4–10%. In this condition, it compacts to a satisfactory high density without the need for quantitative water-content specifications. Since this semi-pervious type of zone 2 is especially sensitive to an excess of water, the specification needs to indicate that water content shall not be so high that the compaction equipment does not operate on firm material. Four pass coverage by a 10 t smooth drum vibratory roller is the suggested method specification, and density tests should be taken for the record.

An important and revolutionary construction procedure has been introduced by CNO contractor Norberto Odebrecht S.A. for the Itá and Itapebi CFRDs.

An extruded concrete wall is built as support for the concrete face. This replaces all the troublesome operations of compaction along the slope of zone 2, prevents erosion, and eliminates the asphalt emulsion spray. After Itá many CFRDs have adopted this system.

The benefits of this approach, named by Cooke as "The Itá Method", were presented by Resende and Materón during the CFRD ICOLD Symposium held in Beijing in 2000.

The extruded concrete works also as a support to the compaction of the cushion material and provides protection for the face against rain erosion. Figures 2.5 and 2.6 show the curb details and curb machine.

Such innovation does not eliminate zone 2, which is necessary for flow control. Since Itá and Itapebi dams the extruded concrete curb has been adopted throughout the world with great success.

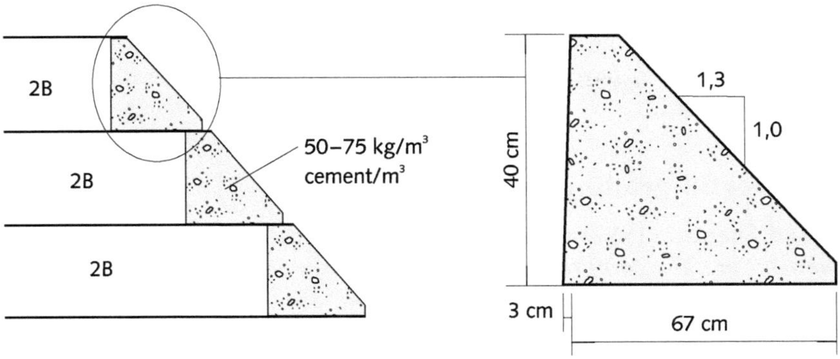

Figure 2.5 Curb detail.

Figure 2.6 Curb machine.

"**Zone 3** – Because most of the water load passes into the foundation through the upstream shell, it is desirable that the compressibility of zone 3B be made as low as practical to minimize slab settlement. Experience has shown that embankments placed in layers about 1 m thick and compacted with four passes of a 10-t (static) smooth steel-drum vibratory roller give satisfactory performance.

The downstream zone 3C takes negligible water load, and its compressibility has little influence on the settlement of the face slab. Consequently, zone 3C is commonly placed satisfactorily in thicker layers, usually about 1.5–2 m, and also compacted with four roller passes. It is desirable to specify thicker layers for zone 3C because it will be highly permeable and there is a substantial cost saving due to less equipment wear. Also, the thicker zone 3C layers provide a location into which larger size rocks can be placed, thickening the layer if necessary."

These specifications regarding layer thicknesses and compaction equipment continue to be applied at dozen of dams around the world, but they have been modified for high dams, such as the Campos Novos Dam (202 m high) which was completed in 2006 and for which Barry Cooke was a consultant. More details are provided in chapters 4, 8 and 12.

2.2.3 Rockfill grading and quality

Cooke and Sherard (1987) state that "*the most important properties of the CFRD embankment are low compressibility and high shear strength*".

Whenever "hard" rock is used in zones 3A, 3B and 3C together with the compaction and grading as specified in Table 2.1, the necessary shear strength and low compressibility are achieved even in high dams.

The implicit hypothesis that zones 3A, 3B and 3C behave independently from the water load, as suggested by Cooke and Sherard, deserves some considerations.

A simple analysis of the stresses in a CFRD is shown in Figures 2.7 and 2.8. This analysis presents the stresses present at the end of construction and after the filling of

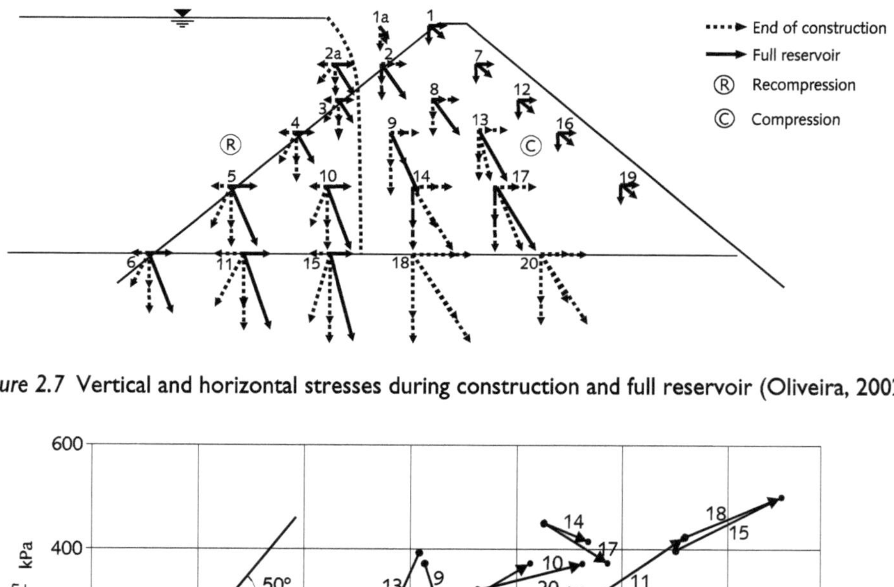

Figure 2.7 Vertical and horizontal stresses during construction and full reservoir (Oliveira, 2002).

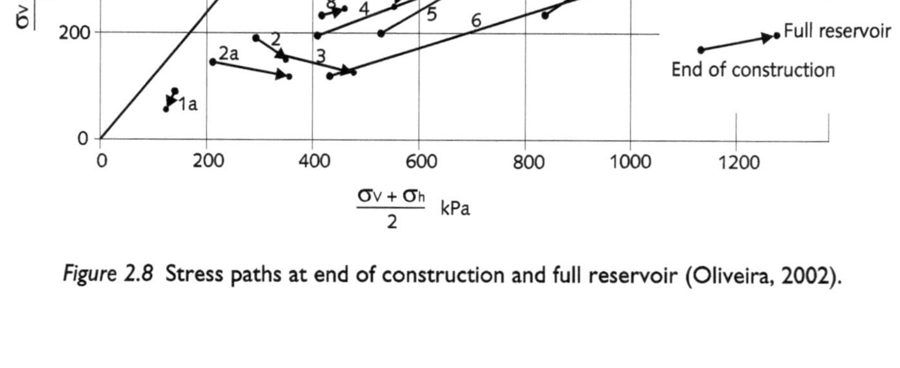

Figure 2.8 Stress paths at end of construction and full reservoir (Oliveira, 2002).

the reservoir at points within the embankment. It is clear that in the upstream shell there is a change in the direction of the stresses from those experienced the end of construction to those present with a full water load. Near the face the shear stress ($\sigma_v - \sigma_h$) reduces and then increases back again. In the downstream shell, the change in the direction of the stresses is practically negligible, and stresses increase with water load, construction and a full reservoir (Oliveira, 2002).

The figures depict clearly how the dam behaves as a whole. On the upstream shell, the rockfill is first unloaded during the impounding of the reservoir and then reloaded again, behaving as a pre-consolidated material.

On the downstream shell, the rockfill is continuously compressed, and if zone 3C is of more compressible material than zone 3B, the displacements may reach the concrete face causing problems.

Whenever different materials are used in zones 3B and 3C, as was the case in the Aguamilpa Dam, the concrete face may bend significantly at its upper end. See Figure 2.9.

With regard to rock quality, it's been recognized by Cooke and Sherard that:

> "For the CFRD (and ECRD) there is no technical need for using rocks with high compressive strength, i.e., compacted embankments from rocks with strengths of 300–400 kg/cm^2 are not more compressible in the completed dam than those of much harder rocks.
>
> Many types of rock with high absorption, or low compressive strength when tested dry, will give considerably lower strengths when tested saturated: saturated strengths of 20–40% of the dry strength are frequently found. When these rocks are compacted with a heavy steel-drum vibratory roller, considerable crushing of the larger rocks occurs. Nevertheless, these materials can be used in CFRD embankments with appropriate zoning and construction methods."

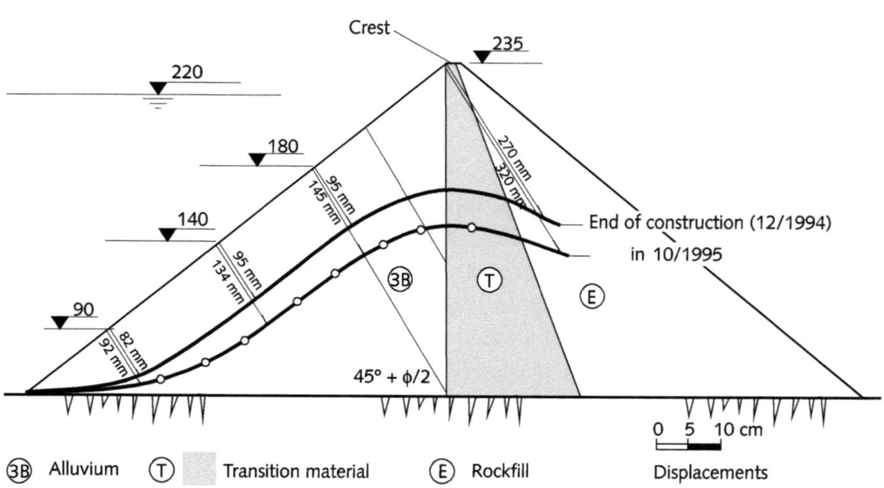

Figure 2.9 Face displacements of Aguamilpa Dam (section in rock foundation) (Cooke, 1999, in Mori, 1999).

In modern high dams, the specifications for material quality and compaction have often been revised and readjusted to reduce the displacements of the concrete face, as discussed in chapter 3.

A final remark on grading and quality concerns a very practical way to evaluate whether the fines in the rockfill are damaging. This was proposed by Cooke and Sherard (1987):

> "When a rockfill contains a fines content exceeding some limits, the final evaluation of suitability can be made on the basis of the trafficability of the rockfill surface when the material is thoroughly wetted.
>
> A stable construction surface under the travel of heavy trucks demonstrates that the wheel loads are being carried by a rockfill skeleton. An unstable construction surface, with springing, rutting, and difficult truck travel, shows that the volume of soil-like fines is sufficient to make the rockfill relatively impervious. Where the surface is unstable, the fines dominate the behavior and the resulting embankment may not have the properties desired for a pervious rockfill zone."

2.2.4 Adding water to rockfill

> "The main objective of adding water is to wet the material to reduce the unconfined compressive strength of the coarser rock particles. The purpose is principally to minimize the post-construction settlement. There is no need (or intention) to use the added water to wash or sluice the fines into the larger rockfill voids. Therefore, it is not necessary that the water be delivered with high-pressure monitors. It is satisfactory that the water be added by any means that thoroughly wets the material, preferably before the material is compacted, though the benefit continues if the rockfill is subject to percolating water and humidity during construction.
>
> The quantity of added water has been commonly 18–20% of the rock fill embankment volume, occasionally 30%. For almost all rocks, 10% (100 l per cubic meter of embankment) is more than adequate. The practice of adding the water to the rock in trucks just before dumping on the construction surface is economical and assures thorough rock wetting."
>
> (Cooke & Sherard, 1987)

The addition of water to rockfill has become a common practice for high dams. This is typically at a minimum rate of 200 liters per cubic meter. When the rock is absorbent, the addition of water brings some benefits. In Shuibuya, China (limestone rockfill) and El Cajón, Mexico (ignimbrite rockfill) a very high modulus of compressibility was achieved after adding a generous amount of water.

Details of the action of water on rockfill are discussed in chapter 4.

2.2.5 Downstream rockfill embankment face

Figure 2.10 shows the downstream rockfill face of Barra Grande Dam, which is made out of large rock blocks. The same practice has been used at Esmeralda Dam (earth core) and Foz do Areia Dam.

Figure 2.10 Barra Grande Dam.

Cooke and Sherard (1987) comment:

"An attractive face is created in a practical and economical manner by dozing selectively handled large rocks to the downstream face, guided by batter boards or laser beams, such that the upper edge of each large rock is between the design line and plus 15 cm. The surface voids should not be filled in. Some stockpiling may be required to have large rocks available for the area near the crest, when quarry production may be low.

In addition to providing an aesthetically pleasing surface, this practice provides a stable surface slope, with minimum excess rockfill, and no need for later dressing of the slope. The practice permits slopes locally steeper than the 1.3(H):1.0(V) slopes of downstream face haul roads. Slopes of 1.3(H):1.0(V)–1 .2(H):1.0(V) have been used.

The practice of dressing up the face by backhoe is sometimes used. It can give the face a satisfactory appearance. It may result, however, in loose rocks rolling down the slope and does not allow steep slopes above the road built on the downstream face."

2.2.6 Temporary construction slopes and ramps

Cooke and Sherard (1987) state that:

"Different embankment sections within zones 3B and 3C can be built to different levels at different times without restriction, using transverse and or longitudinal internal temporary slopes of 1.3(H):1.0(V) between the different sections.

This property of the CFRD gives important flexibility in construction, permitting the maximum use of rock from required excavations to be placed directly in the dam without stockpiling, with important cost and schedule advantages.

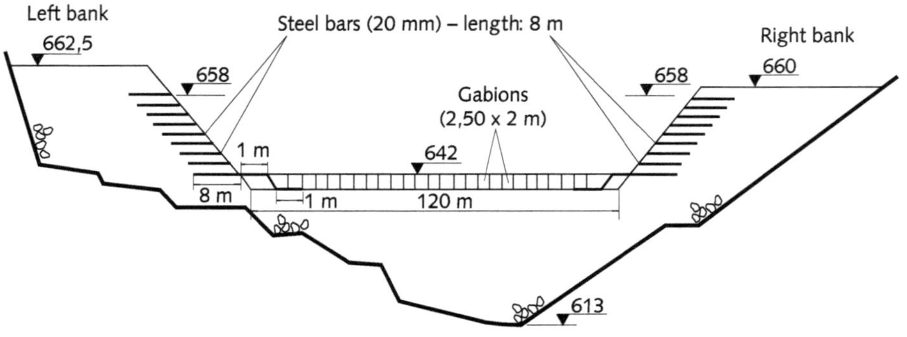

Figure 2.11 Cross section of the canal of TSQ1 dam with anchor bars to protect the rockfill against flood erosion.

A permanent road on the downstream face can be useful during construction. It can help in the optimum planning of internal ramps, and is the most reliable and shortest route between top and bottom of dam. This can be important for access between a powerhouse and power intake and spillway gates."

In fact, this practice is currently used in CFRD construction.

A fine example of such flexibility is the reinforced rockfill canal of TSQ 1 Dam (Fig. 2.11) that allowed water passage during floods that occurred during the first phase of construction of that dam.

For very high dams, however, which may settle down more due to the heavy weight pressure, the proper sequence of construction must be followed in order to minimize post-construction movements of the face slab.

2.2.7 Compaction control tests

According to Cooke and Sherard (1987) compaction control tests are required only at the semi-pervious zone 2:

> "A requirement that the material should be compacted to 98% of the maximum density of the Standard Laboratory Compaction Test, based on the fraction finer than ¾", is a satisfactory and readily achieved result."

For zone 3 rockfill it is recommended that larger pits be excavated to the depth of a layer thickness in zone 3 and with a volume of several cubic meters to determine the density and particle size distribution. This is done for the records, and not for compaction control.

The compaction control should be made as routine procedures, dictated according to layer thickness, water addition, number of passes and type of roller, all governed by experience or as result of field compaction tests.

Fines control should be done by observing the construction surface under the travel of heavy trucks. This surface must be stable, ensuring that loads are transmitted to the rockfill skeleton.

Currently, in situ permeability tests have been carried out in order to provide rockfill compaction control data and monitor performance. Meanwhile, instrumentation records (settlement cells and surface marks) play a key role, complementing rockfill control records during construction and after filling.

Recent experiences have demonstrated that the best control is to have personnel verifying the layer thickness, number of passes, and the roller capacity, which has to apply a force of 5 t/m over the cylinder with vibration frequency of 1400–2000 VM.

The observation of the roller action as described by Cooke and Sherard is an excellent way to evaluate compaction efficiency.

2.3 WATER FLOW THROUGH ROCKFILL AND LEAKAGE

> "Increasing permeability from zone 2 progressively through zones 3A, 3B, and 3C (Fig. 2.4) is desirable during construction in the event of a flood before the concrete face is placed. After the concrete face is placed, there is no credible face problem that could cause more leakage than the rockfill could handle without damage."
>
> (Cooke & Sherard, 1987)

In another part of their first paper, Cooke and Sherard talk about the advantages of a placement that will lead to a stratified rockfill:

> "There are no technical disadvantages to the preferred method of placement in segregated layers. In addition to lower cost, there are several advantages. The stratification assures that any flow through the rockfill embankment will travel much more easily in the horizontal direction than vertically."

The leakage that passes through fissures, cracks, and even fractures on the face of a CFRD is controlled by the sand layer, and it ensures that the flow is considerably less than that required to start a removal process of the downstream rock blocks. Leakages as high as 1000 to 2000 liters/sec have been measured at Barra Grande and Campos Novos CFRDs (2007) without any signs of trouble on the downstream rock slope, confirming Cooke and Sherard's view.

However, if a flood reaches the dam face before the concrete slab placement, the anisotropy of permeability will not be favorable, for it raises the phreatic line and concentrates the flow at the bottom of the segregated layers – as demonstrated by Pinto (1999) in a laboratory experiment with sand and gravel.

From a theoretical point of view, Cruz (2005) demonstrated that the higher the point at which the phreatic line reaches on the downstream shell, the worse is the condition of stability.

The role of zone 2 in controlling the flow through the rockfill in case of a flood during construction is shown in Figure 2.12. It demonstrates that the phreatic line as well as the flow are substantially reduced due the presence of the semi-permeable layer of zone 2.

Details of flow through rockfill are in chapter 6.

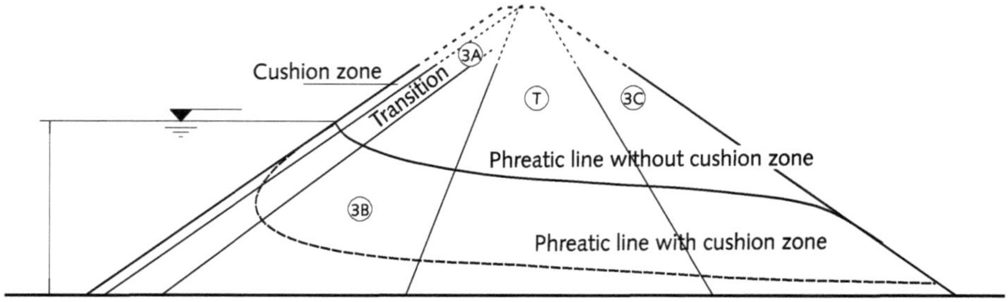

Figure 2.12 Effect of zone 2 in depressing the phreatic line.

2.4 STABILITY

2.4.1 Static stability of the rockfill embankment

The outer slopes of compacted rockfill dams are on average 1.3(H):1.0(V) or 37°, a figure determined by building experience. In case of compacted gravel, they may be 1.5(H):1.0(V), or 33°. When comparing these angles with 45° or more of rockfill shear strength, it becomes clear that dams are safe in relation to a failure parallel to the slope surface.

Cooke and Sherard (1987) are emphatic in affirming that:

> "In conventional stability analysis a trial sliding mass could move by sliding on a curved or plane shearing surface. Shearing forces are calculated and compared to define a safety factor.
>
> For the CFRD, which has no water in the voids and is founded on a sound rock foundation, all experience and theory confirm that such a shear slide cannot occur.
>
> Therefore, conventional stability analysis, when applied to the typical CFRD, calculates the safety factor to prevent a type of movement known to be physically impossible."

In fact, in chapter 5 stability analyses are performed using shear strength of various rockfills and different slope angles, and the resultant safety factor was always above 1.

But whenever the rock foundation has unfavorably located joints or planes of weakness, a sliding block analysis is appropriate and stability analyses are applicable.

The same applies when weathered rock occurs at the foundation. Gravel or alluvial foundations are usually dense and resistant, but occasionaly may require some type of analysis.

2.4.2 Earthquake considerations

Cooke and Sherard (1987) present a long discussion on earthquakes in regard to both the stability and crest settlements of CFRDs. Some of their conclusions are reproduced below.

"Since the entire CFRD embankment is dry, earthquake shaking cannot cause pore pressure in the rockfill voids. The CFRD foundation is rock, which does not magnify the incoming acceleration forces. The embankment is heavily compacted in thin layers to a dense state with vibrating rollers. Earthquakes can only cause small deformations during the short period of strong shaking. After the earthquake is over the CFRD is as stable as before.

In very strong earthquakes the concrete face may be cracked, increasing the leakage. The potential cracking and leakage cannot threaten the overall safety of the dam, because the amount of leakage which can get through the cracks and the zone of small rock under the face slab can easily be passed safely through the main rockfill embankment.

For these reasons the CFRD is considered to have the highest fundamental protection against earthquakes, and the same basic design has generally been used in regions of high seismicity as in non-seismic areas.

For the great majority of sites that may be very strongly shaken, such as those near the epicenter of a magnitude 7.5 quake or at sites with calculated Earthquake Severity Index (28) in the general range of 10–15, the same CFRD design can be used as in non-seismic areas. All present experience of dam behavior combined with the overall results of current dynamic calculations give confidence that the worst earthquake-induced crest settlement will be substantially less than 1% of the dam height. A sudden crest settlement of 1% of the dam height will not threaten the safety of a modern CRFD.

For the small number of sites where the world's strongest ground tremors could occur, within a short distance of major faults capable of generating earthquakes of magnitude 8 or greater, there is very little evidence from any source to guide judgment about the maximum probable earthquake-induced crest settlement in a modern CFRD. At these sites, it is desirable for the designer to make every effort to obtain as much help as is available from the most advanced dynamic analyses. It is essential in such an effort to make parametric studies based on several main assumptions, including the method of feeding the earthquake energy from the foundation to the dam and the shear strength and damping characteristics of the compacted rockfill. We will probably be able to conclude in the future that the maximum possible earthquake-induced crest settlements of a compacted CFRD cannot exceed 1% or 2% of the dam height under the most severe conceivable conditions of earthquake shaking. The provision of a conservative extra freeboard is probably the appropriate and economical seismic design provision."

The recent 8.0 magnitude earthquake (2008) that hit China, had its epicenter 17 km from Zipingpu CFRD, 156 m high. The quake lasted for one minute. A settlement of 74 cm of the crest and an increase in flow from 10 to 19 l/sec were the only consequences in spite of the fissures and cracks suffered both by the concrete slab and the parapet. The dam resisted the earthquake quite well, confirming Cooke and Sherard's predictions.

As mathematical models are becoming more and more accurate in the analysis of CFRDs, there are no reasons not to also consider this type of analysis in the design of high dams. It is, in fact, recommended.

2.5 TOE SLAB OR THE PLINTH

2.5.1 Treatment of the plinth foundation

"The toe slab is usually on hard, non-erodible fresh rock that is groutable. For less favorable foundation rock, after a trench is made to an estimated acceptable foundation, many methods are available to treat local imperfections. The criterion is to eliminate the possibility of erosion or piping in the foundation. Careful excavation is used to minimize fracturing of the rock surface on which the toe slab is placed. Air or air-water cleanup, just prior to placing concrete, is required to obtain a bonded contact of the concrete to the foundation."

(Cooke & Sherard, 1987)

The plinth, usually the name used for the toe slab, has an important role in the performance of CFRDs: it controls the flow control of the foundation because at the upstream side is the reservoir and behind the plinth is the rockfill.

The foundation requirements mentioned by Cooke and Sherard are indispensable.

Whenever the foundation is not made out of sound rock, other treatments are specified.

Good correlations have been obtained between the geomechanical classification of rock foundation and the required gradient to be obtained to prevent rock erosion under the plinth.

Details of foundation treatments are included in chapter 7.

2.5.2 Dimensions of the plinth

Cooke and Sherard (1987) stated that *"for hard and groutable foundation rock, the slab widths should be of the order of 1/20–1/25 of the water depth"*.

An extended width is used for lower quality rock. As far as the length of the plinth is concerned, there is a tendency to define the plinth length as a function of the quality of the rock of the foundation.

Once the rock in the foundation of the plinth is classified according to the usual parameters, the plinth length might be defined by using Table 2.2.

It is a common practice today that an extension of the plinth is made under the dam so that the external length is limited to 3 m or 4 m. Whenever the rock is sound but fractured, an internal filter behind the plinth is included.

Table 2.2 Rock classification and related plinth length.

Rock classification	*Gradient*	*Plinth length*
80–100	18–20	0.053 H
60–80	14–18	0.065 H
40–60	10–14	0.083 H
20–40	4–10	–
<20	*	–

* Necessary to lower the foundation level or use cut-off walls.

As far as the thickness of the plinth is concerned, it seems that the suggestion made by Cooke and Sherard (1987) to make it equal to the face slab thickness remains:

> "The thickness of the toe slab has been frequently made about equal to the face slab thickness. Excavation over break and irregular topography usually provides significantly greater thickness, so that a minimum design thickness of 0.3 to 0.4 m is generally reasonable for most toe slabs."

Details of toe slabs are included in chapters 7 and 8.

2.5.3 Stability of the plinth

> "The toe slab must resist the high horizontal water thrust without support from the rockfill. For a toe slab of normal thickness, there is ample frictional resistance to resist the water thrust, unless the foundation has unfavorably oriented planes of low shearing resistance just below the toe slab. For high toe slabs, and usually in local areas, stability analyses are required. Uplift pressure under the slab is assumed to vary linearly from full reservoir pressure to zero over the width.
>
> Water loading on the face slab opens the perimeter joint so there is no interaction between the toe slab and the face slab. The rockfill may exert a resisting force on the downstream side of the toe slab, but this cannot be relied upon and is neglected in stability computations."
>
> (Cooke & Sherard, 1987)

Figure 2.13 shows the acting forces that should be considered in a stability analysis of the plinth. Anchor bars or tendons may be required, not only to fix the plinth to the foundation, but also to resist part of the upstream water thrust.

2.5.4 Layout of the plinth

"The toe slab is laid out as a series of straight lines. The angle points are selected to suit the foundation conditions and topography, and have no required relation

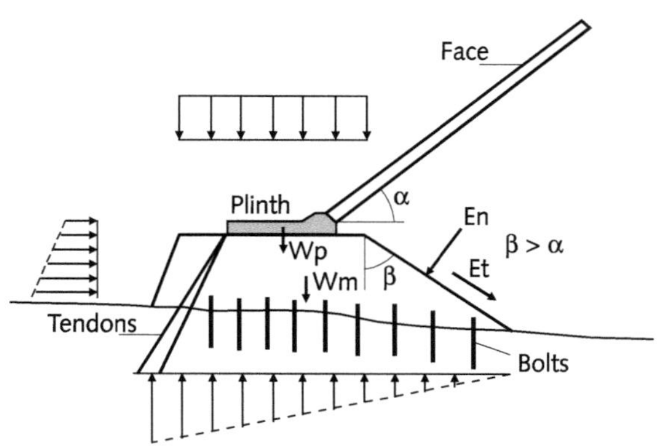

Figure 2.13 Acting forces on the plinth.

to the vertical joints of the face slab. The reference line for the excavation is at the bottom of the toe slab, in the plane of the bottom of the face slab."

(Cooke & Sherard, 1987)

In 2000, Barry Cooke himself recognized that among the many aspects of the construction of CFRDs, the plinth had not received the same attention as other elements (Cooke, 2000b).

The layout of the plinth as revised by Cooke (2000b) is reproduced in chapter 8.

2.5.5 Reinforcing, joints, and anchor bars of the plinth

"The main purpose of the reinforcing in the plinth is the same as in the face slab; i.e., to function as temperature steel, and to spread out and minimize the widths of any cracks that may tend to develop from the small bending strains imposed. While two layers of longitudinal steel have sometimes been used in the past, it is now generally accepted that a single layer should be used. The steel is put 10 to 15 cm clear of the upper surface as temperature steel, where it is hooked by the anchors: 0.3% each way is adequate.

In the past, construction joints with waterstops located at predetermined distances were commonly used. They were needed, since the joints opened. The toe slab waterstop connection to the perimeter waterstop was inconvenient.

Today longitudinal reinforcing is continued through the joints without waterstops. This is considered good practice, is more economical, and has been adopted for more recent dams. With the longitudinal steel passing through the construction joint, there is no need to use waterstops.

The purpose of the anchors is simply to pin the concrete to the rock. The anchors are not to resist any given uplift loads. Lengths, spacing, and bar diameters should be chosen on the basis of precedent and the characteristics of the rock foundation. The anchors and temperature reinforcing improve the slab as a grout cap.

Anchors used in common practice have generally been 25 to 35 mm diameter bars, spaced about 1.0 to 1.5 m in each direction, with lengths usually of 3–5 m."

(Cooke & Sherard, 1987)

Apparently few changes in reinforcing have been introduced in recent projects.

2.5.6 Grouting through toe slab

Grouting has been applied by traditional methods, with mixtures chosen according to the grout penetration and by using the GIN method, where only one mix is selected by using super plasticizers. Both methods have been applied successfully.

Special care has to be taken in areas where the high pressures recommended by the GIN method may lift up the plinth. In karstic zones the GIN method is not applied.

"The toe slab is placed before the adjacent fine-rockfill zone 2 is placed. Grouting is done with the toe slab as a grout cap, and at any time during construction. Both points are important for the shortest schedule and lowest cost.

The independence of grouting from the dam schedule is important. The specifications should not require the grouting to be done in advance of the adjacent rockfill placement. It should always be required that the grouting be done through the toe slab, thus most effectively grouting the upper zone of rock under the toe slab.

For a CFRD, the blanket grouting is of special importance because of the relatively short seepage path through the rock directly under the toe slab. The blanket grouting should be carried to depths sufficient to penetrate any surface zone of open fissures or higher permeability."

(Cooke & Sherard, 1987)

The grouting under the plinth has the purpose of reducing the flow through the rock foundation, for this reason the identification of more permeable layers within the rock foundation is a basic requirement before specifications in regard to depth and grout pressures can be determined.

The traditional three lines of grouting, with a deeper central line deepening from 1/3 H to 2/3 H (H – the water column above the plinth) should be considered only as a basis for the design.

At Alto Anchicayá Dam (Colombia), five lines of grouting were necessary to treat the sedimentary rock of the foundation.

At Foz do Areia (160 m high, Brazil) the outer lines were 20 m deep and the central line 60 m deep, due to the presence of horizontal basalt layers with artesian water pressures.

The GIN method for grout control introduced by Lombardi and Deer in 1993 has become a widely accepted practice. It's been used in Aguamilpa, Mohale and Pichi Picún Leufú. When the GIN method is used special attention must be given to the fact that hydraulic fracturing may occur in some rocks.

Some dams built on alluvium materials, such as mostly compacted gravel as was the case in Santa Juana (110 m high, Chile) and Puclaro (85 m high, Chile), have been provided with a diaphragm wall built at the plinth edge as shown in Figure 2.14.

The performance of the foundations of these dams has been excellent, leading to the construction of other dams using similar foundations.

More details are provided in chapters 7 and 8.

2.6 CONCRETE FACE SLAB

2.6.1 Concrete

"For the concrete, durability and impermeability are more important than strength, a 28 day compressive strength of about 20 MPa (3,000 psi) being adequate. Maximum-size aggregate of 38 mm (1.5″), air entrainment, and use of pozzolan are common features in current practice. However, a maximum size of 64 mm (2.5″) has sometimes been used, and this is satisfactory if special care is taken at construction joints and waterstops."

(Cooke & Sherard, 1987)

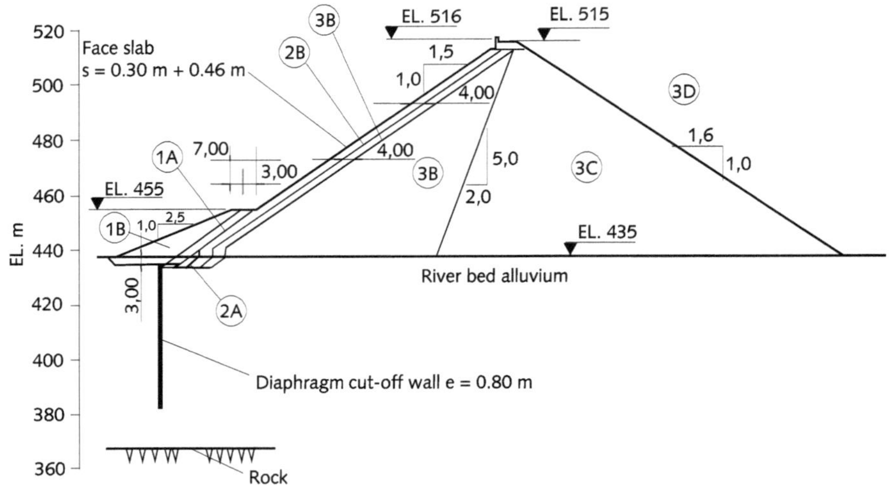

Figure 2.14 Puclaro Dam with diaphragm wall upstream (Nogueira & Vidal, 1999).

Table 2.3 Concrete typical mix.

Pozolan cement	*300 kg*
Sand	720 kg
Aggregate 1½″–¾″	565 kg
¾″–#4	605 kg
Water	157 ℓ
W/C	0,52
Additives	
Air Entrained	4%–6%
Plasticizer	0,5%–1% of cement

As a general rule, concrete mixes are designed according to Cooke and Sherard's specifications. A typical mix is shown in Table 2.3.

The concrete's strength has been specified for either 28 or 90 days in order to take advantage of the pozzolan action throughout that time.

Currently, a 60 to 90 days compressive strength 20–25 MPa is specified. Air entrainment, addictives and slump rates (ranged 3 to 8 cm) must be specified for slip forming concrete. Slump control represents a key factor for face slab pouring and quality control. Both slip forming speed and concrete quality must be controlled by slump management. At batching plants, sand moisture content and concrete water consumption must be controlled in order to ensure that slumps range from 3 to 8 cm.

Slump varies when temperature changes. Concrete curing is important to control cracks which tend to appear with concrete shrinkage and temperature stress. Sprinkling water directly on the concrete surface has been used, and this has achieved effective results thus far. This procedure is considered to be more economical and efficient when compared to chemical treatments.

2.6.2 Thickness of face slab

In their 1987 paper Cooke and Sherard state that:

> "The concrete face has demonstrated its durability under high hydraulic gradients and extreme weather conditions, even for early dams (1920–1930) of dumped rockfill and those constructed before the use of pozzolan and air entrainment.
>
> The thickness of slab on the early, dumped-rockfill dams was traditionally 0.3 m + 0.0067 *H*. The face was underlaid by a layer of large, derrick-placed rocks. For a CFRD with compacted rockfill and a compacted upstream face, the thickness increment was decreased to 0.003 H, and even to 0.002 H or less. These slabs have given satisfactory performance, and there is a current general trend toward thinner slabs.
>
> With modern concrete technology, the quality and durability of concrete is uniformly higher and more reliable than in the past. There is a certain thickness, such as 0.25 or 0.30 m, which would be generally considered as the minimum needed for covering the steel for good construction and for making minimal surface shrinkage, cracking, etc.
>
> Based on the presently available experience and current practice, it is reasonable to use slabs with constant thickness of 0.25–0.30 m for dams of low-to-moderate height (say 75–100 m), and to use an incremental thickness of about 0.002 H for the very important and high dams."

At first it may look surprising that a concrete slab as thin as 0.30 m could be used as a flow barrier for a 70 m high CFRD. But field experience has proved that for compacted rockfill, a concrete slab with a thickness of e = eo+ kH, where eo = 0.30 to 0.35 m and k = 0.002 to 0.0035, has been successful.

The maximum gradient on the slab must be about of 200–220.

When it comes to Campos Novos (202 m high) and Barra Grande (185 m high) both in Brazil, the equation used for the slab thickness was e(m) = 0.30 + 0.0020 H, for H < 100 m, and e(m) = 0.0050 H, for H > 100 m.

For more details see chapter 8.

2.6.3 Reinforcing the slab

The main considerations and recommendations by Cooke and Sherard (1987) are:

> "The use of 0.4% reinforcing in each direction, for compacted rockfill, has been an economical and successful change from the traditional 0.5% used with dumped-rockfill dams.
>
> A trend which appears to be desirable and economical is to carry horizontal reinforcing through the vertical joints, reverting to a practice used successfully on some older CFRDs. The trend is based on the fact that the major and lower area

of the face is under horizontal compression. Several vertical joints near abutments are contraction joints to minimize perimeter joint opening. Where reinforcing passes through vertical joints, a bottom waterstop has sometimes also been used as a carryover from the earlier practice. With the steel passing through, there is little or no more tendency for a crack to open at the construction joint than at other locations in the slab. Bonded joints are assumed in reinforced concrete design. Waterstops are not used in CFRD face slab horizontal construction joints.

A cost advantage of running the steel through the vertical joints is the elimination of the waterstops and anti-spalling steel.

On almost all recent CFRDs, a double row of small bars (anti-spalling reinforcement) has been used at the perimeter joint. All experience shows that there is no possibility of high contact stress and spalling at perimeter joints in dams of low to moderate height, since there is little compression in the slab before the reservoir is filled and the joint opens under the reservoir load. While there has been no trouble with joint spalling in CFRDs of compacted rockfill, it is probably desirable to continue with this edge reinforcement at the perimeter joint for high dams.

Usually ordinary reinforcing steel is used, but on a few recent jobs, including the Foz do Areia Dam, high-yield steel has been used, without changing the amount of steel."

Experience from recent projects has shown the tendency for high stresses to be generated close to abutments. The use of 0.5% reinforcing in a area 25–30 m perpendicular to the plinth alignment has been recommended as it has displayed good results. In high dams located in narrow valleys the compression rebars have been located in two layers increasing the percentage to 0.5% to augment the compressive stress.

Since 1987 some changes in the reinforcing of the slab have been introduced.

- Reinforcing of 0.40% vertically and 0.30% horizontally has been used.
- Steel passing through the vertical joints is not frequently used but vertical joints are included.
- Double rows of small bars (anti-spalling) are still in use.
- Double reinforcing (0.40% in both directions) has been adopted within 10 m to 15 m of the plinth.
- At Tianshengqiao 1 Dam in China (178 m) double reinforcement was adopted in the third stretch of the slab.
- At Campos Novos double reinforcement was adopted up to 15 m from the X line of the plinth along the perimeter joint.

2.7 PERIMETER JOINT

According to Cooke and Sherard (1987):

"The perimeter joint always opens and offsets moderately when the reservoir is filled and is a potential source of leakage if not well designed, inspected, and constructed. For dams of low to moderate height (less than about 75 m), the joint movement has commonly been only a few millimeters, and joints with current waterstop details have usually remained watertight. For some of the higher dams, the joint openings and displacements have been several centimeters. At the 160 m high Foz do Areia Dam, the opening in one area was 25 mm and the offset 50 mm. No joint leakage occurred, but it is probable that the central bulb waterstop was ruptured.

Because of the leakage history at perimeter joints, the trend has been to install first one to two and then three separate waterstops.

An upper mastic waterstop covered with a membrane was first devised for the design of the 160m high Yacambu Dam in Venezuela, but first constructed at the Areia Dam. In principle, the upper mastic should reliably seal the joint and prevent leaks, even with very large joint displacements."

Due to the potential leakage that can occur along the perimeter joint, in addition to the measures discussed by Cooke and Sherard, alternative designs using other materials have been applied to the perimeter joint.

Today, the empirical design based on experience and on observations of CFRD performance still prevails. The perimeter joint is protected by different defenses: copper waterstop, external waterstops, mastic, silt and fly ash covers.

Figures 2.15 and 2.16 show examples of a perimeter joint and a typical vertical compression joint.

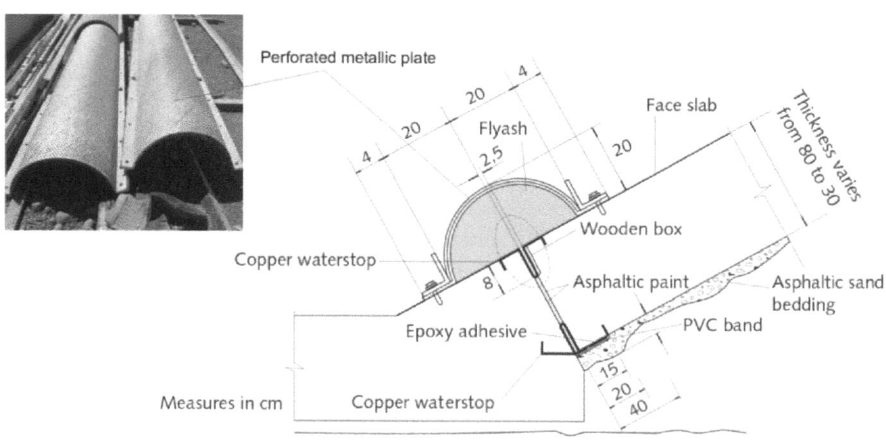

Figure 2.15 Perimeter joint used in El Cajón, México (Mendez, 2005).

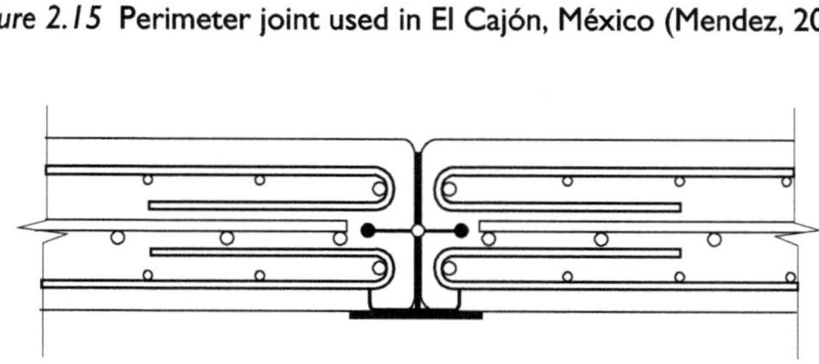

Figure 2.16 Vertical compression joint used in Kárahnjúkar, initial design (Perez, Johannesson & Stefansson, 2007).

Figure 2.17 Parapet wall – Mohale Dam.

2.8 PARAPET WALL AND CAMBER

> "The early dams had a parapet wall of 1.2 m. A higher 3–5 m wall has become the economical and desirable practice. The economy is in saving a slice of upstream rockfill – a saving that exceeds the cost of the wall.
>
> The freeboard for the CFRD is calculated from the top of the parapet wall, rather than from the top of the dam, with the walls being extended into the abutments."
>
> (Cooke & Sherard, 1987)

The compensation for the post-construction settlement (camber) has to take into consideration the amount of rockfill creep that is expected. It depends on zoning, the type of rock that forms the rockfill, settlement due to seismic shakes, and the shape of the valley. Post-construction settlement due to arching relief in narrow valleys may be larger than in open valleys.

See Figure 2.17 for the Mohale Dam parapet (Lesoto). Chapter 8 discusses the use of precast parapets.

2.9 OTHER IMPERVIOUS ALTERNATIVES

2.9.1 Geomembrane

Geomembrane has been used as an impervious solution for repairs on concrete face rockfill dams. Lost Creek (1997), Strawberry (2002), and Salt Springs (2004–2005), all in the USA, are interesting examples of dams where geomembrane has been applied over old concrete slabs as a means of preventing leakage while they were in operation (Larson & Kelly, 2005; Scuero, Vaschetti & Wilkes, 2007).

At Kárahnjúkar CFRD (Iceland, 2007), a geomembrane (CARPI) was placed over the first face slab as an additional protection against cracks. For dams with potential leakage due to high compressibility, the use of a membrane should be considered as a means of preventing excessive leakage. Underwater repairs on the face slab are extremely expensive.

2.9.2 Asphalt concrete

Asphalt concrete has been used in some dams, channels, and pumping store projects in several European countries as well as in Australia. Asphalt concrete has remarkable properties, such as compressive strength, impermeability, and flexibility to support rockfill deformations and stresses.

2.10 CONSTRUCTION

In 1987 Cooke and Sherard stated:

> "Beginning with the Piedras dam in Spain in 1967, all face slabs have been placed in vertical strips by slip form continuously from bottom to top using simple horizontal construction joints, with only a few exceptions. The detail for the horizontal construction joints, with the steel running through and no waterstop, has been used on almost all CFRDs with complete success. Construction joints are used as required by stage construction or by the contractor.
>
> The face slab has usually been placed in 12–18 m wide strips, 15 m being most commonly used width. The actual dimensions should be left to the contractor.
>
> There has been some thinking that the rockfill dam should be completed to its full height before starting the placement of the concrete face. The construction and performance experience at Foz do Areia, Salvajina and Khao Laem dams conclusively shows that the face slabs can be placed in any sequence convenient to the contractor to obtain maximum schedule and cost benefits.
>
> The rockfill upstream slope is built to the design line. By the time the face is compacted and the embankment is completed, the upstream slope will have moved. The concrete slab with the required minimum design thickness is placed on this existing upstream slope. This small difference in the face slab position compared to the design line has no influence in the construction, performance, or appearance of the slab. Measurements show that with careful construction the excess concrete is generally in the range of 5–10 cm more than the design thickness. It is considered good contract practice to acknowledge this by making the pay quantity for concrete about 7.5 cm more than the design thickness of slab."

With the use of the extruded curb, the excess concrete required for face slab construction has been reduced substantially. Today, the payment line is the theoretical one. No excess is taken into consideration.

> "At Foz do Areia Dam – the highest CFRD to date (160 m), and with the highest settlement – the concrete slab was placed on the lower 80 m of the dam before the rest of the embankment was completed. The top of the first-stage face slab in the center of the valley moved downstream normal to the slope by about 0.6 m while the rest of the embankment was being completed, causing no problem."

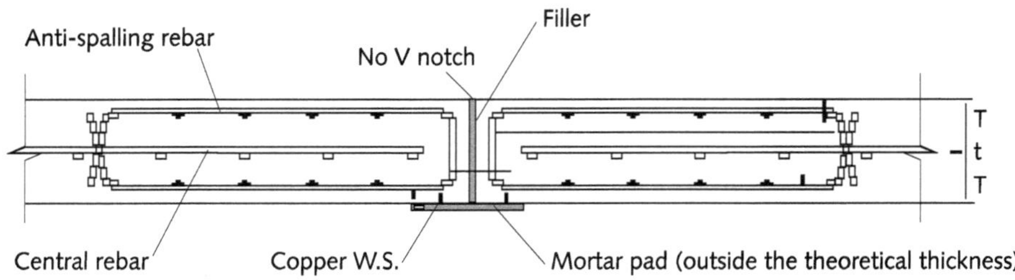

Figure 2.18 Kárahnjúkar face slab joint at central compression zone, modified design (Perez, Johannesson & Stefansson, 2007).

At Campos Novos Dam (202 m, Brazil) the face slab was built in three phases: the first between March 2003 and August 2003, the second from September 2004 to March 2005, and the final stretch (mostly above the maximum water level) at the end of the construction work.

Vertical and horizontal displacements due to the construction loads were still taking place when the first stage of the slab was being built as indicated by measures from the settlement cells.

In high dams, whenever the concrete face is placed before the completion of the embankment for schedule reasons, consideration must be given to possible movements or displacements of the concrete face that can lead to compressive stresses between joints and excessive bending moments. Details of the concrete compressive joints can be modified in order to allow for joint closing without the transmission of compressive stresses that could damage the concrete. See Figure 2.18.

Details of construction sequence for CFRDs are discussed in chapters 8 and 12.

2.11 INSTRUMENTATION

The empirical design of CFRDs that has prevailed for the last 50 years has been improved and modified. This is because CFRDs have been well instrumented and the data can be used to analyze their behavior.

The views and considerations made by Cooke and Sherard in 1987 are relevant today as can be seen from the following two paragraphs:

> "Instrumentation on CFRD dams has been important in gaining knowledge that has led to improvements in joint and face slab design, evaluation of rock and rockfill, and zoning in the rockfill. The results have given confidence in proceeding with higher dams. Instrumentation is not a requirement for safety monitoring. However, a minimum amount should be used.
>
> Foz do Areia would not be a CFRD without the pioneering engineering of the Hydro-Electric Commission of Tasmania, and the publication of instrumentation results. The owners of the Areia (160 m) and Salvajina (148 m) dams have extended the practice. The design of a CFRD is essentially empirical. It is based on experience and judgment. Instrumentation results are a major factor in the

words empirical, experience, and judgment. Instrumentation data from existing and new dams is important to continuing progress."

Nevertheless, the recent incidents at Mohale, Barra Grande, Campos Novos and Tianshengqiao 1, show that the intricate mechanism of stresses transmitted from the rockfill to zone 2, the extruded concrete and the concrete slab are not yet well understood. The instrumentation in those dams did not provide data that could be used to predict what occurred.

The movements of the rockfill towards the valley that resulted in the intricate mechanism of stresses are not measured. We measure settlement and displacements upstream or downstream, but we need to monitor the complete movement.

The benchmarks placed on the crest and on the downstream slope of the dam can give us some information, but only about the external contour of the dam.

A new instrument is necessary in order to give us a view of the complete displacement of the dam body. Mathematical models have been developed, but their results do not match the real displacements.

Attempts to measure stresses in the slab using strain gauges were made at the Mohale Dam, but we still do not have a complete picture of the stresses on the concrete slab. However, improving measures, such as the ones proposed for El Cajón, Shuibuya and Kárahnjúkar, have resulted in good slab performance.

2.12 AN OVERALL CONCLUSION

An imminent challenge for specialists in mathematical modeling and numerical analysis is to estimate the strains, stresses, distortions, and bending moments which take place on concrete slabs and on the joints as a result of the displacements that occur in the transitions and in the rockfill of the embankment during construction and impounding – and, in the long run, as a result of displacements that occur due to the creep of the rockfill.

Today, the empirical design based on experience and on observations of CFRD performance still prevails. Instrumentation should be implemented in dams 170 m or higher, placed mainly on the slab face (ex. electro levels), in order bring about a better comprehension of CFRDs behavior for future projects.

The challenge of building higher and higher dams, in many cases within tighter construction schedules, has forced the specialists in the field to review the basic design criteria proposed by Cooke and Sherard in 1987.

The main purpose of these reviews is to allow better control of the displacements of the concrete face.

Not much has been added to the main principles that underpin the basics of the Cooke and Sherard design criteria. But emphasis has been put on the modification of the central compression joints, introducing flexible materials and increasing the compaction in zones 3B, T and 3C.

Whenever economically feasible, placement of the face slab should be postponed until the rockfill creep develops and reaches values as low as 5–7 mm/month.

In the 20 years since 1987, more than 390 CFRDs over 50 m high have been built or are in planning around the world – most of which have had J. Barry Cooke as a consultant. The two 1987 papers have guided the design and construction of most of those dams.

Chapter 3

Typical cross sections

This chapter presents actual typical cross sections for a selection of existing CFRDs, and reviews their performance and importance for the progress and development of CFRDs structures.

For better identification and to aid comparison, each cross section is presented using the international nomenclature for dam zoning, even if the original papers and drawings were presented using a different nomenclature.

3.1 INTERNATIONAL NOMENCLATURE

For the zoning designation of the dams, and with the purpose of making the case histories in this chapter uniform and comparable, the international nomenclature used is:

- Zone 1A – silt, a low cohesion material.
- Zone 1B – random, material to confine zone 1A.
- Zone 2B – material under the slab or the extruded curb.
- Zone 3A – transition rockfill between zones 2B and 3B.
- Zone 3B – main upstream rockfill located downstream of 3A zone.
- Zone T – central rockfill placed between 3B and 3C zones.
- Zone 3C – downstream rockfill, placed after T or 3B zones.
- Zone 3D – material placed close to the downstream slope.
- Zone 4 – protection material of downstream slope.

3.2 EVOLUTION OF COMPACTED CFRDs

Table 3.1 charts the evolution of the highest CFRDs and their main characteristics, as well as of other dams in which the authors of this book were involved either in the design or construction.

Table 3.1 Height progression of CFRDs.

Dam	*Height (m)*	*Country*	*Face slab area m²*	*Year of completion*	*Highest period remark*
Cethana	110 m	Australia	30,000	1971	1971–1974
Alto Anchicayá	140 m	Colombia	22,300	1974	1974–1980
Mohale	145 m	Lesotho	87,000	2002	Highest in Africa
Salvajina	148 m	Colombia	57,500	1983	Highest in gravels
Xingó	150 m	Brazil	135,000	1994	–
Messochora	150 m	Greece	51,000	1995	Highest in Europe 1995–2007
Porce III	155 m	Colombia	57,000	2010	U.C.
Foz do Areia	160 m	Brazil	139,000	1980	1980–1993
Tianshengqiao	178 m	China	173,000	1999	–
Barra Grande	185 m	Brazil	108,000	2006	–
Mazar	166 m	Ecuador	45,000	2008	U.C.
Aguamilpa	187 m	Mexico	137,000	1993	1993–2006
El Cajón	188 m	Mexico	113,300	2006	Complete
Kárahnjúkar	196 m	Iceland	93,000	2007	Highest in Europe
Campos Novos	202 m	Brazil	106,000	2006	Highest until 2006
Bakún	205 m	Malaysia	127,000	2007	U.C.
La Yesca	210 m	Mexico	129,000	2010	U.C.
Shuibuya	233 m	China	120,000	2008	Highest in the world

U.C. – Under construction.

3.3 CASE HISTORIES

3.3.1 Cethana (Australia, 1971)

Figure 3.1 shows the position of each zone of Cethana Dam. Table 3.2 shows the materials used in the construction of the dam.

Main characteristics: H – 110 m; L – 213 m; L/H – 1.94; A/H^2 – 2.48; type of material – quartzite. External slopes are 1.3(H):1.0(V) upstream and downstream. The volume is 1.400.000 m³ (Fitzpatrick et al., 1973).

Cethana Dam, Australia, was built on the Forth River, north of Tasmania.

Completed in 1971, this dam was the highest in the world from 1971 to 1974. The transition zone 2B under the slab was built with 22.5 cm $D_{max.}$ quartzite in compacted layers of 45 cm, by four passes of a 10 t vibratory roller and adding water. The transition was a well-graded material with a coefficient of non-uniformity equal to 19. Due to over break, the concrete was reported to be 12.5 cm thicker than specified by the design (Fitzpatrick et al., 1985). Thus, the real expression for the slab was:

$$T = 0.30 + 0.002\ \mathrm{H} + \text{over break of } 0.125\ \mathrm{m} \qquad (3.1)$$

It is important to mention that in order to calculate the slab reinforcement, designers considered the theoretical slab thickness plus 10 cm, recognizing that excess concrete would be used in construction.

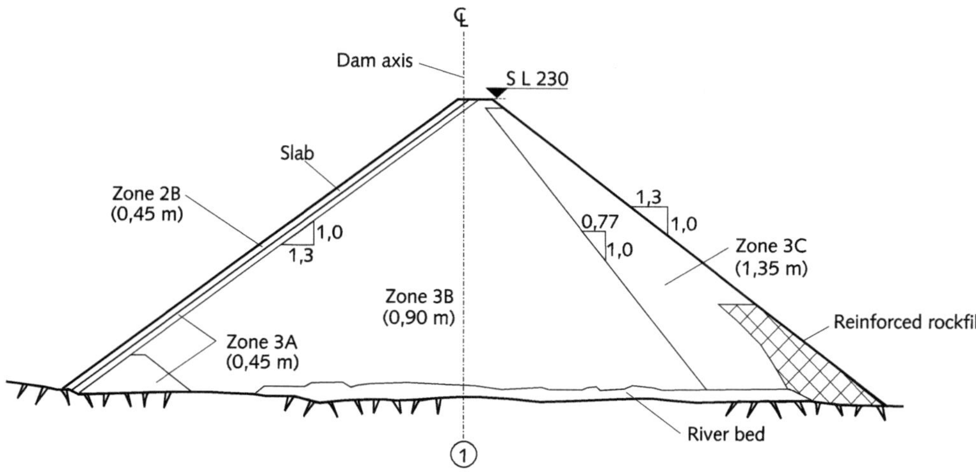

Figure 3.1 Cethana Dam.

Table 3.2 Cethana Dam materials.

Description	*Zone*	*Placing*	*Compaction*
Quartzite, max. size 22.5 cm	2B	Compacted in 0.45 m layers	10 t vibratory roller
Quartzite, max. size 22.5 cm	3A	Compacted in 0.45 m layers	compacted by
Rockfill, well graded $Ø_{max.}$ 60 cm	3B	Compacted in 0.90 m layers	4 passes
Rockfill, $Ø_{max.}$ 120 cm	3C	Compacted in 1.35 m layers	
Reinforced rockfill	D	Compacted in 0.45 m layers	

The main rockfill 3B was compacted in layers of 90 cm, with four passes of a 10 t vibratory roller and watering at a rate of 150 ℓ/m³. The maximum rockfill size of zone 3B was 60 cm, the C_u non-uniformity coefficient was 25 and the average compression modulus was 140 MPa.

The slab characteristics were width 12 m, face area 30,000 m², valley shape ratio $A/H^2 = 2.48$ (narrow valley). The perimetric joint and the tension joints had two waterstops, one made of copper and a central one made of rubber. The compression joints had one copper waterstop, without V notch, but with double anti-spalling reinforcement.

On the abutments, the slabs were divided in widths of 6 m. The slab was built in a single stage, and the maximum deflection recorded after reservoir filling was 11.6 cm.

The slab displayed retraction cracks at every 7 m, but it behaved excellently, with very little leakage at around 35 ℓ/s.

3.3.2 Alto Anchicayá (Colombia, 1974)

Figure 3.2 shows the position of each zone of Alto Anchicayá Dam. Table 3.3 presents the materials used in the construction of the dam.

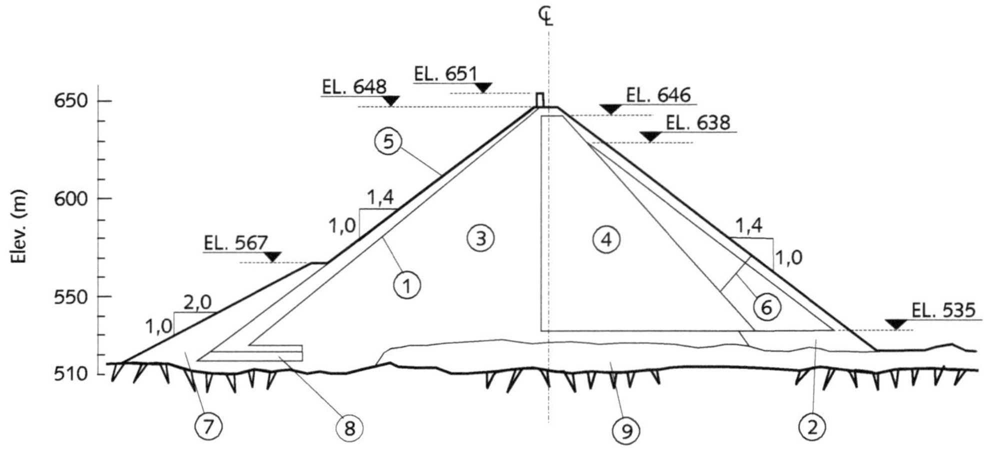

Figure 3.2 Alto Anchicayá Dam, cross-section.

Table 3.3 Alto Anchicayá Dam materials.

N°	*Zone*	*Placing*	*Compaction*
1	2B	Compacted in 0.50 m layers	Vibratory roller 10 t 4 passes horizontally and 8 in the upslope direction
2	Drain	1.0 m layers	Vibratory roller 10 t
3	3B	Compacted in 0.60 m layers	Vibratory roller 10 t 4 passes – water 200 ℓ/m³
4	3C	Fine rockfill 0.60 m layers	Vibratory roller 10 t 4 passes – water 200 ℓ/m³
5	Slab	Concrete	Face slab – variable thickness
6	3D	Over sizes	Vibratory roller 10 t 4 passes
7	1	Clayed silt	Compacted with construction equipment
8	Filters	Sand and coarse filter	Sand filters and coarse filter 2 passes of 10 t vibratory roller
9	River bed	Gravels	Dense natural gravels

Main characteristics: H – 140 m; L – 260 m; L/H – 1.86; A/H^2 – 1.14; type of material – hornfels.

Alto Anchicayá CFRD (Fig. 3.2) was built in Colombia, on the Anchicayá River, in western Colombia. Upstream and downstream slopes are 1.4(H):1.0(V) and its volume amounts to 2,400,000 m³ (Materón et al., 1982).

The dam was completed in 1974, and remained the world's highest dam until 1980. The 2B zone transition was built with blocks of 30 cm maximum size in 50 cm layers compacted by four passes of a 10 t vibratory roller with the addition of water. The transition material was well-graded, with a coefficient of non-uniformity equal to 17. However, the over excavation and consequent addition of concrete was very high, resulting in much larger thicknesses than those calculated using the theoretical formula:

$$T = 0.30 + 0.003\,H\ (\text{m}) \tag{3.2}$$

Figure 3.3 Alto Anchicayá Dam.

To calculate the slab reinforcement, 0.5% of the theoretical section was considered for the horizontal and vertical reinforcements placed on one or two layers.

The 3B main rockfill was compacted into 60 cm layers with four passes of a 10 t vibratory roller and addition of 200 ℓ/m³ of water.

The 3B material had a maximum size of 60 cm, and a coefficient of non-uniformity of 16. It was well-graded, resulting in a dense rockfill with a compressibility modulus of 135 MPa.

The main characteristics of the slab area are 15 m in width, area of 22,300 m², and a valley shape factor of $A/H^2 = 1.14$ (very narrow). The perimetric joint and the tension joints had one central rubber joint waterstop only. The compression joints also had just one rubber joint waterstop without a V notch, but with anti-spalling rebars.

At Alto Anchicayá, sub-parallel joints were placed by the perimetric joint in order to distribute potential movements due to the highly steep abutments (Sigvaldason et al., 1975).

During the filling of the reservoir, a high leakage volume was observed (1,800 ℓ/s), particularly from concentrated spots near the abutments. The flow had greater intensity from the right abutment, and this was caused by the detachment of the waterstop, which was found loose. After a fast treatment of mastic, leakage was reduced to 180 ℓ/s, and has remained almost constant throughout the project's life. The slab was built in two stages, and its maximum displacement after reservoir filling was 12 cm.

The slab suffered only minor cracks in the central portion. An inspection conducted after lowering the reservoir revealed excellent slab behavior. Figure 3.4 illustrates slab construction aspects, with four slip forms.

Alto Anchicayá has demonstrated the importance of considering several protection lines in the perimetric joint (Materón, 1985).

Figure 3.4 Alto Anchicayá slab construction. (See colour plate section).

3.3.3 Foz do Areia (Brazil, 1980)

Figure 3.5 shows the position of each zone of Foz do Areia Dam. Table 3.4 presents the materials used in the construction of the dam.

Main characteristics: H – 160 m; L – 828 m; L/H – 5.18; A/H^2 – 5.43; type of material – basalt. The slopes on both sides are 1.40(H):1.00(V), the volume is 14,000,000 m^3 (Pinto, Materón & Marques Filho, 1982).

The Foz do Areia CFRD (Figs. 3.6 and 3.7) was built on the Iguaçu River, in the state of Paraná, in southern Brazil.

Completed in 1980, this was the world's highest dam from 1980 until 1993. The 2B transition zone was built with 7.5 cm maximum grain size, although specifications allowed maximum sizes up to 15 cm. This material was compacted into 40 cm layers with four passes of a 10 t vibratory roller and the addition of water. The transition material was uniform, with a coefficient of non-uniformity equal to 10 due to the lack of fines in the crushed basalts. Similar to Cethana Dam, the additional concrete was reported to be 12.5 cm, which means the dam had a greater concrete thickness than that specified by the theoretical formula used, which was:

$$T = 0{,}30 + 0{,}00357\ \mathrm{H}\ (\mathrm{m}) \tag{3.3}$$

To calculate the slab reinforcement, 0.4% of the theoretical slab section was considered for both directions (Pinto, Materón & Marques Filho, 1982).

The main 3B rockfill was compacted into 80 cm layers with four passes of the 10 t vibratory roller and the addition of 250 ℓ/m^3 of water.

The 3B rockfill, with maximum size of 80 cm, had a coefficient of non-uniformity of 6 and a average compressibility modulus of 40 MPa.

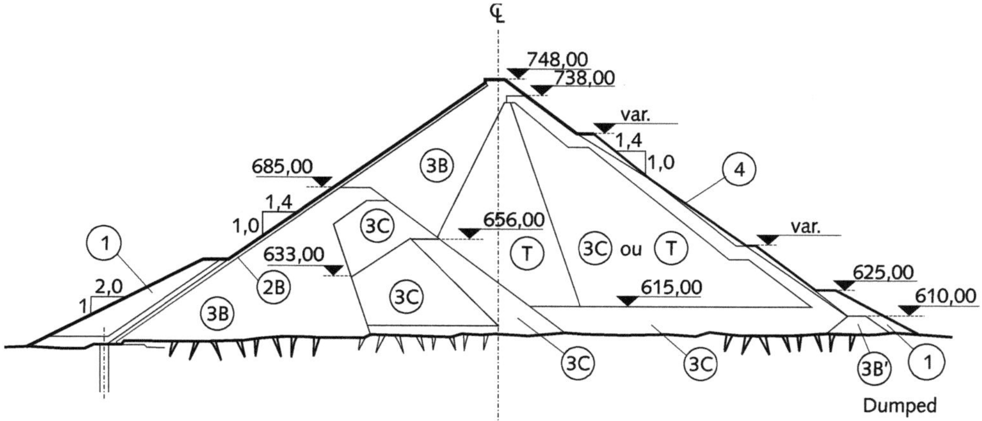

Figure 3.5 Foz do Areia Dam.

Table 3.4 Foz do Areia Dam materials.

Material	*Description*	*Zone*	*Placing*	*Compaction*
Rockfill	Sound basalt with up to 25% of basaltic breccia	3B′	Dumped	–
		3B	Compacted in 0.80 m layers	4 passes of 10 t. vibratory roller; 25% of water
		3C	Compacted in 1.60 m layers	
	Basalt and breccia inter-bedded	T	Compacted in 0.80 m layers	
	Sound basalt selected rock – 0,80 m ($Ø_{min.}$)	4	Equipment placed rocks	–
Transition II	Crushed rock from sound basalt	2B	Crushed basalt, 6″ max. size compacted in 0.40 m layers	10 t vibratory roller, 4 passes horizontally and 6 passes up in the slope direction
Clay fill	Impervious material	I	Max. size ¾″ compacted in 0.30 m layers	Construction equipment or pneumatic roller

The main characteristics of the slab are 16 m in width, area of 139,000 m^2, and a valley shape factor of $A/H^2 = 5.43$ (wide valley). The perimetric joint had two waterstops, one made of copper and a central one made of PVC, as well as a mastic cover. The tension joints had two waterstops, one made of copper and an upper one made of mastic. The compression joints had a copper waterstop with one small V notch and an anti-spalling reinforcement. The slab was built in two stages, and the maximum displacement after filling was relatively high (69.2 cm).

The initial leakage was 236 ℓ/s, and it decreased over time. Figure 3.8 shows the slab built up to the parapet level. The slab behavior has been satisfactory.

Figure 3.6 Foz do Areia Dam – aerial view.

Figure 3.7 Foz do Areia: lateral view of downstream slope.

3.3.4 Aguamilpa (Mexico, 1993)

Figure 3.9 shows the position of each zone of Aguamilpa Dam. Table 3.5 presents the materials used in its construction.

Main characteristics: H – 187 m; L – 642 m; L/H – 3.43; A/H^2 – 3.92; type of material – gravels and ignimbrite. Upstream slope is 1.5(H):1(V) (gravels) and

Figure 3.8 Face slab built up to the parapet level.

downstream slope is 1.4(H):1(V) (rockfill); its volume is 13,000,000 m^3 (Montañez, Hacelas & Castro, 1993).

The Aguamilpa dam was built on the Santiago River, in the state of Nayarit, western Mexico.

Completed in 1993, this was the world's highest dam up until 2006. The 2B zone transition was built with processed gravels with a maximum size of 7.5 cm and compacted into 30 cm layers with four passes of a 10 t vibratory roller. The transition material was well-graded, with a coefficient of non-uniformity in excess of 100. We have no information on over excavation and additional concrete, but the slab thickness was calculated using the formula:

$$T = 0{,}30 + 0{,}003\ \mathrm{H}\ (\mathrm{m}) \tag{3.4}$$

The upstream portion under the slab consisted of natural gravel from the Santiago River, followed by a transition of alluvium and rockfill. The downstream portion consisted of rockfill using rock from the structure's excavation (Gómez, 1999).

To calculate the reinforcement, the theoretical slab thickness was considered at a rate ranging from 0.3% to 0.5% depending on the location. The reinforcement was placed at the central part of the slab, with anti-spalling reinforcement near the perimetric joint and the tension joints.

The 3B main rockfill (gravel) was compacted into 60 cm thick layers with four passes of a 10 t vibratory roller, a maximum size of 60 cm, a coefficient of non-uniformity over 100, and an average compressibility modulus of 250 MPa. Downstream rockfill was compacted into 1.20 m thick layers, with a maximum size of 1 m and an average compressibility modulus of 50MPa.

The main characteristics of the slab are 15 m in width, area of 137,000 m^2, and a valley shape factor of $A/H^2 = 3.92$.

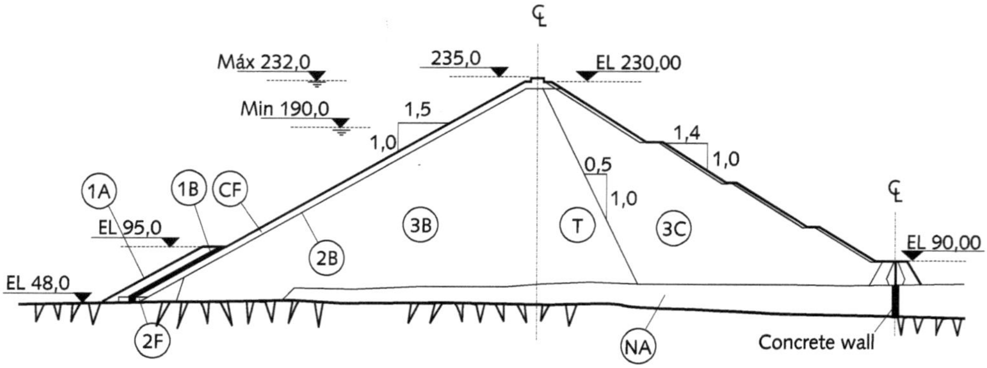

Figure 3.9 Aguamilpa Dam.

Table 3.5 Aguamilpa Dam materials.

Material	*Specification*	*Zone*	*Placing*	*Compaction*
Soil	Random	1A	Just placed in 80 cm thick layers	Construction equipment
Soil	Fine silty sand $Ø_{max.}$ 0.2 cm	1B	Just placed in 30 cm thick layers	Construction equipment
Filter 2F	Alluvial gravel and silty sand mix. $Ø_{max.}$ 3.8 cm	2A	Compacted in 30 cm layer	Layers 4 passes 100 kN vibratory roller
Gravels	Crushed alluvial gravel and sand mix. $Ø_{max.}$ 7.5 cm	2B	Compacted in 30 cm layer	Layers 4 passes 100 kN vibratory roller Face: 6 passes of 40 kN or 130 kN PC
	Alluvium, $Ø_{max.}$ 40 cm	3B	Compacted in 60 cm layer	4 passes 100 kN vibratory roller
Rockfill	Rockfill 3C with reduced $Ø_{max.}$ 50 cm	T	Compacted in 60 cm layer	4 passes 100 kN vibratory roller
Rockfill d/s	Rockfill ignimbrite $Ø_{max.}$ 100 cm	3C	Compacted in 120 cm layer	4 passes 100 kN vibratory roller
Slab	Concrete face	CF	–	–
River bed	Natural alluvium	NA	–	–

The perimetric joint and the tension joints had two joint waterstops, one made of copper and a central one made of PVC, where the upper portion was covered with fly ash. The compression joints had one copper water stop, with small upper V notch, but without any anti-spalling rebars.

An important aspect of the central compression joints was the introduction of a 2 cm thick wood board placed alternately at each fifth joint to mitigate potential compression stresses.

The slab was built in three stages (Fig. 3.10), and its the maximum displacement after filling was relatively small, 15 cm, but its crest moved downstream by twice as much.

The slab displayed a series of retraction cracks and a predominant high horizontal crack due to the difference in compressibility modulus between the alluvial gravel and the rockfill.

Figure 3.10 Aguamilpa: main slab at third phase construction. (See colour plate section).

The behavior of the dam is good with initial maximum leakage of 258 ℓ/s, a rate that has been decreasing with the passing of time.

3.3.5 Campos Novos (Brazil, 2006)

Figure 3.11 shows the position of each zone of Campos Novos Dam. Table 3.6 presents the materials used in the construction of the dam.

Main characteristics: H – 202 m; L –592 m; A/H^2 – 2.60; type of material: basalt. Upstream slope is 1.3(H):1.0(V) and downstream slope is 1.4(H):1.0(V), the volume is 12,100,000 m^3 (Xavier et al., 2007a).

Campos Novos Dam (Fig. 3.12) was built on the Canoas River, in the state of Santa Catarina, southern Brazil.

- Slab panels: 16 m wide
- Variable thickness according to: e = 0,30 + 0,002 H (for $H < 100$ m) and e = 0,005 H (for $H > 100$ m)
- Concrete face area: 106,000 m^2
- Reinforcing: 0.4% in both directions.

Completed in 2006, this was world's highest dam in operation for a short period. The 2B zone transition under the slab was built with maximum sizes of 7.5 cm. Although the specification allowed for maximum sizes up to 10 cm, this material was compacted into 50 cm layers with six passes of a 12 t vibratory roller and the addition of water. The transition material was uniform, typical of crushed basalts, with a coefficient of non-uniformity of 5 due to the lack of fines. The theoretical formula for the slab design was:

$$T = 0{,}30 + 0{,}002\ \text{H (m) up to 100 m in depth} \tag{3.4}$$

For depths over 100 m high, the slab was calculated as $T = 0.005\ H$. This increase of slab thickness was specified to prevent the hydraulic gradients from exceeding 200, which had proved to be appropriate for other dams.

In order to calculate the reinforcement (Fig. 3.13), 0.3% of the theoretical slab thickness was considered horizontally, and 0.4% was considered vertically. In areas near the abutments, reinforcement was increased to 0.5% and placed in a double layer.

The 3B rockfill was compacted into 1 m thick layers with six passes of a 12 t vibratory roller and the addition of 200 ℓ/m³ of water. This 3B rockfill, with a maximum size of 80 cm and uniform grading, had a coefficient of non-uniformity of 6 and an average compressibility modulus of 55 MPa.

The main characteristics of the slab are panels 16 m wide, area of 106,000 m², and a valley shape factor of $A/H^2 = 2.60$, typical of a narrow valley.

The compression joints had only one copper waterstop without any anti-spalling reinforcement. The slab was built in two stages, and the maximum slab displacement after filling was 86 cm.

The slab displayed failures in the vertical compression joints, as well as inclined cracks, which are discussed in chapter 8.

In October 2005 the central compression joint failed, accompanied by shifiting and superposition of the lateral slabs by 12 cm to 15 cm and with distortions of the reinforcement, rupture of cooper waterstop and increasing in leakages (Antunes Sobrinho et al., 2007).

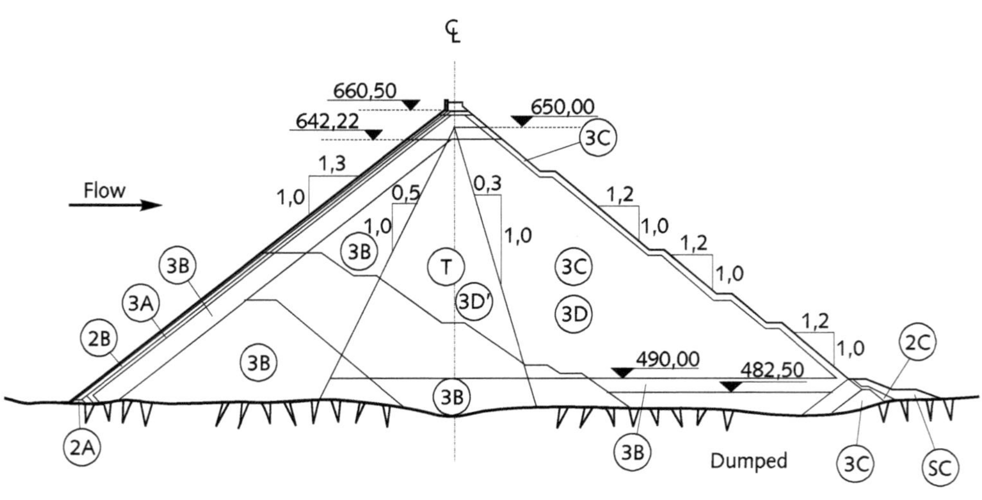

Figure 3.11 Campos Novos Dam.

Table 3.6 Campos Novos Dam materials.

Material	*Designation*	*Zone*	*Placing*	*Compaction*	*Strength specification*
Fine filter D/S the perimetric joint	$Ø_{max.}$ 25 mm	2 A	Compacted in 0.50 m layers	12 t vibratory roller	–
Fine transition	Sound processed basalt $Ø_{max.}$ 100 mm	2B	Compacted in 0.50 m layers	6 passes of 12 t vibratory roller	–
Coarse transition	Sound basalt $Ø_{max.}$ < 0.50 m	3A	Compacted in 0.50 m layers	6 passes of 12 t vibratory roller	–
Upstream rockfill	$Ø_{max.}$ 100 cm 70% min. of sound basalt or rhyodacite	3B	Compacted in 1.0 m layers	6 passes of 12 t vibratory roller water 200 ℓ/m³	At least 70% with unconfined compression strength above 50 MPa
Downstream rockfill	$Ø_{max.}$ 1.60 m	3C	Compacted in 1.60 m layers	6 passes of 12 t vibratory roller	At least 70% with unconfined compression strength above 40 MPa
Downstream rockfill	$Ø_{max.}$ 1.60 m	3D	Compacted in 1.60 m layers	6 passes of 12 t vibratory roller	At least 70% with unconfined compression strength above 25 MPa
Central rockfill	$Ø_{max.}$ 1.0 m with at least 70% sound basalt or dense rhyodacite	3D′ T	Compacted in 1.0 m layers	6 passes of 12 t vibratory roller water (200 ℓ/m³)	At least 70% with unconfined compression strength above 25 MPa
Soil	Saprolitic soil	SC	Dumped	–	–
Soil	Saprolitic soil	2C	Compacted by construction equipment in 0.40 m layers	–	–

Due to an accident in one of the diversion tunnels, the Campos Novos reservoir water level completely lowered from June 18 to 22, 2006. After the drawdown, it was possible to see that besides the rupture of the central joints, there was a transversal surface failure at approximately 30% to 40% of the dam height, and about 300 m long. Figures 3.14 and 3.15 show other aspects of the slab failure and distortions of reinforcement steel.

It is important to mention that Campos Novos Dam was built in a narrow valley making use of a basaltic rockfill which has an uniform grading. Compression modulus was within 50–60 MPa. Leakage reached values of 1,500 ℓ/s when the reservoir was at 93% of its height. Some remedial treatments reduced the leakage slightly. The behavior of the dam is good.

Figure 3.12 Campos Novos Dam – aerial view.

Figure 3.13 Campos Novos Dam: double reinforcement layer close to abutments. (See colour plate section).

3.3.6 Shuibuya (China, 2009)

Figure 3.16 shows the position of each zone of Shuibuya Dam. Table 3.7 presents the materials used in the construction of the dam.

Main characteristics: H – 233 m; L – 660 m; L/H – 2.83; A/H^2 – 2.21; type of material – limestone. Upstream and downstream slopes are 1.4(H):1.0(V), and its volume is 15,640,000 m^3 (Sun & Yang, 2005).

Figure 3.14 Campos Novos Dam: slab failure. (See colour plate section).

Figure 3.15 Campos Novos Dam: central compressive joints failure. (See colour plate section).

The Shuibuya CFRD was built on the Qingjian River, a tributary of the Yangtze, in the Hubei province, central China.

Shuibuya Dam was completed in 2008. In April of that year the reservoir was only 9 m below its maximum level. The compaction of the dam was quite rigorous, as shown in Table 3.7. Fibers were incorporated in the concrete in order to reduce the fissure frequency on the slab. During construction, the frequency of fissure was considered high.

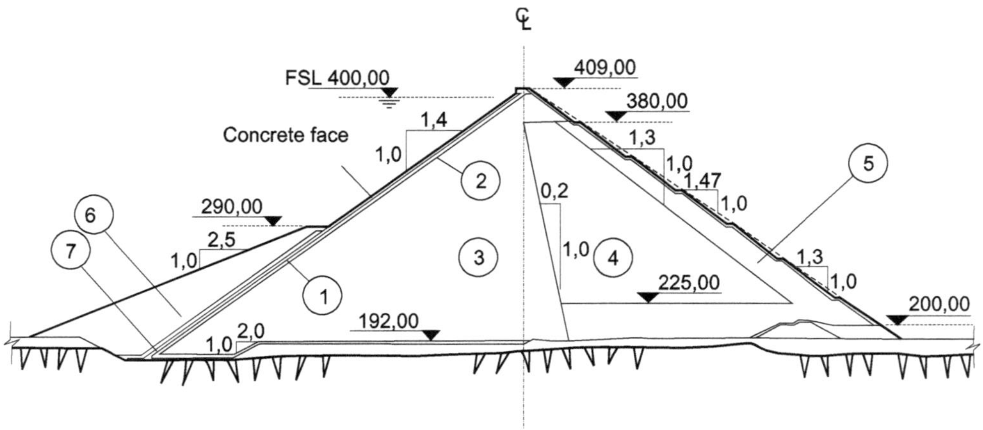

Figure 3.16 Shuibuya Dam.

Table 3.7 Shuibuya Dam materials.

Material	*Description*	*Zone*	*Compaction procedure*	*Compaction specification*
1	Filter, processed limestone γ_d 2.25 t/m^3	2B	Compacted in 0.40 m layers	Compacted by 8 passes of a 18 t vibratory roller, and water
2	Transition processed limestone γ_d 2.20 t/m^3	3A	Compacted in 0.40 m layers	Compacted by 8 passes of a 18 t vibratory roller, and 15% water
3	Limestone, blasting, excavation γ_d 2.18 t/m^3	3B	Compacted in 0.80 m layers	Compacted by 8 passes of a 25 t vibratory roller, and 15% water
4	Limestone, blasting, excavation γ_d 2.15 t/m^3	3C	Compacted in 0.80 m layers	Compacted by 8 passes of a 25 t vibratory roller, and 10% water
5	Limestone, blasting, excavation γ_d 2.15 t/m^3	3D	Compacted in 1.20 m layers	Compacted by 8 passes of a 25 t vibratory roller
6	Soil	1B	Placed	Construction equipment
7	Soil	1A	Placed	Construction equipment

An important feature of this dam is the use of a special type of waterstop, along with mastic (called GB) developed by the China Institute of Water Resources and Hydropower Research (IWHR). This mastic is protected by a band of a resistant material manufactured with EPDM. All joints (either tension or compression) are protected with this same mastic. After the performance of Campos Novos and Barra Grande dams in Brazil, at Shuibuya Dam some compressive materials were included in the central compression joints to avoid the concentration of efforts.

The number of joints was increased close to the abutments, dividing the slab width with intermediate joints. Shuibuya today is the highest dam in the world, with an excellent performance.

3.3.7 Tianshengqiao I (China, 1999)

Figure 3.17 shows the position of each zone of Tianshengqiao 1 Dam. Table 3.8 presents the materials used in the construction of the dam.

Main characteristics: H – 178 m; L – 1.104 m; L/H – 6.2; A/H^2 – 5.68; type of material – limestone, mudstone. Both external slopes are 1.40(H):1.0(V) and its volume is 17,700,000 m^3 (Wu et al., 2000b).

Tianshengqiao 1 Dam (TSQ1) was built on the Nanpanjiang River, at the border of Guizhou province and Guangxi autonomous province, in southeastern China.

Completed in 2000, this dam is currently one of the highest in China and the largest commercially operating dam. The 2B zone transition under the slab was built

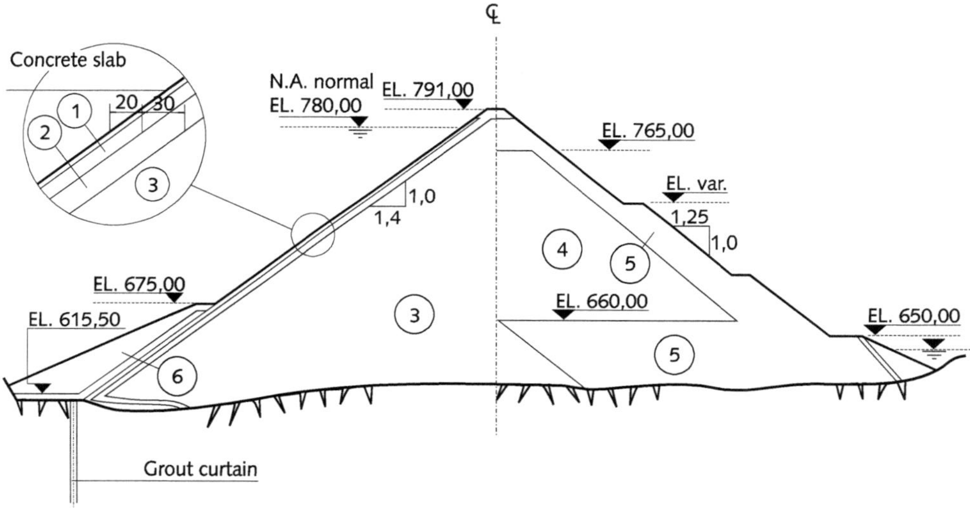

Figure 3.17 Tianshengqiao I Dam.

Table 3.8 Tianshengqiao I Dam materials.

Material	*Description*	*Zone*	*Placing*	*Specifications*
1	Filter processed limestone $Ø_{max.}$ 7,5 cm	2B	0.40 m layers	Compacted by 6 passes of a 9 t in the vibratory roller and 8 passes upstream direction
2	Transition processed limestone $Ø_{max.}$ 0.30 m	3A	0.40 m layers	Compacted by 6 passes of a 9 t vibratory roller
3	Limestone, blasting excavation $Ø_{max.}$ 0.80 m	3B	0.80 m layers	Compacted by 6 passes of a 18 t vibratory roller
4	Mudstone, blasting excavation $Ø_{max.}$ 0.80 m	3C	0.80 m layers	Compacted by 6 passes of a 18 t vibratory roller
5	Limestone, blasting excavation $Ø_{max.}$ 1.0 m	3D	1.0 m layers	Compacted by 6 passes of a 18 t vibratory roller
6	Clay	I	0.30 m layers	Compacted by construction equipment

with maximum sizes of 7.5 cm, and obtained by processing limestone. The transition material was well-graded, with a coefficient of non-uniformity equal to 33. The theoretical formula used for designing the slab was:

$$T = 0{,}30 + 0{,}0035 \text{ H (m)} \quad (3.5)$$

To compute the reinforcement, designers considered 0.3% of the theoretical slab section horizontally and 0.4% vertically, and it was located at the central part of the slab. The presence of cracks on the slab due to the multiple stages of the fill construction indicated the need for placing reinforcement at the upper and lower parts during the slab's third phase. Those cracks were dutifully treated before reservoir filling.

The 3B main rockfill was compacted with the addition of water. This 3B rockfill, with maximum size of 80 cm, was well-graded and had a coefficient of non-uniformity over 17, and an average compressibility modulus of 45 MPa.

The main characteristics of the slab are 16 m in width, area of 173,000 m^2, and a valley shape factor of $A/H^2 = 5.46$, typical of a wide valley as Foz do Areia. The performance of the dam was good. Maximum displacement was 2.92 m at the central portion of the dam and a measured leakage of 55 ℓ/s, which is relatively low considering the size of the dam.

The compression joints had one copper waterstop only, with an anti-spalling reinforcement and V notch in the upper part. The slab was built in three stages.

Tianshengqiao 1 Dam suffered rupture in the central compression joint due to a concentration of stresses. This happened three years after the first filling of the reservoir. Joints have been repaired and the performance is good and shows no increase in leakage.

3.3.8 Mohale (Lesotho, 2006)

Figure 3.18 shows the position of each zone of Mohale Dam. Table 3.9 presents the materials used in the construction of the dam.

Main characteristics: H – 145 m; L – 600 m; L/H – 4.14; A/H^2 – 4.14; type of material – basalt, doleritic basalt. Upstream slope is 1.40(H):1.0(V) and downstream slope is 1.45(H):1.0(V). The volume is 7,800,000 m^3 (Johannesson & Tohlang, 2007b).

Mohale Dam (Fig. 3.19) was built on the Senqunyane River, close to the confluence with Likalaneng River, East Maseru, near the capital of Lesotho, Africa.

This dam had its reservoir completely full in February 2006. It is the highest CFRD in Africa. The 2B transition zone under the slab was built with maximum sizes of 7.5–10 cm with the addition of water. The transition 2B material is uniform and made from processed basalt, and it has a coefficient of non-uniformity equal to 10. The slab was designed using the formula:

$$T = 0{,}30 + 0{,}003 \text{ H (m)} \quad (3.6)$$

To calculate the reinforcement, 0.35% of the section was considered in both directions. However, near the abutments and at the starter slabs, a rate of 0.4% was used.

The 3B main rockfill was compacted into 1 m layers with the addition of 150 ℓ/m^3 of water. With a maximum size of 1 m, the 3B rockfill has a coefficient of non-uniformity lower than 15 and low compressibility modulus values of around 32 MPa.

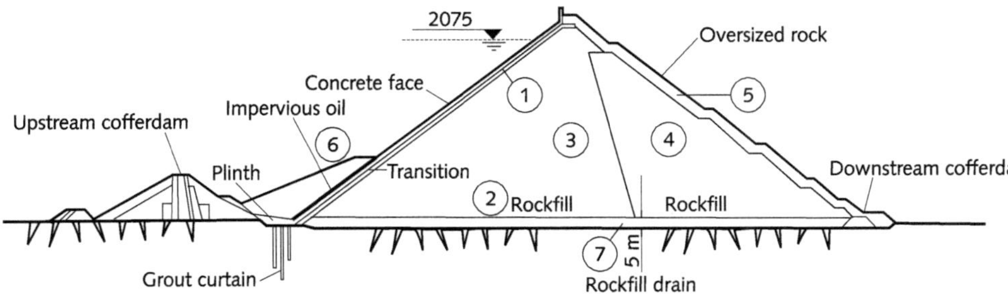

Figure 3.18 Mohale Dam, cross-section.

Table 3.9 Mohale Dam materials.

Number	*Description*	*Zone*	*Placing*	*Compaction*
1	Processed doleritic basalt max size 75 mm	2B	Placed in 0.40 m layers	12 t vibratory roller, 4 passes
2	Selected rockfill, $Ø_{max.}$ 400 mm	3A	Placed in 0.40 m layers	12 t vibratory roller, 6 passes
3	Rockfill $Ø_{max.}$ 1.0 m	3B	Placed in 1.0 m layers	12 t vibratory roller, 6 passes
4	Rockfill $Ø_{max.}$ 2.0 m	3C	Placed in 2.0 m layers	12 t vibratory roller, 6 passes
5	Downstream slope protection	4	Backhoe placing	Placed
6	Soil	1	Placed in 0.30 m – 0.60 m layers	Compacted by construction equipment
7	Drain	Drain	Doleritic basalt over 1.0 m	Placed with dozer without compaction

Figure 3.19 Mohale Dam. (See colour plate section).

The main characteristics of the slab are 15 m in width, area of 87,000 m², and a valley shape factor of $A/H^2 = 4.14$. The perimetric joint and the tension joints had two joint waterstops, one made of copper and a central one made of PVC, as well as a cover consisting of a non-cohesive, fine material. The compression joints had one copper waterstop and a V notch, but no anti-spalling reinforcement. The rebars were placed overtapping the joint.

The slab was built in two stages. The slab displayed retraction fissures, which were treated prior to reservoir filling. The behavior was appropriate, but when the reservoir reached 90% of its total height, ruptures occured on the central compression joints, similar to those that occurred at Campos Novos and Barra Grande dams, with leakage increasing to 600 ℓ/s (Johannesson & Tohlang, 2007a). The dam has been performing well.

3.3.9 Messochora (Greece, 1996)

Figure 3.20 shows the position of each zone of Messochora Dam. Table 3.10 presents the materials used in the construction of the dam.

Main characteristics: *H* – 150 m; *L* – 337 m; *L/H* – 2.25; A/H^2 – 2.27; type of material – limestone. The slopes are 1.4(H):1.0(V) both upstream and downstream, although the very upper downstream slope is 1.55(H):1.0(V) (Materón, 2006).

The Messochora Dam (Fig. 3.20) was built on the Acheloos River, in northeastern Greece.

The dam was completed in 1995, and was the highest in Europe at that time. It was built in a narrow valley with A/H^2 ratio of 2.27.

The transition 2B was built with processed material of rounded gravel with a maximum size of 7.5 cm and with 38% medium sand content, and less than 5% of the material passed a number 200 sieve. The material is well graded, with densities of 2.25 to 2.30 t/m³ and a void ratio of 0.16.

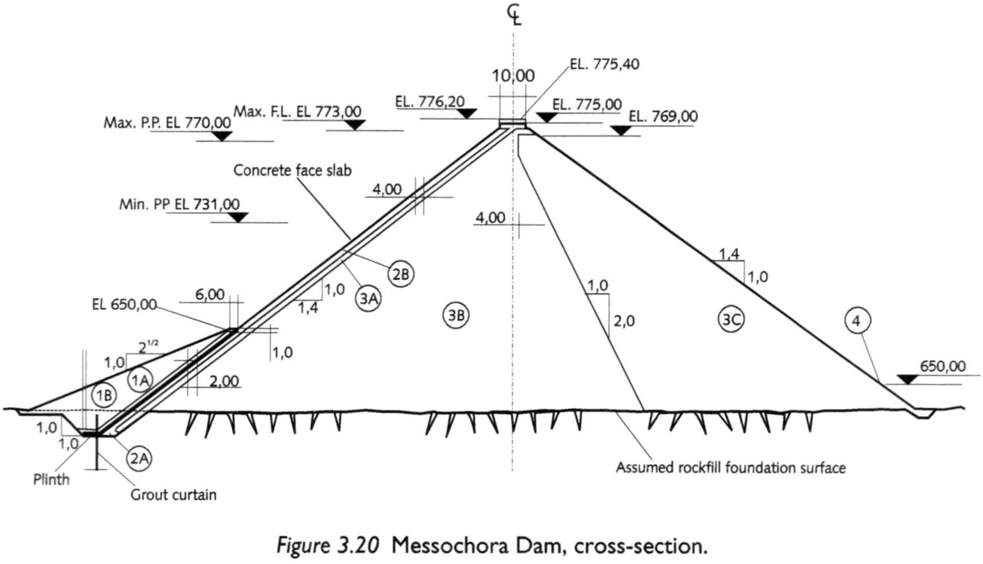

Figure 3.20 Messochora Dam, cross-section.

Table 3.10 Messochora Dam materials.

Material	*Zone*	*Placing*	*Compaction*
Impervious material 80%<n°4 sieve	1A	0.30 m layers	Construction equipment
Random fill	1B	0.30 m layers	4 passes of vibratory roller
Special processed filter from gravels	2A	0.40 m layers	4 passes of vibratory roller
Processed gravels $Ø_{max.}$ 75 mm	2B	0.40 m layers	4 passes of vibratory roller
Selected rockfill from quarries $Ø_{max.}$ 400 mm	3A	0.40 m layers	4 passes of vibratory roller water 10%
Main rockfill $Ø_{max.}$ 1,000	3B	1.00 m layers	4 passes of vibratory roller water 10%
Downstream rockfill $Ø_{max.}$ 1,500 mm	3C	1.50 m layers	4 passes of vibratory roller water 10%
Selected rockfill $Ø_{max.}$ 1,500 mm	4	Placed over the downstream slope	–

Figure 3.21 Messochora Dam. (See colour plate section).

Zones 3B and 3C were built with rockfill from limestone quarries near the dam site and had a density of 1.9–2.0 t/m³. The compressibility modulus is in the order of 44 MPa and corresponds to a relatively high void ratio of 0.30 to 0.36. To compute the slab reinforcement this formula used was:

$$T = 0.30 + 0.003\ H\ (m) \tag{3.7}$$

The reinforcement was conventional, with an increasing reinforcement rate close to the abutments.

The perimetric joint was designed with anti-spalling reinforcement and a filling of mastic with a copper waterstop at the slab bottom; on the upper part, the mastic was protected by a band of neoprene.

The compression joints had top *V* notches with mastic, protected by a 3 mm neoprene band.

The Messochora reservoir has not been filled so far, due to social and ecological concerns, but these issues are close to being resolved. A recent inspection showed that it will be necessary to improve the central compression joints to avoid problems similar to the ones that occurred at Mohale, Tianshengqiao 1, Barra Grande and Campos Novos CFRDs.

To analyze the problem the Public Power Corporation of Greece (PPC), the project owner, developed a methodology that allowed analysis of non linear behavior by a 3D program. The conclusions were that in narrow valleys with low rockfill compressibility modulus, high tensions will occur close to the plinth and high compression will occur in the central third of the dam. See also chapter 11 (Dakoulas, Thanopoulos & Anastassopoulos, 2008).

3.3.10 El Cajón (Mexico, 2007)

Figure 3.22 shows the position of each zone of El Cajón Dam. Table 3.11 presents the materials used in the construction of the dam.

Main characteristics: H – 188 m; L – 550 m; L/H – 2.93; A/H^2 – 3.21; type of material – ignimbrite. Upstream and downstream slopes are 1.40(H):1.0(V) and its volume is of 10,900,000 m^3 (Marengo-Mogollón & Aguirre-Tello, 2007).

El Cajón Dam was built on the Santiago River, upstream of Aguamilpa's reservoir, in the state of Nayarit, Mexico.

El Cajón Dam was completed in 2007, it is one of the highest in the world and the highest of this type in Mexico.

The transition 2B was built in layers of 0.30 m, with processed material of 7.5 cm maximum diameter. The material is well graded. The coefficient of non-uniformity of the material was over 100, with 40% to 55% sand content and about 10% fine material (<200 sieve).

The void ratio was ≤0.22 and the average permeability coefficient was 5×10^{-3} cm/s.

An interesting aspect of this dam is that 2B material was placed using paving equipment, leading to nicely conformed layers as it can be seen in Figure 3.23.

The 3A material was processed with a maximum size of 0.15 m, and was well graded and compacted, in a similar way to the 2B material. The coefficient of non-uniformity was over 60.

Materials in zones 3B, T and 3C were properly compacted. The moduli were high. The rockfill was of ignimbrite, a rock with a relatively low specific gravity. The reported densities were relatively low when compared with other rockfills.

Quality control values are in Table 3.12.

The volume of water used during compaction was higher than the 200 ℓ/m^3 originally specified. The effect was good, due to the high absorption of the rock (5%) (Mena Sandoval et al., 2007b).

An important feature of the construction of these zones was the use of 10–12 t compactors (rollers) with transmitted pressures over 5 t/m on the vibratory cylinder.

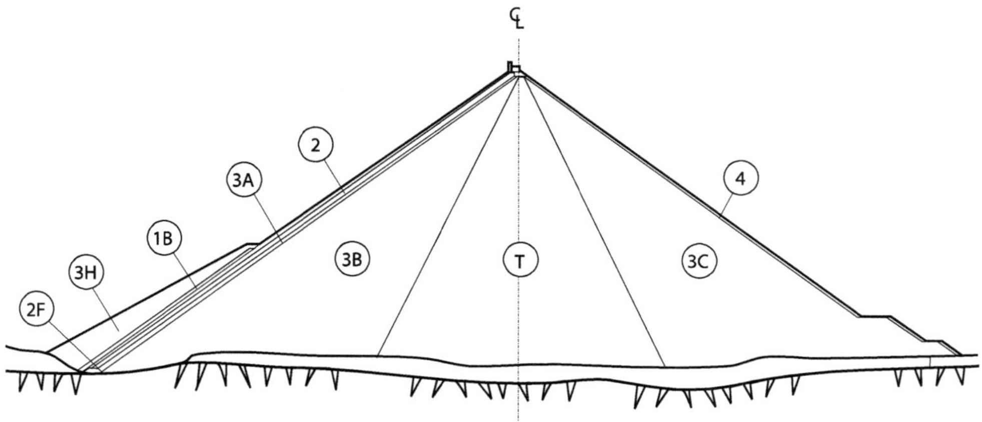

Figure 3.22 El Cajón Dam.

Table 3.11 El Cajón Dam materials.

Material	*Zone*	*Placing*	*Compaction*
1B	1B	0.3 m layers	Compacted by Dozer
2F	2A	0.3 m layers	10 t vibratory roller, 6 passes
2	2B	0.3 m layers	10 t vibratory roller, 8 passes
3A	3A	0.3 m layers	10 t vibratory roller, 8 passes
3B	3B	0.8 m layers	12 t vibratory roller, 6 passes and water > 200 ℓ/m^3
T	T	1.0 m layers	12 t vibratory roller, 6 passes and water > 200 ℓ/m^3
3C	3C	1.4 m layers	12 t vibratory roller, 6 passes and water > 200 ℓ/m^3
3H	Random	0.4 m layers	Compacted by dozer or construction equipment
4	4	Placed as slope protection	Placed by backhoe

The face slab has an area of 113.300 m^2 (Mendez et al., 2007) and the thickness is:

- 0.30–0.50 m (0–100 m deep)
- 0.50–0.80 m (>100 m deep)

The slab was built in four stages, with a variable production between 3.0–5.50 m/h.

Before the reservoir filling, the maximum settlement was 0.85 m, which corresponds to 0.45% of the dam height, relatively smaller than that observed in other CFRDs.

The maximum displacement of the slab was 0.18 m at a height of 0.54 *H*.

The displacements of the perimetric joints in various places were: settlement – 24.4 mm; opening – 8.8 mm; shear displacement – 3.4 mm; these are consistent with

Figure 3.23 Equipment used in El Cajón Dam to build 2B transition. (See colour plate section).

Table 3.12 El Cajón quality control.

Zone	*Density (t/m³)*	*Compressibility modulus (MPa)*
3B	1.8	110
T	1.8	125
3C	1.78	75

the predicted values and much lower than the deformation capacity of the waterstops used in the joint.

The performance of the rockfill and the slab of El Cajón is a clear demonstration that well-compacted rockfill, with high compressibility modulus, has low displacements and the slab compression stresses are within design compatible values.

The performance of the dam is excellent, for a high dam in a narrow valley:

$$A/H^2 = \frac{113.300}{188^2} = 3.21 \tag{3.8}$$

The maximum leakage is 150 ℓ/s, quite low when compared with leakages from other dams of similar size.

3.3.11 Kárahnjúkar (Iceland, 2007)

Figure 3.24 shows the position of each zone of Kárahnjúkar Dam. Table 3.13 presents the materials used in the construction of the dam.

Main characteristics: H – 196 m; L – 700 m; L/H – 3.57; A/H^2 – 2.42; type of material – basalt. Uupstream and downstream slopes are 1.3H:1.0V; the volume is 8,500,000 m³ (Perez, Johannesson & Stefansson, 2007).

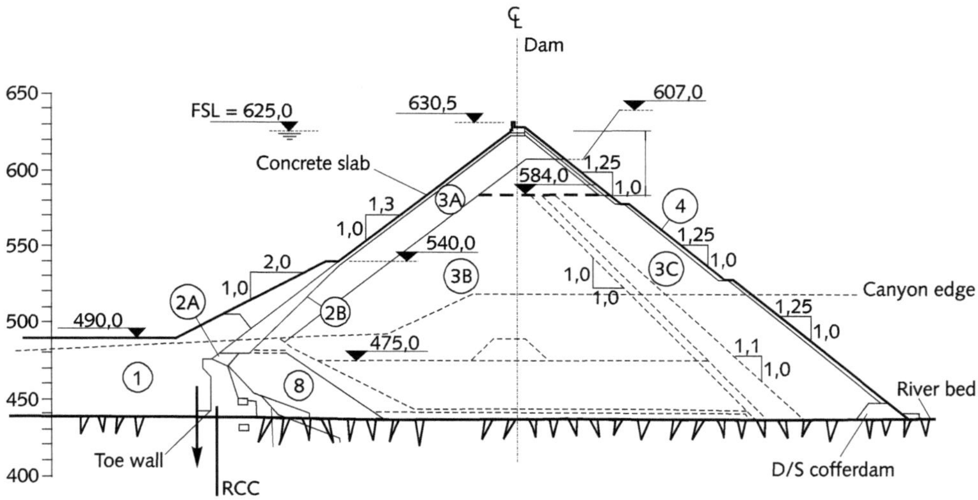

Figure 3.24 Kárahnjúkar Dam.

Table 3.13 Kárahnjúkar Dam materials.

Description	*Zone*	*Placing*	*Compaction specification*
Glacial Silt, fine sand	1	Compacted in 50 cm layers	Compacted by 6 passes of 350 kN
Fine filter processed pillow lava	2A	Compacted in 20 cm layers	Compacted with vibratory plate and water (50 kN)
Transition processed pillow lava	2B	Compacted in 40 cm layers	Compacted by 4 passes in summer with water; 6 passes in winter without water (350 kN)
Pillow lava, or Moberg	3A	Compacted in 40 cm layers	
Natural gravel	8	Compacted in 40 cm layers	
Sandy gravel, Moberg and pillow lava	3B	Compacted in 40, 60, 80 cm layers	
Durable free draining basalt	3C	Compacted in 160 cm layers	
Selected durable slope protection	4	Placed by backhoe in compact array	–

Kárahnjúkar Dam, Iceland, was built on the Jökulsá River, fed by the largest glacier of Europe, the Vatnajôkull Glacier.

Kárahnjúkar Dam was completed in 2007, and the reservoir was completely full in the same year. Today it is the highest dam in Europe.

The dam is located in a narrow valley with a 45 m deep canyon, and vertical walls at its base. There is a very steep slope on the right margin, and an extended left margin with an average slope of 2.5H:1.0V.

The 2B material was processed from lava flow with a maximum size of 7.5 cm, a sand content of 15% to 40%, and less than 5% fines. The material did not conform to Sherard's proposed grain sizes, but is a well graded material, with a non-uniformity coefficient C_u of 20.

Zones 3B and 3C were compacted in layers of 0.60 and 1.20 m, and the downstream material in layers of 1.60 m.

In the original project it was considered that there should be a higher percentage of reinforcement, above that which would be normally used, in order to serve the particular characteristics of a narrow valley in one of the abutments and the presence of a deep canyon where the dam was supported by a RCC (roller compacted concrete) structure.

When the dam was under construction, ruptures in the central joints of basaltic rockfill dams with similar compression modulus, as at Barra Grande, Campos Novos and Mohale, were being reported.

The designers, along with a group of international consultants (Perez, Johannesson & Stefansson, 2007), required the following adaptations to be implemented:

- Reducing the compacted layer to 0.40 m between El. 584–625.
- Placement of a 3 mm bond breaker film over the extruded curb.
- Increasing the thickness of the 10 central slabs by 10 cm above El. 535, with a transition between the old and new slabs.
- Reduction of the mortar pad and elimination of the top V notch to increase the contact area betwen the slabs.
- Use of spacers, of 15 mm, between compression joints.
- Use of spacers, of 25 mm, between the joints in the parapet wall.
- Improvement of the anti-spalling reinforcement as shown in Figure 2.18.
- Raising up of the non-cohesive material placed over the slab to El. 540.

The performance of the dam after filling the reservoir has been excellent with low settlement and leakage below 200 ℓ/s. This leakage value is considered normal in glacial areas.

3.3.12 Bakún (Malaysia, 2008)

Figure 3.25 shows the position of each zone of Bakún Dam. Table 3.14 presents the materials used in the construction of the dam.

Main characteristics: H – 205 m; L – 750 m; L/H – 3.66; A/H^2 – 3.02; type of material – graywacke, shale. With an upstream slope of 1.4(H):1.0(V) and a downstream slope of 1.5(H):1.0(V) due to the access road built on the slope, the dam volume is 16,200,000 m^3 (Long et al., 2005).

Bakún Dam is under construction on the Balui River. The dam was completed in 2008, the parapet wall is under construction at the time of writing and the reservoir filling is scheduled for April/May 2009.

The 2B transition material was compacted in 0.40 m layers, with a maximum size of 80 mm and well graded, and with 4–6 passes of a 12 t vibratory roller. The material was processed from a sound graywacke, which was the same material used in zone 3A.

Although the grain size distribution of the 3B material was specified within the limits of Sherard, the average material had a sand content between 18% and 46% (average of 32%) and a coefficient of non-uniformity $C_u = 23$.

Materials 3B and 3C, comprising high percentages of graywacke (sandstone), had variable percentages of shale or mudstone. An important feature of this dam is the decision to mix different materials in the quarry. A mixture of sound

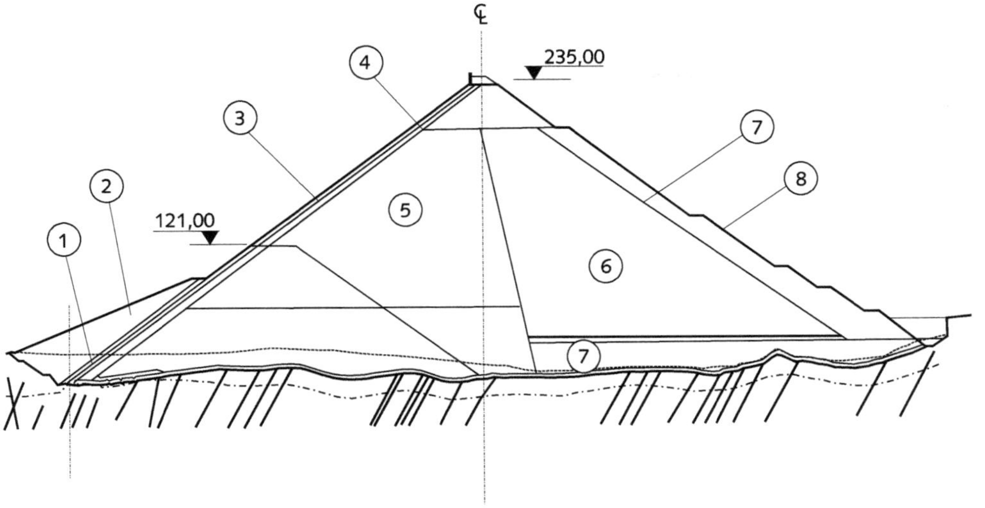

Figure 3.25 Bakún Dam, cross-section.

Table 3.14 Bakún Dam materials.

Material	*Description*	*Zone*	*Placing*	*Compaction*
1	Compacted silt	1A	0.20 m layers	Vibratory roller 2–3 passes
2	Random material	1B	0.30 m layers	Construction equipment
3	Processed fresh graywacke, $Ø_{max.}$ 80 mm	2B	0.40 m layers	8 passes of a 12 t vibratory roller
4	Processed fresh graywacke, $Ø_{max.}$ 300 mm	3A	0.40 m layers	8 passes of a 12 t vibratory roller
5	Fresh to slightly weathered graywacke, $Ø_{max.}$ 800 mm	3B	0.80 m layers	8 passes of a 12 t vibratory roller, water 150 ℓ/m³
6	Graywacke and mudstone moderately weathered, $Ø_{max.}$ 800 mm	3C	0.80 m layers	8 passes of a 12 t vibratory roller, water 150 ℓ/m³
7	Fresh graywacke, $Ø_{max.}$ 1.600 mm	3D	1.60 m layers	8 passes of a 12 t vibratory roller, water 150 ℓ/m³
8	Selected material	4	With equipment	–

graywacke with a maximum of 30% of shale mudstone was considered adequate for the dam.

The dam characteristics are slab area of 127,000 m², height of 205 m, and a valley ratio below 4. These characteristics have guided the designers to adopt a similar joint treatment to that used in Kárahnjúkar at the central compression joints.

Beyond these treatments, the central area of the slab was built only when the rockfill deformation was below 7 mm per month. Figure 3.26 shows major aspects of the dam.

Figure 3.26 Bakún Dam – aerial view. (See colour plate section).

3.3.13 Golillas (Colombia, 1978)

Figure 3.27 shows the position of each zone of Golillas Dam. Table 3.15 presents the materials used in the construction of the dam.

Main characteristics: H – 125 m; L – 108 m; L/H – 0.86; A/H^2 – 0.92; type of material – dirty gravels. Upstream and downstream slopes are 1.6H:1.0V. One of the main characteristics of the dam is the narrow valley leading to an arching effect between abutments (Amaya & Marulanda, 1985).

Golillas Dam (Fig. 3.28) was completed in 1978 on the Chuza River, as part of the Chingaza project to convey potable water to Bogotá in Colombia.

The transition 2B material was processed to a maximum diameter of 0.15 m and it was compacted in 0.60 m layers with four passes of a 10 t vibratory roller. In addition, four passes were applied in the direction of the upstream slope. The densities reported were 2.18 t/m^3 with a void ratio of 0.25.

The available material in the region was dirty gravels with inadequate permeability. Therefore, the dam has an inclined internal drain built with clean gravel, with a maximum size of 0.30 m.

Zones 3B and 3C were built in layers of 0.60 m with dirty gravel, and four passes of a 10 t vibratory roller. Zone 4 at downstream slope was built with large size blocks in 1.20 m thick layers.

During the filling of the reservoir, the movements close to the abutments were similar to those in the dam central area leading to failure of the copper and PVC waterstops in the perimetric joint. Leakage increased.

These leakages reached values over 1 m^3/s when the reservoir's water level reached its maximum limit. It was also observed that some rock joints filled with clay were

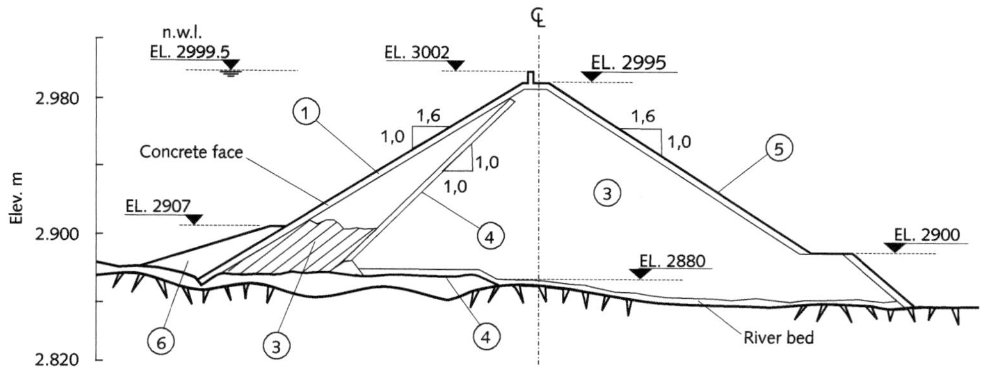

Figure 3.27 Golillas Dam, cross-section.

Table 3.15 Golillas Dam materials.

Material	*Specification*	*Zone*	*Placing*	*Compaction*
1	Processed dirty gravels, with $Ø_{max.}$ 0.15 m	2B	0.6 m layers	Vibratory roller 10 t, 4 passes horizontally Face: 4 passes upslope
3	Dirty gravels	3B	0.6 m layers	4 passes of 10 t vibratory roller
4	Clean gravels $Ø_{max.}$ 0.30 m	Drain	Layers of 0.6 m	2 passes of 10 t vibratory roller
5	Donwnstream slope $Ø_{max.}$ 1.20 m	4	Larger blocks	–
6	Silty clay	1	0.3 m layers	Compacted with bulldozer

Figure 3.28 Golillas Dam. (See colour plate section).

being eroded by the high gradients. Golillas shows how important it is to build protective filters over the potentially erodible zones of the dam.

The perimetric joints in narrow valleys ought to be provided with several lines of defense to avoid the rupture of the waterstops.

After the repair, the leakage was reduced to 650 ℓ/s, and it has decreased further since then. The performance of the dam is good.

3.3.14 Segredo (Brazil, 1992)

Figure 3.29 shows the position of each zone of Segredo Dam. Table 3.16 presents the materials used in the construction of the dam.

Main characteristics: *H* – 145 m with a 6 m high parapet wall; *L* – 720 m; *L*/*H* – 4.97; A/H^2 – 4.14; type of material – basalt. With an upstream slope of 1.3H:1.0V

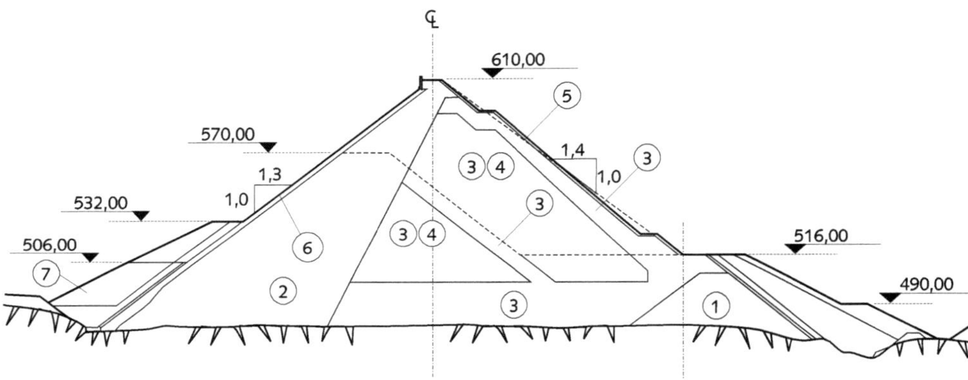

Figure 3.29 Segredo Dam, cross-section.

Table 3.16 Segredo Dam materials.

Designation	*Material*	*Description*	*Zone*	*Placing*	*Compaction*
1	Rockfill	Sound basalt (up to 25% breccia)	3B′	Dumped	–
2	Rockfill	Sound basalt (up to 25% breccia)	3B	0.80 m layers	4 passes of 10 t vibratory roller, water 25%
3	Rockfill	Sound basalt (up to 25% breccia)	3C	1.60 m layers	4 passes of 10 t vibratory roller, water 25%
4	Rockfill	Basalt and intercalated breccia	3D	0.80 m layers	4 passes of 10 t vibratory roller, water 25%
5	Rockfill	Sound basalt $\varnothing_{max.}$ 0.80 m	4	Blocks arranged by equipment	–
6	Processed rockfill	Well graded crushed Basalt $\varnothing_{max.}$ 7.5 cm	2B	0.40 m layers	6 passes of 10 t vibratory roller and 6 passes upslope
7	Earthfill	Impervious soil Max. size ¾	1	0.30 m layers	Tire compactor or construction equipment

and a downstream a slope of 1.4H:1.0V (average slope between berms 1.2H:1.0V), its rockfill volume is 7,200,000 m^3.

- Concrete face area: 87,000 m^2
- Variable thickness: min. 0.30 m to max. 0.70 m
- Theoretical slab thickness varies per: $e = 0.30 + 0.003\ H$ (m)
- Used concrete: 260 kg/m^3 of cement
- Water/cement ratio: 0.65
- Air entrainement: $4 \pm 0.5\%$
- Built in air additive: 1 kg/m^3 and 7 ± 1 cm slump
- Concrete strength: 16 MPa at 90th day
- Reinforcement: 0.3%
- Rockfill settlement: 2.22 m at the centre of its vertical axis during construction

Segredo Dam (Fig. 3.30) was built on the Iguaçu River, in the state of Paraná, southern Brazil, from 1987 to 1992 by Copel, Parana's energy company and also owner of Foz do Areia Dam (Pinto, Blinder & Toniatti, 1993).

The transition material 2B was processed and had a maximum size of 7.5 cm. Width deposition was constant and equal to 5 m. Near the basalt foundation it was enlarged to 10 m.

Zones 3B and 3C were compacted in layers of 0.80 m and 1.60 m with six passes of a 10 t vibratory roller and watering in the upstream zone at a rate of 250 ℓ/m^3.

At Segredo Dam sound basalt was more predominant than breccia or amygdaloidal basalts, both of which were very abundant in Foz do Areia Dam.

The compressibility modulus measured in the dam varies from 20 MPa (downstream) to 70 MPa (upstream) in the lower layers. The deformations were typical of basalts with a few fines, and were similar to those observed in Foz do Areia.

Figure 3.30 Segredo Dam. Courtesy Copel. (See colour plate section).

The slab maximum displacement was 34 cm. Slab displacements measured at the joints were small, varying from 6 mm opening to 2 mm settlement.

An interesting incident occurred before the closure of the diversion tunnels. A flood in the Iguaçu River increased the arrival flow to 7,000 m^3/s, which impounded the water and raised its level to El. 580, i.e., nearly 115 m above the plinth foundation on the river bed located at El. 465.

This flow generated high water velocities of above 20 m/s in the tunnels without revetment, but there was no erosion of the sound basalt.

The maximum leakage measured during the impounding was 390 ℓ/s. This value dropped to 50 ℓ/s, although an additional leakage of 100 ℓ/s is assumed to flow through the downstream cofferdam.

A total leakage of 150 ℓ/s is normal for a dam of this magnitude. The performance of Segredo is within expectations, with displacements smaller than Foz do Areia and compatible with its height of 145 m.

3.3.15 Xingó (Brazil, 1994)

Figure 3.31 shows the position of each zone of Xingó Dam. Table 3.17 presents the materials used in the construction of the dam.

Main characteristics: H – 150 m; L – 850 m; L/H – 5.67; A/H^2 – 6.0; type of material – gneiss. The upstream slope has an inclination of 1.4H:1.0V as does the downstream slope, which incorporates two access berms, 12 m wide each.

Xingó Dam (Fig. 3.32) was built on the São Francisco River, in northeastern Brazil. The dam foundation was gneissic rock exhibiting either a schist or granite structure. The dam was completed in 1994, with a rockfill volume of about 13,000,000 m^3. The gneissic granite rockfill came from the structure's excavations (Eigenheer & Mori, 1993).

The transition 2B was a sound granite material processed by using a grizzly. The grizzly allowed a maximum size of 0.10 m, with a sand content of ~45%, and 12% of fines passing a Nr. 200 sieve. This grading was much finer than the usual transitions applied in other Brazilian dams, but followed Sherard's recommendations (Table 3.18).

The material was compacted in 0.40 m layers with six passes of a 10 t vibratory roller. Additionally the material was compacted on the slope direction, with four passes of a 6 t vibratory roller without vibration, followed by another eight passes with full vibration (Souza et al., 1999).

The 2B material was placed beyond the upstream alignment and removed with a Gradall bucket, as shown schematically in Figure 3.33.

The material compaction was controlled as a dam core, using the Hilf method to calculate the degree of compaction and the water content deviation from optimun (Koch et al., 1993).

During the placement in some zones the material showed "weaving" due to an excess of water, but performed well when the excess water was drained.

Zones 3B and 3C of the dam were compacted in layers of 1.0 m and 2.0 m respectively with four passes of a 10 t vibratory roller; water was added to zone 3B (150 ℓ/m^3).

- Average secant deformability modulus at Zone I (2B): 47 and 68 MPa
- Deformability modulus at Zone I (2B): 59 MPa

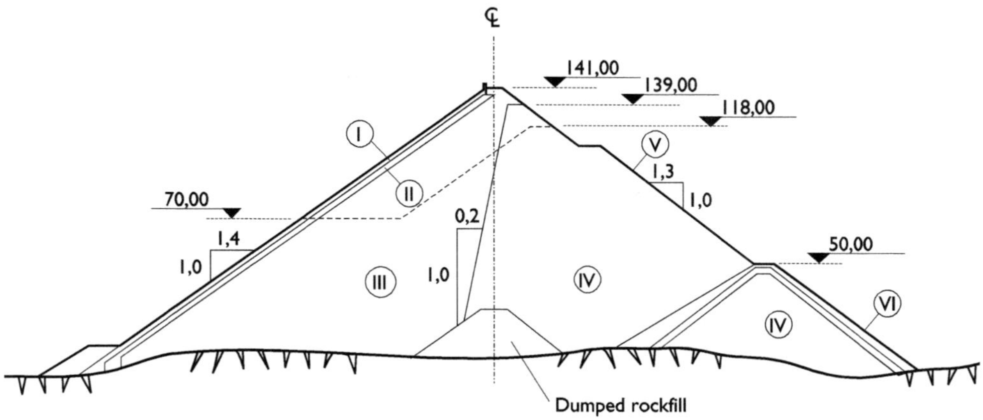

Figure 3.31 Xingó Dam, cross-section.

Table 3.17 Xingó Dam materials.

Material	*Description*	*Zone*	*Placing*	*Compaction*
I	Rock and saprolite with fines	2B	0.40 m layers, $\varnothing_{max.}$ 0.10 m	6 passes of 10 t vibratory roller Face compaction (Gradall use)
II	Coarse transition $\varnothing_{max.}$ 0.40	3A	0.40 m layers	6 passes of 10 t vibratory roller
III	Sound rockfill $\varnothing_{max.}$ 1.0 m	3B	1.0 m layers	6 passes of 10 t vibratory roller with water addition 150 ℓ/m³
IV	Downstream rockfill	3C	2.0 m layers	4 passes of 10 t vibratory roller, without water
V	Downstream slope protection	4	Arranged 1.0 m block	–
VI	CCR protection	–	0.30 m layers	4 passes of 10 t vibratory roller

- Deformability modulus at Zone II (3A): 40 MPa
- Settlement at Zone III (3B): 170 cm (at higher section of dam)
- Settlement at Zone IV (3C): 290 cm (at center point)
- Vertical compressibility modulus at Zone III (3B): 32 MPa
- Vertical compressibility modulus at Zone IV (3C): 20 MPa

During construction some fissures appeared in the proximity of the left abutment, and were treated before the construction of the slab.

The reservoir filling started in June 10, 1994 at El. 40 m, reaching El. 120 m with a very high average speed of above 0.50 m/hour.

The maximum displacement of the slab was 0.30 m and the largest settlement of the benchmarks on the crest surface reached a similar value.

Figure 3.32 Xingó Dam. (See colour plate section).

Table 3.18 Materials grading and in situ density.

Zone	*% passing #4*	*% passing #1″*	*% passing #200*	*tf/m³*
I	35–60		4–12	2.3
II		35–70	3–8	2.25
III	up to 23	up to 40	3	2.15
IV	7–38	15–60	2–7	2.1

- Concrete face: 16 panels, 16 m wide (135,000 sq. m)
- Variable thickness per: 0.3 + 0.002 H
- Min. thickness: 0.30 next to parapet wall 0.70 next to plinth at max height section
- Slab reinforcement ratio: 0,4% in both directions
- Concrete strength: fck = 15 MPa at 28 days
- Cement consumption 204 kg/cu. m
- Pozzolan consumption: 43 kg/cu. m
- Slipping molds: 1.6 m/hour

During operation, the displacements developed in a normal pattern until September 1995, when the measured settlement cells indicated accelerated deformation.

The leakage observed on the left abutment might have apparently saturated the rockfill causing the increase in settlement.

On the slab at left the abutment, fissures and cracks were observed during underwater investigations.

The treatment consisted of placing fine silty sand in bags over the places where suction was detected.

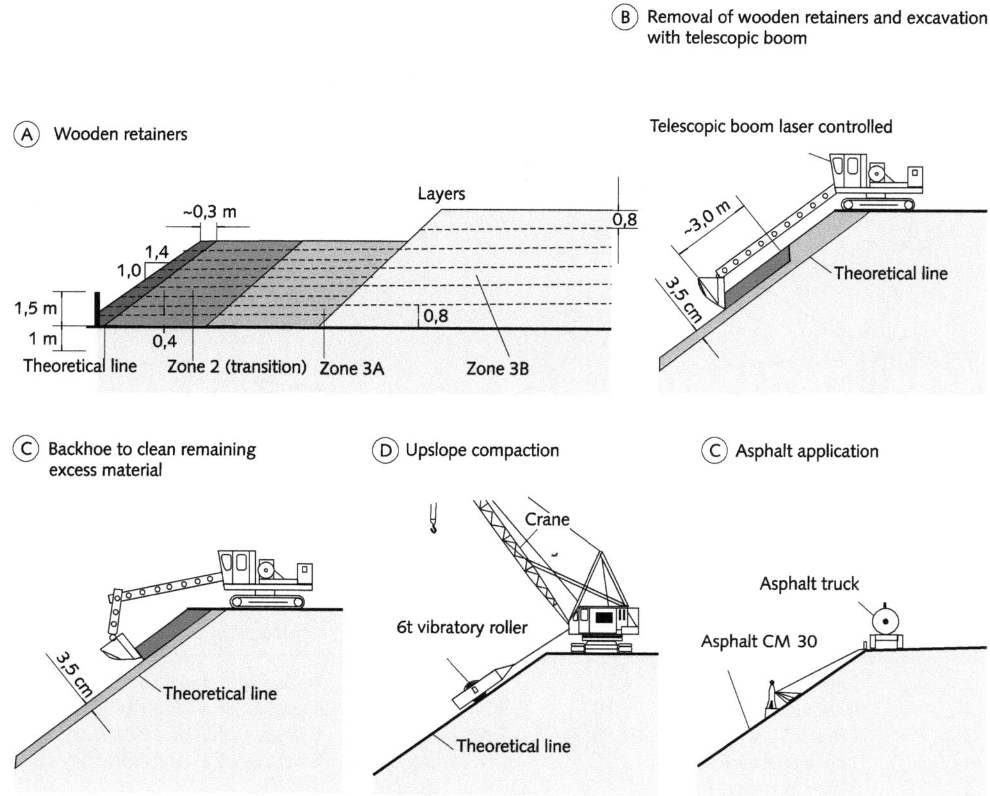

Figure 3.33 Material placing using a Gradall excavator.

The initial leakage of 200 ℓ/s was reduced to 135 ℓ/s as a result of this treatment. In 2003 the leakage increased again, stabilizing eventually at an ultimate value of 175 ℓ/s in 2005. The performance of the dam is satisfactory (Silva, Casarin & Souza, 1999).

3.3.16 Pichi Picún Leufú (Argentina, 1995)

Figure 3.34 shows the position of each zone of Pichi Picún Leufú Dam. Table 3.19 presents the materials used in the construction of the dam.

Main characteristics: H – 50 m; L – 1.100 m; L/H – 22; A/H^2 – 36 aprox.; type of material – gravels; rockfill volume – 1,400,000 m^3.

Pichi Picún Leufú (PPL) Dam was built on the Limay River, in Patagonia, Argentina, in 1999. It's a dam of compacted gravels with an upstream slope of 1.5(H):1.0(V) and a similar slope downstream. It was built over a berm of gravel that was already in place. As indicated on the original design, this berm has been conceived as the dam's central core (Machado et al., 1993).

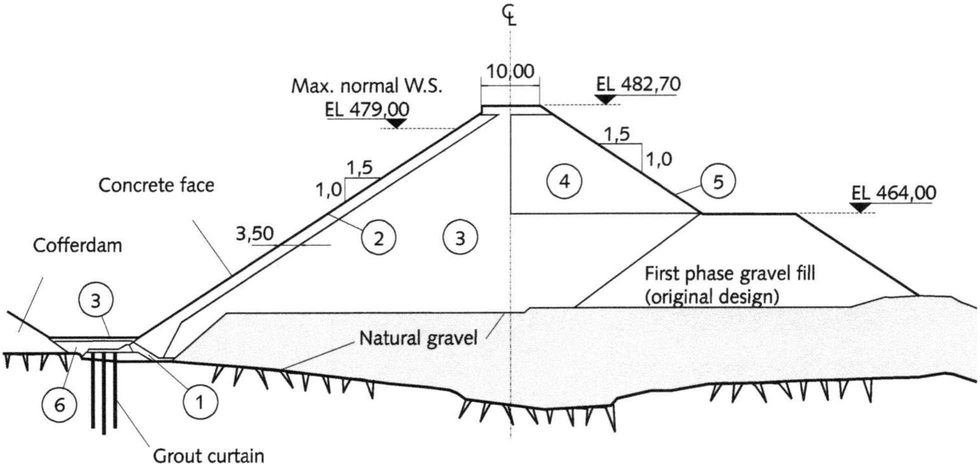

Figure 3.34 Pichi Picún Leufú Dam.

Table 3.19 Pichi Picún Leufú Dam materials.

Material	*Designation*	*Zone*	*Placing*	*Compaction*
1	Fine filter	2A	0.20 m layers	Manually compacted
2	Processed transition	2B	0.30 m layers	6 passes of a 10 t vibratory roller
3	Natural gravels	3B	0.6 m layers	6 passes of a 10 t vibratory roller
4	Natural gravels	3C	1.2 m layers	6 passes of a 10 t vibratory roller
5	Slope protection	4	Coarse boulders	–
6	Silt		0.2 m layers	–

The 2B material was processed from natural gravels with a 5 cm maximum size, 45% sand content and with 5% of fines passing a Nr. 200 sieve. The material was placed in 0.30 m layers and compacted with six passes of a 10 t vibratory roller after being slightly watered.

In addition, the upstream slope was compacted in the up slope direction with six passes of a 6 t vibratory roller, stabilizing the face with cutback asphalt.

The 3B and 3C materials were placed in 0.60 m layers, and the 3C material was compacted every two layers (1.20 m) with six passes of a 10 t vibratory roller after wetting.

The 3B and 3C materials were natural gravels with a maximum size of 25 cm, 33% sand content and a low percentage of fines passing Nr. 200 sieve (<2%).

The compactness of these materials was excellent, with densities ranging from 2.23 to 2.38 t/m^3 and a relative density of 102%.

The plinth had a width of 6 m reduced to 4 m on the left abutment. The criteria used to calculate the plinth dimensions were based on an evaluation of the foundation with regards to the different degrees of erodibility, as shown in Table 3.20.

The reservoir filling was slow and the measured leakage very low, at about 13.5 ℓ/s. The slab maximum displacement was less than 2 cm. The dam stands as an example of the excellent performance that can be achieved with compacted gravels.

Table 3.20 Design critera for the plinth (Marques Filho et al., 1999).

A	*B*	*C*	*D (%)*	*E*	*F*	*G*	*H*
I	Non erodible	1/18	>70	I–II	1–2	<1	1
II	Low erodibility	1/12	50–70	II–III	2–3	1–2	2
III	Moderate erodibility	1/6	30–50	III–IV	3–4	2–4	3
IV	Very high erodibility	1/3	0–30	IV–V	4–5	>4	4

A – Foundation type (I–IV); B – Erodibility; C – Gradient; D – RQD (%); E – Weathering: I – sound rock; IV – decomposed rock; F – Consistency: 1 – very hard rock; 6 – friable rock; G – Number of weathered discontinuities along 10 cm of length; H – Excavation classes: 1 – Blasting; 2 – Heavy ripper and blasting; 3 – Excavation with light ripper; 4 – Excavated with a very heavy bulldozer blade.

3.3.17 Itá (Brazil, 1999)

Figure 3.35 shows the position of each zone of Itá Dam. Table 3.21 presents the materials used in the construction of the dam.

Main characteristics: H – 125 m; L – 881 m; L/H – 7.05; A/H^2 – 7.04; type of material – basalt. The proposed inclinations were 1.3(H):1.0(V) for both upstream and downstream slopes with 1.2H:1.0V between berms on the downstream slope (access berm incorporated). The rockfill volume is 8,900,000 m^3 of compacted basalt.

Itá Dam (Fig. 3.36) was built on the Uruguai River, between the states of Santa Catarina and Rio Grande do Sul in southern Brazil, in 2000 (Antunes Sobrinho et al., 2000).

- Concrete face area: 110,000 m^2
- Concrete volume: 46,000 m^3
- Concrete face: 57 panels 16 m wide
- Variable thickness of: e = 0.30 + 0.002 H
- Min. thickness: 0.30 m
- Reinforcement ratio: 0.40% in vertically and 0.30% horizontally
- Concrete strength: 21 MPa at 90th day

The 2B material was processed to a maximum size of 0.10 m, the sand content was 25% and the percentage of fines passing Nr. 200 sieve was 1%. These transitions processed from basalt with sand content below Sherard's limits are typical of basalts that do not produce fines.

The 3A material was obtained from the quarries. It was more fragmented rock and did not require any special processing, although processing might be needed at another dam.

Zone 2B was compacted in 0.40 m layers with four passes of a 9 t vibratory roller. Zone 3B was compacted in 0.80 m layers, with 200 ℓ/m^3 of water and with four passes of the same roller.

Zone 3C was spread in 1.60 m layers and compacted like zone 3B, but with no watering.

The foundation of the dam was severely affected by the presence of very thick soil on the abutments. It was decided that the plinth should be founded on rock, which required deep excavation in places. High walls were founded and stabilized with anchors to the rock.

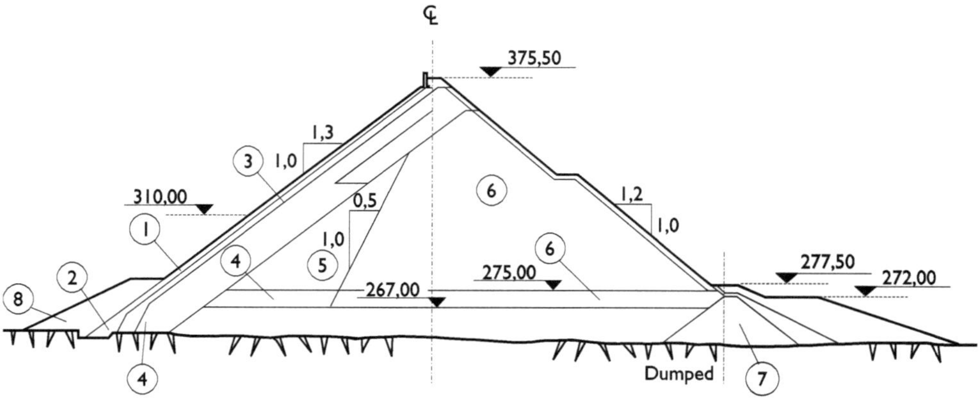

Figure 3.35 Itá Dam.

Table 3.21 Itá Dam materials.

Material	*Designation*	*Description*	*Zone*	*Placing*	*Compaction*
1	Transition	Max. size 0.10 m	2B	0.40 m layers	4 passes of 9 t vibratory roller
2		Under perimetric joint	2A	0.20 m layers	Vibratory plate
3	Transition	Fine rockfill	3A	0.40 m layers	4 passes of 9 t vibratory roller
4	Main rockfill	70% sound basalt	3B	0.80 m layers	4 passes of 9 t vibratory roller and 200 ℓ/m³ water
5	Rockfill	Vesicular basalt, breccia	3B′	0.80 m layers	4 passes of 9 t vibratory roller and 200 ℓ/m³ water
6	Downstream rockfill	Breccia, vesicular and sound basalt	3C	1.60 m layers	4 passes of 9 t vibratory roller no water
7	Rockfill	70% sound basalt	3B″	Dumped	–
8	Soil	Soil and random	1A Random	Placed	Construction equipment

The upstream zone (one third of the base) was cleaned down to the weathered rock in 50% of its surface, exposing portions of soil or saprolite that were left in place. In the remaining two-thirds of the base, the dam was founded on saprolite with SPT above 15 blows.

A distinctive feature of Itá when compared to other dams of its kind was the construction of a priority section downstream. This provides as a protection from floods with a recurrence interval of 500 years and it was built to circumvent the obstacles forseen in completing the plinth before the diversion.

Itá Dam was the first dam to feature an extruded curb that simplified the construction of zone 2B material and protects of the upstream slope, avoiding the traditional compaction of this slope.

Figure 3.37 shows the extruded curb.

Figure 3.36 Itá Dam – aerial view. (See colour plate section).

Figure 3.37 View of upstream slope protected by the extruded curb, Itá Dam. (See colour plate section).

During the building of the dam some improved construction methods were introduced in comparison those used in the construction of earlier Brazilian dams (Tsunoda et al., 1999):

- The confinement of the transition zone, provided by the extruded curb and by placing material with the aid of a metallic mold, reduced segregation along the slope.

- The performance of the extruded curb was excellent. It served as an adequate protection during heavy rains and provided additional security to labor and equipment involved in the executive phases execution. Upstream displacements of 3 to 8 cm were observed at the lower part of the dam, but these did not affect the construction of the slab. On the abutments, tension fissures, typical of CFRDs, confirmed the appropriateness of the tension joint selection.
- The slipping forms pulled up by jacks in cables, similar to post-tension cables, led to good productivity results. This process, used for the first time at Itá, is comparable to the method of hydraulic jacks or synchronized winches.

The characteristics of the extruded curb, which Barry Cooke called in one of his classic memoranda the "Itá Method", has been copied around the world (Resende & Materón, 2000).

The impounding of the reservoir started in February 2000, and it reached its maximum level in May of that year.

Leakages occurred on both margins, reaching values of 1,700 ℓ/s. Underwater investigations revealed cracks on the slab at a depth of 90 m; sub-parallel fissures along the plinth were also observed. They were treated by dumping sand mixed with fines, after which the leakage dropped to 380 ℓ/s.

The slab had a deflection of 45 cm, and the crest had a downward displacement of 55 cm.

Leakage at the right abutment probably reached and weakened the downstream rockfill, which suffered new settlement after saturation and ruptures of the slab, as happened at Xingó.

Last measurements (2008) indicated a remaining leakage of 80 ℓ/s and the performance of the dam can be considered normal tending towards to stabilization.

3.3.18 Machadinho (Brazil, 2002)

Figure 3.38 shows the position of each zone of Machadinho Dam. Table 3.22 presents the materials used in the construction of the dam.

Main characteristics: H – 125 m; L – 700 m; L/H – 5.6; A/H^2 – 4.93; type of material – basalt. Upstream and downstream average slopes were 1.3H:1.0V, with 1.2H:1.0V between berms with incorporated access berm. Rockfill volume is 6,500,000 m^3 (Mauro et al., 1999).

Machadinho Dam (Fig. 3.39) was built on the Pelotas River, on the border between the states of Santa Catarina and Rio Grande do Sul in southern Brazil, with a design similar to Itá Dam. The works ended in 2002.

To support a 500 years flood, at the beginning of the rainy season after diversion, the priority section was built downstream and the rockfill was covered with a fine rockfill equivalent to 2B material to break the saturation line.

- Concrete face area: 77,000 m^2
- Concrete face: panels 16 m wide
- Variable thickness of: e = 0.30 + 0.002 H
- Min. thickness: 0.30 m
- Reinforcement ratio: 0.40% to 0.60% in both directions

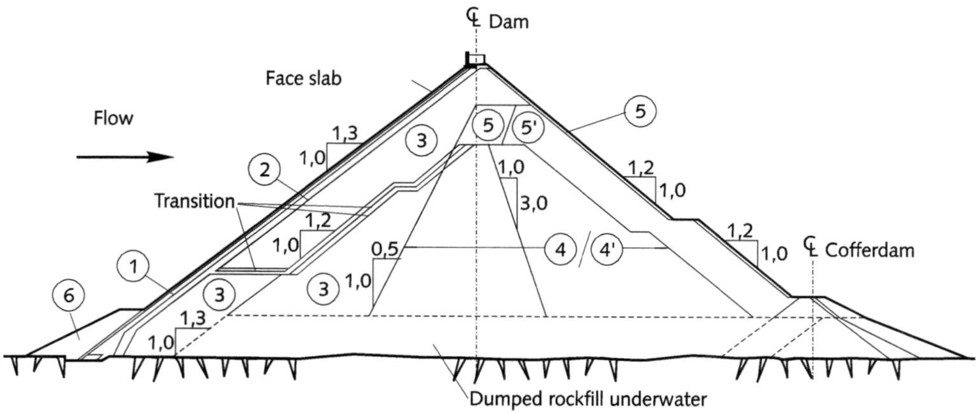

Figure 3.38 Machadinho Dam.

Table 3.22 Machadinho Dam materials.

Material	*Designation*	*Specification*	*Zone*	*Placing*	*Compaction*	*Unconfined Compressive Strength (UCS)*
1	Fine transition	Processed basalt $Ø_{max.}$ 100 mm	2B	0.40 m layers	Vibratory roller 9 t, 4 passes	UCS over 50 MPa
2	Coarse transition	Basalt $Ø_{max.}$ 400 mm	3A	0.40 m layers	Vibratory roller 9 t, 4 passes	70% having UCS over 50 MPa
3	Main upstream rockfill	Basalt $Ø_{max.}$ 800 mm	3B	0.80 m layers	6 passes of 12 t vibratory roller – water 200 ℓ/m³	70% having UCS over 50 MPa
4–4′	Downstream rockfill	Basalt or breccias	3C 3C′	1.60 m layers (4) 1.20 m layers (4′)	Vibratory roller 9 t, 4 passes	UCS over 25 MPa
5–5′		Basalt or breccias	3D 3D′	1.60 m layers (5) 1.20 m layers (5′)	Vibratory roller 9 t, 4 passes	70% having UCS over 40 MPa
6	Soil	Saprolite soil	I	0.20 m e 0.30 m layers	Construction equipment	–

The transition 2B material was processed and compacted in 0.40 m layers with the use of the extruded curb developed earlier at Itá. This material was compacted with four passes of a 9 t vibratory roller.

The 3B material was spread in 0.80 m layers with six passes of a vibratory roller of 9 t, and the 3C material was spread with the same intensity of compaction but in layers of 1.60 m.

The plinth construction in both abutments presented problems with foundation stability. The rhyodacite rock has acidic composition and has vertical and inclined

Figure 3.39 Machadinho Dam works before diversion. (See colour plate section).

fractures. It is present in blocks surrounded by soil. The weathering process of this rock was enhanced by water infiltration through vertical fractures to the basalt.

The highly weathered rock with soil patches was inadequate to support the foundation to the plinth. Removing the material made it necessary to build high walls to support the plinth (Mauro et al., 2007). The stability of these walls up to 17 m in height was guaranteed by anchors, tendons and internal drainage to reduce the uplift pressure induced by the reservoir.

The dam performance is adequate, but leakage is relatively high (>600 ℓ/s) probably due to the drains within the wall foundations in the abutments that were built to reduce uplift pressures.

3.3.19 Antamina (Peru, 2002)

Figure 3.40 shows the position of each zone of Antamina Dam. Table 3.23 presents the materials used in the construction of the dam.

Main characteristics: H – 109 m (starter dam) up to 210 m (following completion of the mine exploration); L – 1.030 m; L/H – 4.90; A/H^2 – 5.08; type of material – limestone.

Antamina Dam (Fig. 3.41) will be 210 m high at its completion. The dam is being built to retain the tailings of a copper mine excavation in Peru. Initially a CFRD was built as a starter dam to retain the water from the mining process, and afterwards the dam will be increased up to a final height of 210 m (Marulanda, Amaya & Milian, 2000).

The slopes are 1.4(H):1.0(V), but on the downstream area, access roads to the trucks are being built and, as a result of the berms, the average slope is flatter.

The 2B material for this dam was specially processed with a maximum size of 7.5 cm, a sand content ranging between 40–55% and 8% of fines passing

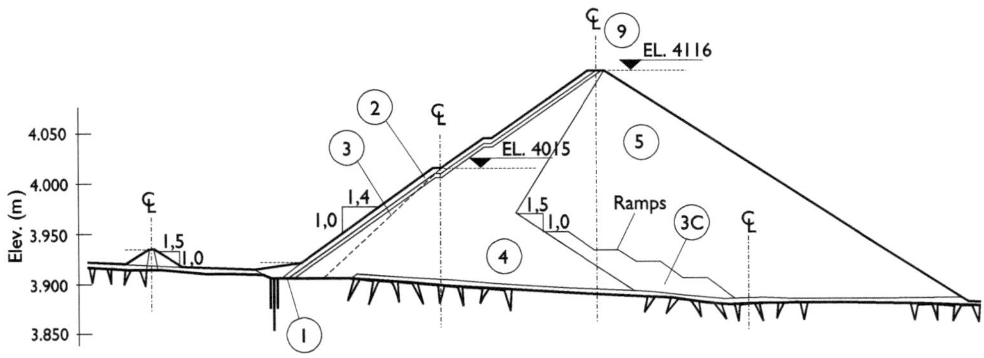

Figure 3.40 Antamina Dam, cross-section.

Table 3.23 Antamina Dam materials.

Material	*Designation*	*Zone*	*Placing*	*Compaction*
1	Processed sand and gravel, $\varnothing_{max.}$ 9 mm, under perimetric joint	2A	50 cm layers	2 passes of 5 t vibratory roller
2	Processed gravel and sand, $\varnothing_{max.}$ 75 mm	2B	50 cm layers	6 passes of 10 t vibratory roller
3 Transition zone	Processed boulders, cobbles, gravel and sand, $\varnothing_{max.}$ 300 mm	3A	50 cm layers	6 passes of 10 t vibratory roller
4	Rock from mine pit	3B	1.0 m layers thickness	6 passes of 10 t vibratory roller – water 250 ℓ/m³
5	Rock from mine pit	3C	2.0 m layers	6 passes of 10 t vibratory roller – water 250 ℓ/m³

Figure 3.41 Antamina Dam – detail. (See colour plate section).

Nr. 200 sieve. To avoid fines migration into the tailings, the width of this cushion zone (8 m) is double that usually applied in this type of dam.

An extruded curb 0.50 m in height was used to confine the 2B material, which was compacted in 0.50 m layers with six passes of a 10 t vibratory roller.

Materials 3A, 3B and 3C were compacted respectively in 0.50 m, 1.0 m and 2.0 m layers with six passes of a vibratory roller. The slab was built with a constant width of 0.30 m, increasing to 0.45 m close to the plinth. Reinforcing was 0.35% in both directions. The design took into account that the slab could fail, since its purpose was to work only for the first two years. Afterwards, the tailings deposition and zone 2B would be enough to control potential water flows.

Above the crest of the starter dam, it was considered that the extruded curb plus the 8 m of zone 2B would be sufficient to control the water flow through the dam.

During the process of raising the dam and filling up the reservoir the deformations were considered adequate, but the leakage increased to values above 425 ℓ/s. This was treated by the application of a membrane. When the dam was 160 m high, leakage had dropped to 250 ℓ/s.

The performance of the structure is good, particularly as far as deformations are concerned (Marulanda, Amaya & Millan, 2000).

3.3.20 Itapebi (Brazil, 2003)

Figure 3.42 shows the position of each zone of Itapebi Dam. Table 3.24 presents the materials used in the construction of the dam.

Main characteristics: H – 120 m; L – 583 m; L/H – 4.86; A/H^2 – 4.65; type of material – gneiss, micaschist. The upstream slope is 1.25(H):1.0(V) and the downstream slope is 1.3(H):1.0(V) with an incorporated access road with slope of 1.2(H):1.0(V) above the horizontal berm of access. The dam has a rockfill volume of 3,900,000 m^3 (Fernandez et al., 2007).

Itapebi Dam (Fig. 3.43) was built on Jequitinhonha River, in the state of Bahia, northeast Brazil.

- Concrete face area: 67,000 m^2
- Variable thickness of: $e = 0.30 + 0.002$ H
- Max. thickness: 0.51 m
- Concrete face: 35 panels 16 m

The transition 2B material was processed from gneiss, with a maximum size of 10 cm, sand content of 35–55% and fines below 7% in the Nr. 200 sieve.

This material was compacted with four passes of a 10 t vibratory roller in 0.40 m layers.

The extruded curb was similar to that at Itá, with a height of 0.40 m.

Materials 3B and 3C were compacted with the same vibratory roller with six and four passes on layers of 0.80 m and 1.60 m respectively.

In the central zone of the dam, the rockfill was weathered rock placed in layers of 1.60 m and compacted with four passes of a 10 t roller.

The face slab was calculated using the formula:

$$e = 0.30 + 0.002 \text{ H (m)} \quad (3.9)$$

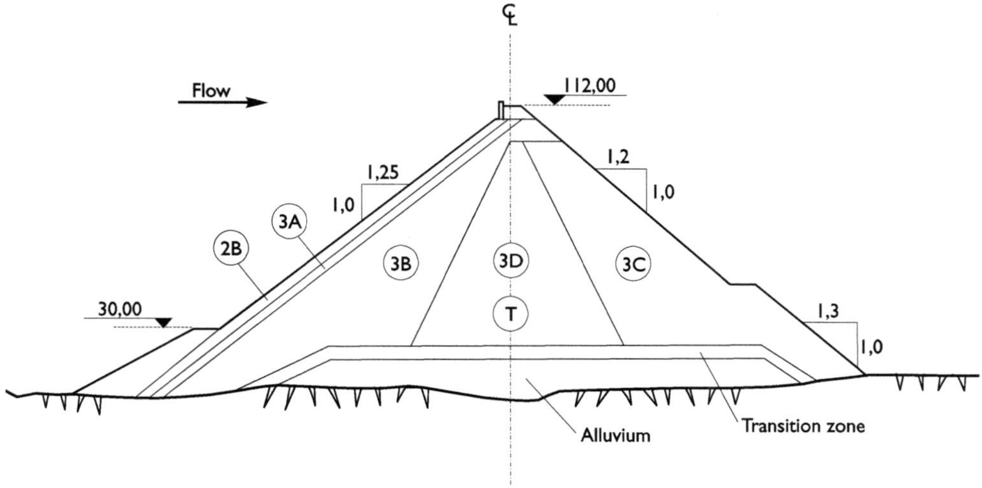

Figure 3.42 Itapebi Dam, cross-section.

Table 3.24 Itapebi Dam materials.

Material	*Specification*	*Zone*	*Placing*	*Compaction*	*Density*
Transition	Processed material $\varnothing_{max.}$ 100 mm	2B	0.40 m layers	4 passes of 9 t vibratory roller	2.15 t/m^3
Coarse transition	$\varnothing_{max.}$ 400 mm	3A	0.40 m layers	4 passes of 9 t vibratory roller	2.15 t/m^3
Upstream rockfill	Sound rock $\varnothing_{max.}$ 0.80	3B	0.80 m layers	6 passes of 9 t vibratory roller – water	2.10 t/m^3
Downstream rockfill	Weathered rock, high fines content (20%)	3C	1.60 m layers	4 passes of 9 t vibratory roller	2.00 t/m^3
Central rockfill	Weathered rock, high fines content (20%)	3D T	1.60 m layers	4 passes of 9 t vibratory roller	1.95 t/m^3

The dam foundation on the abutments was in rock. On the river bed, a 15 m deep sand deposit was left in place, properly protected with filters. Only the first 40 m downstream from the plinth and the 30 m at the downstream slope toe (base) were removed for stability purposes.

The rock mass at Itapebi is basically composed of gneiss granite layers, interbedded with biotite schist and amphibolite of low strength (bx/af).

In early July 2001, a slope failure occurred on the left abutment, along some 200 m through a bx/af layer. It was necessary to build walls to support the plinth. In addition, new galleries were built and filled with concrete (shear keys) to guarantee the stability of the left abutment and the spillway.

Back analyses of stability revealed that the bx/af layers had a shear strength with friction angle below 12° and Nr. cohesion as could be observed on unstable blocks.

Figure 3.43 Itapebi Dam. (See colour plate section).

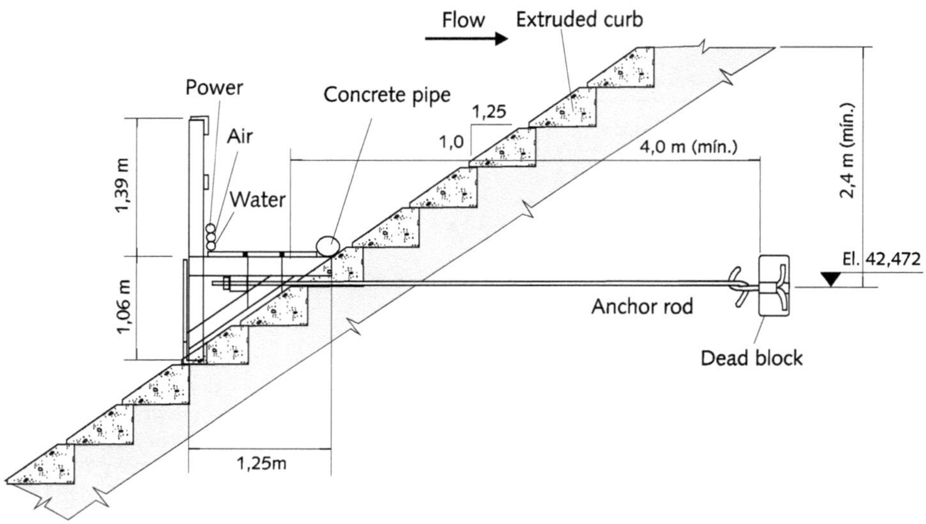

Figure 3.44 Construction method allowed the slab to be constructed simultaneously with the upstream rockfill.

From a construction point of view, at Itapebi a new construction method was used for the first time, which allowed simultaneous construction of the concrete upstream slab together with the rockfill raising as can be seen in Figures 3.44 and 3.45 (Materón & Resende, 2001).

A platform was built. This platform allowed the lower part of the slab to be constructed at the same time as the rockfill was being placed, so keeping the project up with the schedule (Fig. 3.46).

The reservoir filling started on December 10, 2002, and maximum water level was reached on January 28, 2003, only 49 days later.

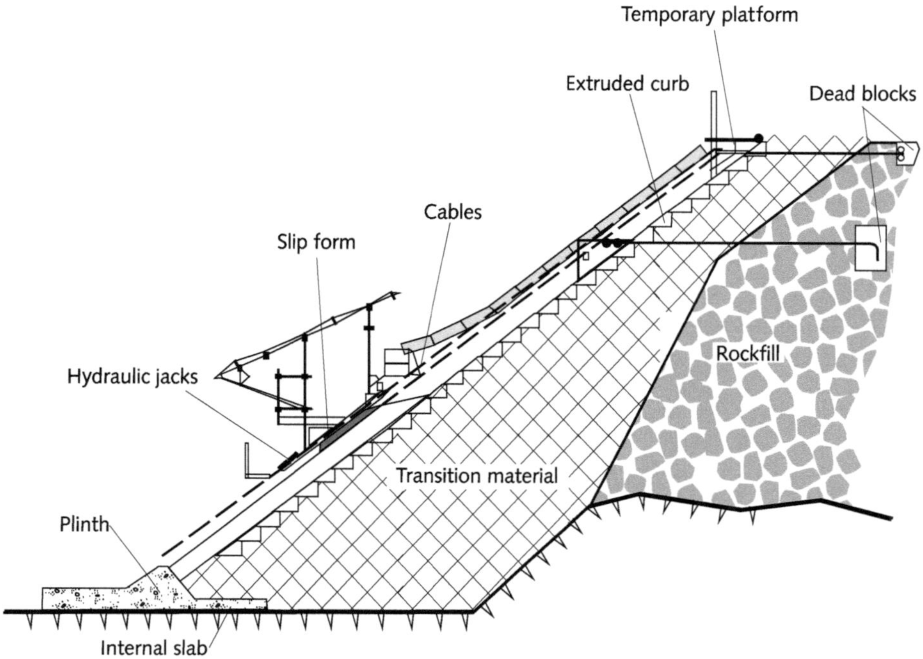

Figure 3.45 Itapebi Dam: the construction method allowed the construction of the slab simultaneously with the upstream rockfill.

Figure 3.46 Itapebi Dam: slab partial view, built with the platform incorporated on the rockfill. (See colour plate section).

When the reservoir was at El. 105 (95% H), the settlement cells showed an increasing rate of settlement. Apparently fissures in the slab increased the degree of rockfill saturation, leading to new fissures and cracks in the slab.

Underwater observations detected openings in the slab. Leakage reached values close to 1,000 ℓ/s. After treatment with fine soil, the leakage dropped to 127 ℓ/s. Today, the actual leakage is 50 ℓ/s. The performance of the dam is normal.

3.3.21 Quebra-Queixo (Brazil, 2003)

Figure 3.47 shows the position of each zone of Quebra-Queixo Dam. Table 3.25 presents the materials used in the construction of the dam.

Main characteristics: *H* – 75 m; *L* – 672 m; *L*/*H* – 9; *A*/H^2; type of material – basalt. Upstream slope is 1.25(H):1.0(V) and the downstream slope is 1.20(H):1.0(V). The volume is 1200,000 m^3.

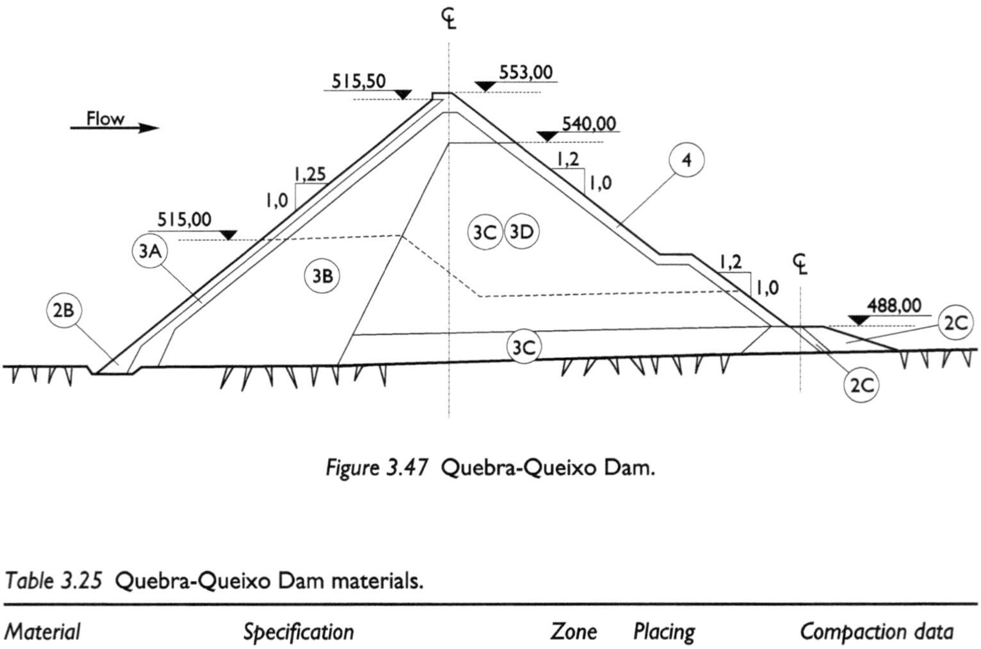

Figure 3.47 Quebra-Queixo Dam.

Table 3.25 Quebra-Queixo Dam materials.

Material	*Specification*	*Zone*	*Placing*	*Compaction data*
Transition	Processed basalt $Ø_{max.}$ 100 mm	2B	0,40 m layers	6 passes of 9 t, vibratory roller
Fine rockfill	$Ø_{max.}$ 400 mm, sound basalt	3A	0,40 m layers	6 passes of 9 t, vibratory roller
Upstream rockfill	$Ø_{max.}$ 0.80 m, at least 70% basalt or dense dacite	3B	0,80 m layers	6 passes of 9 t, vibratory roller
Downstream rockfill	Basalt, breccias, $Ø_{max.}$ 1.6 m	3D	1,60 m layers	6 passes of 9 t, vibratory roller
Downstream slope	Sound basalt rodacite, at least 70%, $Ø_{max.}$ 1.60 m	4	1,60 m layers	–

Quebra-Queixo Dam (Fig. 3.48) was built on the Chapecó River, a tributary of the Uruguai River in the state of Santa Catarina, southern Brazil.

- Concrete face: panels 16 m wide
- Variable thickness of: e = 0.30 + 0.002 H (m)
- Max. thickness: 0.45 m.
- Min. thickness: 0.30 m.

The 2B material was processed from sound basalt to a maximum size of 10 cm, compacted in 0.40 m layers. Compaction of material 3A was also in 0.40m layers, with a maximum block size of 40 cm of sound basalt.

Material 3B was compacted in 0.80 m layers with six passes of a 9 t vibratory roller. The void ratio was of the order of 0.30. The material 3C was compacted in 1.60 m layers with the same roller and same number of passes.

The upstream face was protected by an extruded curb (similar to Itá) of 0.40 m in height, with a cement content of 55 kg/m^3. The rate of transition placement, together with the curb element, was about 1 m/day or two layers/day.

On the abutments the rhyodacite presented a thick soil layer rich in decametric boulders, which affected the excavation and the alignment of the plinth. This resulted in the construction of walls up to 8 m in height to support the plinth.

The excavation of the dam foundation down to the rock surface was extended along one third of the dam base in the upstream zone. The remaining two-thirds of the foundations were considered adequate, except for a small strip close to the downstream toe that was also excavated to the rock.

The proposed layout allowed a gross head of 122 m, almost 50 m more than the dam height.

The dam plant was built in a record time of 26 months.

The performance is quite good and there is practically no leakage.

Figure 3.48 Quebra-Queixo Dam – lateral view. (See colour plate section).

3.3.22 Barra Grande (Brazil, 2005)

Figure 3.49 shows the position of each zone of Barra Grande Dam. Table 3.26 presents the materials used in the construction of the dam.

Main characteristics: *H* – 185 m; *L* – 665 m; *L*/*H* – 3.59; *A*/H^2 – 3.16; type of material – basalt. The slopes are 1.30(H):1.0(V), both upstream and downstream, and the dam has a volume of 12,000,000 m^3.

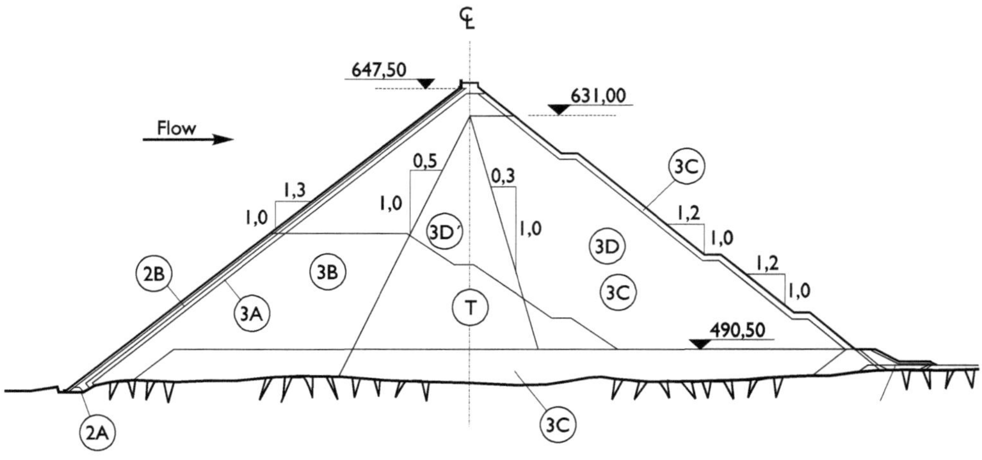

Figure 3.49 Barra Grande Dam, cross-section.

Table 3.26 Barra Grande Dam materials.

Material	*Classification*	*Zone*	*Placing*	*Compaction*	*Unconfined Compressive Strength (UCS)*
Fine transition under perimetric joint	Sound basalt $Ø_{max.}$ 25 mm	2A	0.50 m layers	Compacted by 9 t vibratory roller	
Fine transition	Dense basalt $Ø_{max.}$ 100 mm	2B	0.50 m layers	Compacted by 9 t vibratory roller	
Fine rockfill	Dense basalt $Ø_{max.}$ 500 mm	3A	0.50 m layers	Compacted by 9 t vibratory roller	–
Upstream rockfill	$Ø_{max.}$ 1.0 m min. 70% of dense basalt or rhyodacite	3B	1.0 m layers	Compacted by 12 t vibratory roller – water 200 ℓ/m^3 – 6 passes	70% with UCS over 50 MPa
Downstream rockfill and river bed	$Ø_{max.}$ 1.60 m	3C	1.60 m layers	Compacted by 12 t vibratory roller 4 passes	70% with UCS over 40 MPa
Downstream rockfill		3D			70% with UCS over 25 MPa
Central zone rockfill	$Ø_{max.}$ 1.0 m min. 70% of dense basalt or rhyodacite	3D′ T	1.0 m layers	Compacted by 12 t vibratory roller – water 200 ℓ/m^3 – 6 passes	70% with UCS over 25 MPa

Barra Grande Dam (Fig. 3.50) was built on the Pelotas River, at the border of the States Rio Grande do Sul and Santa Catarina, southern Brazil.

- Concrete face: panels 16 m wide
- Variable thickness of: e = 0.30 + 0.002 H (for H < 100 m) and e = 0.005 H (for H > 100 m)
- Concrete face area: 108,000 m^2
- Reinforcement ratio: 0.4% in both directions

The transition material 2B was processed from basalt to a maximum size of 10 cm and had a low percentage of sand, typical of basalts.

Transition 2B was compacted in 0.50 m layers, with four passes of a 9 t vibratory roller.

Basalt material 3B had a maximum size of 0.80 m and was compacted in 1.0 m layers with six passes of a 12 t vibratory roller with 200 ℓ/m^3 of water.

Material 3C had a maximum size of 1.60 m and was compacted in 1.60 m layers with four passes of a 12 t vibratory roller.

The slab width was determined by the formula:

$$e = 0.30 + 0.002\ H\ (m) \quad \text{for H up to 100 m}$$
$$e = 0.005\ H\ (m) \quad \text{for } H \geq 100 \text{ m} \tag{3.10}$$

A double reinforcement was adopted, with a strip up to 20 m from the plinth with 0.5% of the area in both directions.

Figure 3.50 Barra Grande Dam. (See colour plate section).

In other areas of the slab, the reinforcement was 0.4% in the vertical direction and 0.3% in the horizontal direction.

When the reservoir was 93% full on September 22, 2005, a failure was observed at the central compressive joints and a leakage of the order of 1,000 ℓ/s increased to 1,300 ℓ/s. Treatment with fine material reduced the leakage to 500 ℓ/s, but later increased again to approximately 900 ℓ/s.

Central joints 19 and 20 failed. Underwater investigations have shown that the failure went down to a depth of 90~100 m deep (Borges, Pereira & Antunes, 2007).

A space was opened in the central slabs above El. 630 to release the compression pressure.

The performance of the dam, from the point of view of internal deformation (displacements), is normal and stabilized, but the leakage has not been treated.

The maximum displacement measured at the slab is 50 cm, and the leakage remains of the order of 935 ℓ/s without further treatment .

3.3.23 Hengshan (China, 1992)

Figure 3.51 shows the position of each zone of Hengshan Dam. Table 3.27 presents the materials used in the construction of the dam.

Main characteristics: H – 70.2 m; a CFRD over an old gravel dam with impervious core.

Hengshan Dam has historical interest because it is a rockfill dam with an impervious core which has been raised, with a concrete face extended over the original dam (Hong Tao, 1993).

The dam was built on the Xianjlang River, in Fenhua, China, with the raised section holds the water supply for the district of Ningbo. The upstream slope is 1.4(H):1.0(V) and the downstream slope is 1.3(H):1.0(V).

The reservoir of Hengsban was built in 1966 with the lower dam, with an impervious core 49 m in height and rockfill shoulders. The reservoir had a capacity of 50,000,000 m^3.

In 1987 the dam was raised and rebuilt as a concrete face rockfill dam by installing a diaphragm wall inside the core, 72 m in depth and 0.80 m wide, to form a new reservoir with a capacity of 112,000,000 m^3.

In 2007, during a visit to the site we were informed of the characteristics of the dam (Materón, 2007).

- The plinth is 4.40 m wide, and was built over the crest of the old dam after the conclusion of the diaphragm wall. The concrete in the diaphragm wall (0.80 m thick) was not plastic concrete, but one with a low resistance (10 MPa).
- The plinth was connected to the diaphragm using cooper waterstops. The 2B material was processed and spread in 0.40 m layers compacted by a 13.5 t vibratory roller with eight passes.
- Materials 3B and 3C were compacted in layers of 0.80 m with the same vibratory roller and the same number of passes.

The rockfill was controlled by density tests, with values in the range of 2.10–2.13 t/m^3.

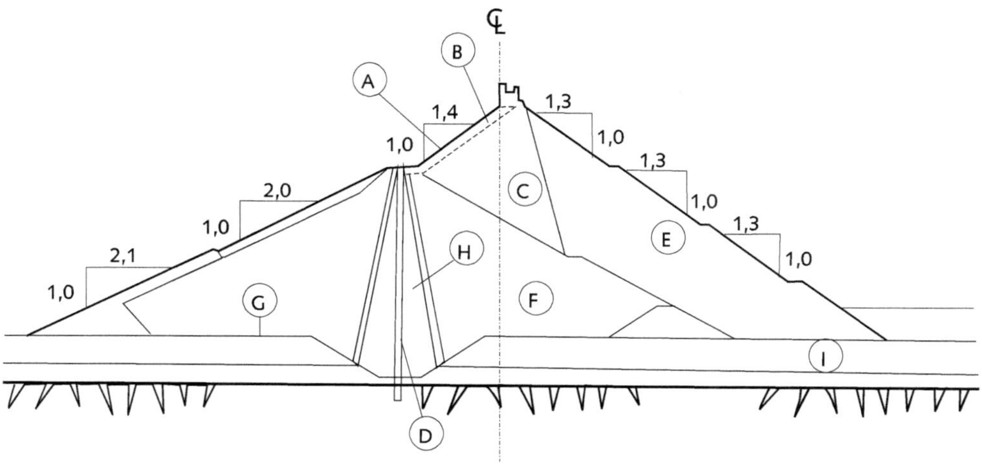

Figure 3.51 Hengshan Dam, cross-section.

Table 3.27 Hengshan Dam materials.

Material	*Description*	*Zone*	*Placing*	*Compaction*
B	Transition	2B	0.40 m layers	8 passes of 13.5 t vibratory roller
C	Rockfill	3B	0.80 m layers	8 passes of 13.5 t vibratory roller
D	Concrete diaphragm	–	Concrete 0.80 m	10 MPa low strength concrete
E	Downstream rockfill	3C	0.80 m layers	8 passes of 13.5 t vibratory roller
F	Gravels	–	Existing fill	–
G	Gravels	–	Existing fill	–
H	Impervious core	–	0.30 m layers	Compacted
I	River bed	–	Natural gravel	–

Figure 3.52 Hengshan Dam. (See colour plate section).

The concrete face was built with a constant thickness of 0.30 m, with a specified strength of 25 MPa at 28 days.

The reinforcement was 0.45% in the vertical direction and 0.35% in the horizontal direction. Joints were placed at every 12 m with copper waterstops.

The plinth on the abutments had a constant width of 4 m and was treated by grouting. The grouting had primary holes drilled at every 8 m, secondary ones at every 4 m, and tertiary ones at every 2 m.

Two rolls of holes at every 2 m were built as consolidation grouting.

The reservoir water level fluctuates every year, from a level below the base of the plinth to one that covers the concrete face and reaches the parapet wall. The performance is excellent

Figure 3.52 shows Hengsham Dam.

3.3.24 Salvajina (Colombia, 1983)

Figure 3.53 shows the position of each zone of Salvajina Dam. Table 3.28 presents the materials used in the construction of the dam.

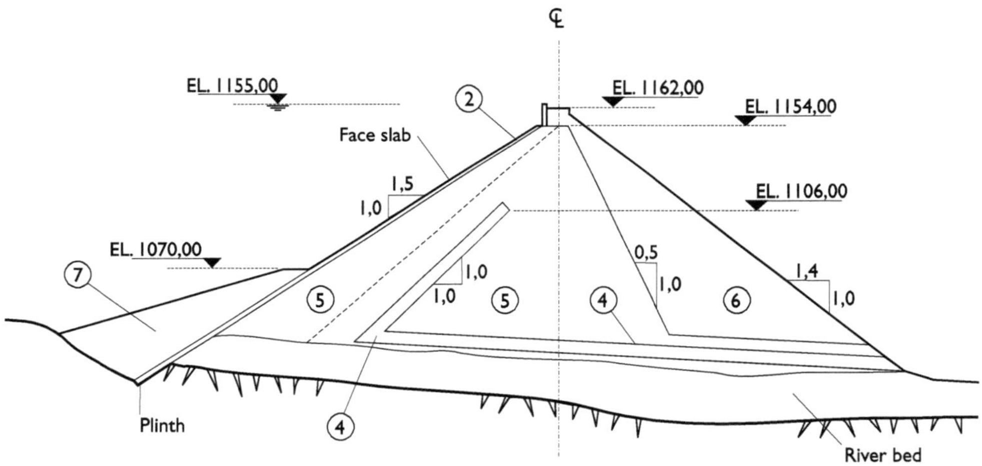

Figure 3.53 Salvajina Dam, cross-section.

Table 3.28 Salvajina Dam materials.

Material	*Specification*	*Zone*	*Placing*	*Compaction specification*
2	$\varnothing_{max.}$ 10 cm	2B	45 cm layers	4 passes of 10 t vibratory roller, 8 passes 5 t on the upstream slope
4	$\varnothing_{max.}$ 40 cm	Drain	60 cm layers	4 passes of 10 t vibratory roller
5	Gravel, $\varnothing_{max.}$ 30 cm	3B	60 cm layers	4 passes of 10 t vibratory roller
6	Rockfill, $\varnothing_{max.}$ 60 cm	3C	90 cm layers	6 passes of 10 t vibratory roller
7	Impervious, $\varnothing_{max.}$ 30 cm	I	30 cm layers	Compacted by construction equipment

Main characteristics: H – 148 m; L – 362 m; L/H – 2.44; A/H^2 – 2.62; type of material – gravels and siltstones. The slopes were 1.5(H):1.0(V) upstream (gravels) and 1.3–1.4(H):1.0(V) downstream (rockfill) and the volume is 3,395,000 m^3 (Sierra, Ramirez & Hacelas, 1985).

Salvajina Dam (Fig. 3.54) was built on the Cauca River and is part of the regulation project of the same river under the Corporación Autónoma Regional Del Cauca (CVC) southwest Colombia.

Transition material 2B, with a maximum size of 10 cm, was processed from gravels, with an average sand content of 35%, and less than 5% passing Nr. 200 sieve.

During the placement, weaving of the fill occurred during the passage of the trucks, due to the high water content, phenomenon that had already been seen at the Messochora and Xingó dams (Materón, 1998).

The 2B material was compacted in 45 cm layers, with four passes of a 10 t vibratory roller. The upstream slope was also compacted with eight passes of a 5 t vibratory roller.

In Salvajina a chimney drain was included with uniform grain sizing, with a maximum size of 0.40 m and minimum size of 1″, with practically no fines.

Zones 3B and 3C, with maximum sizes of 0.30 m and 0.60 m, were compacted in layers of 0.60 m and 0.90 m respectively with four passes of a 10 t vibratory roller.

Zone 3C was built with rock from the spillway excavation.

The plinth foundation was determined by the rock classification according to safe hydraulic gradients as set out in Table 3.29.

An interesting feature noted during the construction of the plinth was the foundation on residual soil (saprolites) that developed on the right abutment due to hydrothermal weathering. The plinth was designed with transversal joints at every 8 m to promote an articulated behavior and to avoid differential settlement that could cause ruptures in a conventional plinth (Sierra, Ramirez & Hacelas, 1985).

Figure 3.54 Salvajina Dam.

Table 3.29 Plinth dimensions (Sierra et al., 1985).

Foundation	*Designation*	*Gradient* *Max.*	*Gradient* *Present*	*Width (m)*
Original design	Hard groutable rock	18	–	4–8
I	Competent rock	18	17,5	6–8
II	Highly fractured	9	6,2	15–23
III	Intensely weakened sedimentary	6	3,1	15–18
III	Residual soil saprolite	6	1,3	13–14

Figure 3.55 Salvajina Dam face slab.

During the construction of Salvajina, the 2B material was protected with shotcrete that showed fissures and crushing during the raising of the dam. However, during a period of rain, May 3–4, 1963, the 2B material on the right abutment was eroded, forming a canal 8 m wide and down to a depth of 10 m between El. 1015–1037 (Sierra, Ramirez & Hacelas, 1985), was rebuilt with the dam already high.

This shows the importance of having the constant protection provided by the extruded curb, called the Itá Method by Barry Cooke.

The behavior of Salvajina was excellent, and the maximum face displacement was 10 cm, the maximum settlement of the gravels was 30 cm. The rockfill, with a lower modulus, settled 60 cm. The leakage was 60 ℓ/s.

3.3.25 Puclaro (Chile, 2000)

Figure 3.56 shows the position of each zone of Puclaro Dam. Table 3.30 presents the materials used in the construction of the dam.

Main characteristics: H – 80 m; L – 640 m; L/H – 8; A/H^2 – 10.62; type of material – gravels.

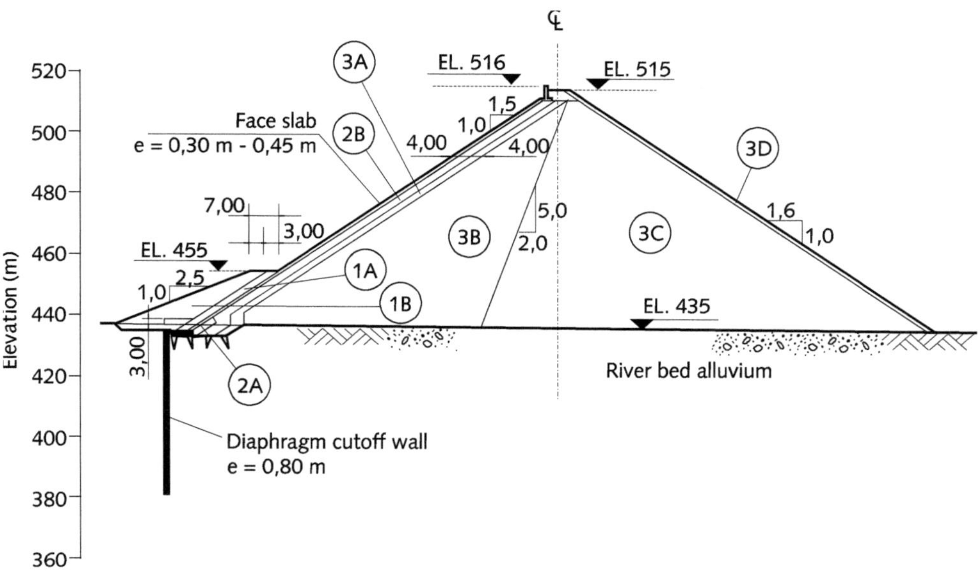

Figure 3.56 Puclaro Dam.

Table 3.30 Puclaro Dam materials.

Material	*Specification*	*Zone*	*Compaction procedure*	*Compaction specification*
Silt	Non cohesive soil, $\varnothing_{max.}$ Nr. 8	I A	20 cm layers	Construction equipment
Random	Random $\varnothing_{max.}$ 12″	I B	40 cm layers	Construction equipment
Fine filter	Sand, $\varnothing_{max.}$ Nr. 4″	2A	20 cm layers	Manual vibratory compactor
Processed transition	Gravel, $\varnothing_{max.}$ 1½″	2B	30 cm layers	4 passes of a 10 t vibratory roller and upstream compaction
Coarse gravels transition	Gravel, $\varnothing_{max.}$ 6″	3A	30 cm layers	4 passes of a 10 t vibratory roller
Upstream gravel	Gravel, $\varnothing_{max.}$ 24″	3B	60 cm layers	4 passes of a 10 t vibratory roller
Downstream gravel	Gravel $\varnothing_{max.}$ 24″	3C	90 cm layers	4 passes of a 10 t vibratory roller
Oversize blocks		3D	–	–

Puclaro Dam was built on the Elqui River, 40 km from La Serena and 500 km to the north of Santiago in Chile. Puclaro has slopes of 1.5(H):1.0(V) upstream and 1.6(H):1.0(V) downstream and a volume of 5,000,000 m^3 (Anguita, Alvarez & Vidal, 1993).

An important characteristic of this dam is that the foundation is on thick alluvial deposits with depths down to 113 m. The plinth is articulated and seepage is controlled by diaphragm walls as have been used on compressible foundations for over 50 years.

Technical references to dams with compressible foundations include Campo Moro II Dam, built with an articulated plinth, in Italy in 1958. Another example, also in Italy, is Zoccolo Dam, completed in 1964 with an asphalt face.

In Puclaro, the diaphragm was built 60 m deep after analysis had shown that coping with water losses was more economical than extending the diaphragm to the rock.

The diaphragm wall was connected to a horizontal articulated plinth that allowed differential movements between this wall and the concrete face.

The transition material 2B was processed from natural gravels with a maximum size of 4.0 cm, in accordance with the grain sizes proposed by Sherard.

Compaction was carried out with a 10 t vibratory roller and four passes on layers of 0.30, 0.60 and 0.90 m for the 2B, 3B and 3C material respectively. The 3A material was also compacted in 0.30 m layers.

Puclaro Dam is located in a seismic zone, where the plates of Nazca and South American converge, accumulating (compressive) efforts generating seismic events in the area.

The maximum probable earthquake was 0.25 g and the maximum credible earthquake acceleration was 0.54 g (Noguera, Pinilla & San Martin, 2002).

Stability analysis showed that slope inclinations were adequate for such seismic conditions. The diaphragm wall was built with conventional concrete with 20 MPa at 28 days and a width of 0.80 m. The last 6 m were reinforced to withstand differential efforts during construction.

Puclaro plinth has three slabs of approximately 2.0 m that add to a 6.50 m plinth slab. The dam was built with a mini plinth that afterwards was connected with the diaphragm when the dam was completed.

The performance of the dam is quite good, and deformations and leakages are nominal and within the design requirements.

3.3.26 Santa Juana (Chile, 1995)

Figure 3.57 shows the position of each zone of Santa Juana Dam. Table 3.31 presents the materials used in the construction of the dam.

Main characteristics: H – 113 m; L – 360 m; L/H – 3.19; A/H^2 – 3.05; type of material – gravels.

Santa Juana Dam was built on the Huasco River, 20 km from Vallenar and 660 km north of Santiago, Chile.

The dam has an upstream slope, which is 1.5(H):1.0(V) and the downstream slope is 1.6(H):1.0(V), and a volume of 2,700,000 m^3 (Anguita, Alvarez & Vidal, 1993).

The dam is built on a gravel deposit with a depth of 35 m, approximately.

As in the case of Puclaro, the plinth is articulated over the gravels, and is connected with a 35 m diaphragm wall that penetrates into the foundation rock.

The crest width is 6 m, with a parapet wall 6 m high. The slab was designed using the same formula that was applied at Pulclaro:

$$e = 0.30 + 0.002 \text{ H (m)} \quad (3.12)$$

The concrete was pozzolanic, with a strength of 21 MPa at 28 days.

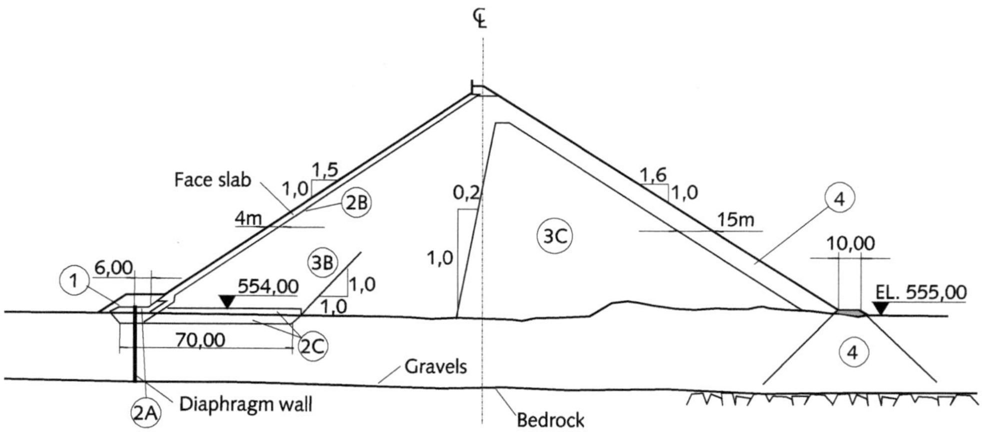

Figure 3.57 Santa Juana Dam.

Table 3.31 Santa Juana Dam materials.

Material	*Zone*	*Placing*	*Compaction*
Silt	1	0.20 m layers	Construction equipment
Fine filter	2A	0.20 m layers	Manual vibrators
Processed transition max. size 100 mm	2B	0.30 m layers	4 passes of 10 t vibratory roller and on upstream direction
Processed filter	2C	0.30 m layers	4 passes of 10 t vibratory roller
Main fill	3B	0.60 m layers	4 passes of 10 t vibratory roller
Rockfill or gravels	3C	0.90 m layers	4 passes of 10 t vibratory roller
Oversize blocks	4	–	–

The transition 2B material was built with processed gravel, with a maximum size of 4.0″, with 35–55% sand content and a Nr. 200 sieve fraction of 4–12%. The material was compacted in 0.30 m layers, with four passes of a 10 t vibratory roller and compaction along the slope.

The 3B material was compacted in layers of 0.60 m with four passes of the vibratory roller, and 3C the material was in 0.90 m layers with natural gravel alternated with rockfill from the spillway excavations. The maximum size of the gravel was 0.60 m in general.

Zone 4 was done with oversized blocks of the rockfill.

The dam is located within a zone of probable seismic activity that led the designers to consider the following parameters in the stability analysis:

- maximum probable earthquake 0.27 g
- maximum credible earthquake 0.56 g.

The articulate plinth design and its connections with the diaphragm wall were computed using fine element analysis that determined two slabs of 3.0 m. As at

Puclaro, the dam was built with the mini plinth and, once completed, the connecting slab was built between the diaphragm and the plinth.

During construction of the articulated plinth in an area of one third of the dam upstream zone, at 5 m down into the earth a layer of clay was reached, which had to be removed downstream (Noguera, Pinilla & San Martin, 2000).

The reservoir of Santa Juana remained low during the first years due to low precipitation but in 1997 during a heavy rainy season the reservoir became completely full. The displacements were small, as is usual for these fills built with gravels of high compressibility modulus.

In October 1997, an earthquake hit the dam. The epicenter was 250 km away from the site and 39 km below ground, with magnitude 6.80 (Richter scale). The parapet wall settled 9.7 cm without causing any strutural consequence to the dam (Noguera, Pinilla & San Martin, 2000).

3.3.27 Mazar (Ecuador, 2008)

Figure 3.58 shows the position of each zone of Mazar Dam. Table 3.32 presents the materials used in the construction of the dam.

Main characteristics: H – 166 m; L – 340 m; L/H – 2.05; A/H^2 – 1.7; type of material – quartzitic, chloritic, sencitic schists.

Mazar Dam is located in southeast Ecuador, 100 km from Cuenca on the Paute River. It has an upstream slope of 1.4(H):1.0(V) and a downstream slope of 1.5(H):1.0(V), and a volume of 5,000,000 m^3. The asymmetric valley has a steep right abutment with a slope of 0.60(H):1.0(V). The narrow valley has a ratio of $A/H^2 = 1.7$ (Ramirez, 2007).

The 2B material was processed from the rock of the Paute and Negro rivers. It is well graded and has a maximum size of 7.5 m, sand content of 35–60%, and 0–8% fines passing the Nr. 200 sieve.

The density is very high, 2.25 t/m^3, because it was compacted in 0.20 m layers, with six passes of a 10 t vibratory roller.

The 3B material is being compacted in 0.50 m layers with six passes of a 13.6 t vibratory roller. The rock comes from a quarry of quartzitic, chloritic and sericitic schists.

The 3C material is being quarried from the structure's excavation area. It is being placed in 0.80 m layers with six passes of the 13.6 t vibratory roller.

The rockfills are watered with a volume equivalent to 300 ℓ/m^3. The project has some non-conventional characteristics on the slab.

- On the right abutment sub-parallel slabs to the plinth have been included to reduce differential settlements of the perimetric joint.
- The sub-parallel slabs (hinge slabs) are thicker and defined by the formula:

$$e = 0.30 \text{ m} + 0.006 \text{ H (m)} \tag{3.13}$$

- The central compressive joints are spaced at every 7.5 m with compressive elements of 3.2 cm in thickness.
- The reinforcement of the slab is 0.5% in both directions.
- The perimetric joint at the right abutment has double corrugated waterstops similar to those used in China.

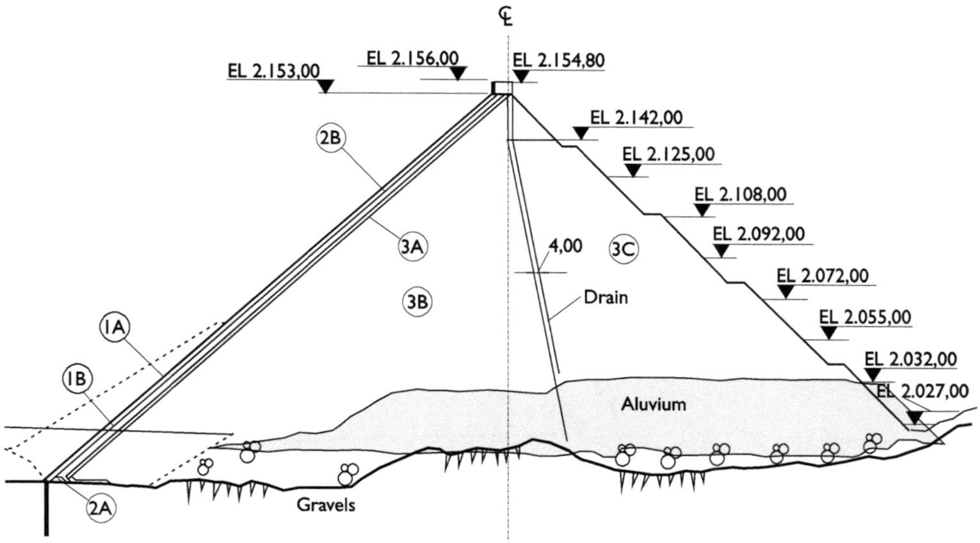

Figure 3.58 Mazar Dam.

Table 3.32 Mazar Dam materials.

Designation	*Zone*	*Placing*	*Compaction*
Silt	1A	0.30 m layers	Construction equipment
Selected material	1B	0.60 m layers	Construction equipment
Fine transition $\varnothing_{max.}$ 30 mm	2A	0.40 m layers	Vibratory roller
Processed gravels $\varnothing_{max.}$ 75 mm, well graded	2B	0.50 m layers	10 t vibratory roller – 6 passes
Rockfill transition $\varnothing_{max.}$ 0.40 m	3A	0.50 m layers	10 t vibratory roller – 6 passes
Quarry rockfill	3B	0.50 m layers	13.6 t vibratory roller – 6 passes, water 300 ℓ/m³
Rockfill from excavations	3C	0.80 m layers	13.6 t vibratory roller – 6 passes water 300 ℓ/m³

The river was diverted on December 1, 2006 and the placement of the materials progressed well. At the time when these comments were written, the volume of rockfill was 4,900,000 m³ and the quality control indicated a compressibility modulus between 46 MPa and 85 MPa, measured in 0.50 m diameter plates in zone 3B, with void ratios of 0.16 to 0.22 and permeability above 10^{-1} cm/s. The compressibility modulus evaluated from settlement cells showed values between 90 and 140 MPa.

3.3.28 Merowe (Sudan, 2008)

Figure 3.59 shows the position of each zone of Merowe Dam. Table 3.33 presents the materials used in the construction of the right margin of the dam.

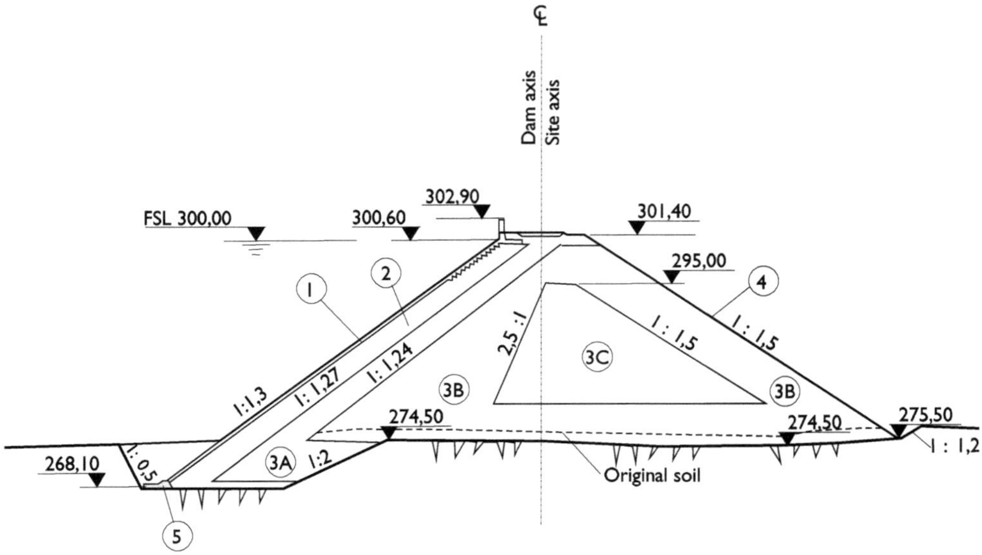

Figure 3.59 Merowe Dam.

Table 3.33 Materials of the right abutment of Merowe Dam.

Material N°.	*Designation*	*Zone*	*Placing*	*Compaction*
1	Concrete face	–	–	–
2	Processed rockfill $Ø_{max.}$ 10 mm well graded	2B	0.40 m layers	10 t vibratory roller – 4 passes
3A	Coarse transition from quarry	3A	0.40 m layers	10 t vibratory roller – 4 passes
3B	Main rockfill $Ø_{max.}$ <1.20 m CUn >70 MPa	3B	1.20 m layers	10 t vibratory roller – 6 passes-water 150 ℓ/m³
3C	Rockfill with fines – CUn <70 MPa	3C	1.20 m layers	10 t vibratory roller – 6 passes-water 150 ℓ/m³
4	Oversized to protect downstrean slope	4	Equipment	–
5	Plinth	–	–	–

Main characteristics: *H* – 53 m; *L* – 4.364 m; *L*/*H* – 8.4; A/H^2 – 50; type of material – granitic gneiss.

Merowe Dam is located in Sudan, at the margins of the Nile River. With a length of 5,800 m, it is one of the longest dams in the world. The slopes are 1.3(H):1.0(V) upstream and 1.5(H):1.0(V) downstream. The dam joins with a central core dam on the river bed 67 m high and with a diaphragm wall to control the flow through the alluvium of the river area (Schewe & El Tayeb, 2005).

The rockfill on both margins is founded on moderately weathered rock, with an unconfined compression strength higher than 25 MPa after the removal of the loose material. The rockfill is being built with rock from a quarry of granitic gneiss with interbedded biotite. All materials have an unconfined compression strength over 70 MPa.

The 2B material is being processed with a maximum size of 0.10 m, a sand content between 35–55% and fines below Nr. 200 sieve of 5%. This material is being compacted in 0.40 m layers and six passes of a 12 t vibratory roller; the 3A material is being compacted in a similar way.

The 3B material is being spread in 1.2 m layers and compacted with eight passes of a 12 t vibratory roller.

The 3C material is also being compacted in layers of 1.2 m, with the same number of passes of the vibratory roller, but with rock of lower quality.

As watering using monitors becomes difficult, because the distance from the river to the compaction site increases, the addition of water is done by trucks at a rate of 150 ℓ/m³ with showers being installed outside of the dam.

In this dam the plinth was built with slipping forms. Dimensions were defined using Bieniawski's rock classifications by geomechanical methods (see Table 3.34).

The plinth followed the concept of external and internal slabs, keeping an external slab of 3.5 m.

The slipping forms at the plinth were moved by cables installed in a winch inside the same form as shown in Figure 3.60 (Materón, 2006).

Table 3.34 Rock mass classification by Bieniawski.

RMR	*Gradient*
>80	20
80–60	18
60–40	16
40–20	10
<20	deepen the excavation

Figure 3.60 Slipping forms for plinth construction. (See colour plate section).

The performance of the dam during the impounding of the reservoir has been excellent. The leakage is very low and is in the order of 30 ℓ/s in each abutment.

3.4 CONCLUSIONS

From this analysis of the dams and from their performance to date, the following conclusions can be drawn.

- It is desirable to process 2B material with a maximum grain size of 10 cm and a sand content between 35% to 55% as recommended by the consultant James L. Sherard, limiting the fines passing Nr. 200 sieve to 8%. However, different grain sizes have been used with adequate results. There are materials such as the Brazilian and African basalts, for which the production of a grain size as suggested by Sherard becomes expensive. For these materials, intermediate solutions with sand content up to 20% have shown satisfactory results.
- The 3A materials have to be designed as a transition between the 2B material and the main rockfill 3B. It is not necessary to process the material of zone 3A. Practical experience has shown that it is possible, with a bigger fragmentation, to produce the 3A material in the quarry, so that it can be placed in layer similar to the 2B material.
- The 3B and 3C materials should be spread out and compacted in such a way that their compressibility moduli should not be too different from each other. The performance of dams in which the difference in the thicknesses of 3B and 3C is too high have resulted in undesirable displacements of the upper part of the dam, which can affect the slab. A classical example is Aguamilpa Dam, built with gravel upstream and thick rockfill downstream.
- Zone T at the central portion of the dam has been used to accommodate materials of lower quality or mixtures of gravel and rockfill. These are practical and economical solutions.
- The erosions experienced at Alto Anchicayá, Salvajina, Messochora and other dams in Asia, and described in the technical literature, have shown the importance of the inclusion of a supportive element to the upstream slope. The Brazilian innovation of the extruded concrete curb, first used at Itá, has been a construction solution displaying excellent results.
- The evolution of compacted dams of CFRD type within the last 35 years (Cethana 1970 to Shuibuya 2008) has shown the importance of undertaking compaction with vibratory rollers and the addition of water, as dams get higher and higher. The vibratory rollers ratio must have a weight/width ratio of the roller of values that exceed 5 t/m to achieve a high density and an acceptable deformability.
- Alto Anchicayá, in Colombia, has shown the importance of providing several defense lines in the perimetric joint. The dams built after Alto Anchicayá have included various types of waterstops, both internal and external, with good results.
- The design of the slab follows empirical procedures in the majority of the dams and are determined according to the formula $T = (0{,}30 + KH)$ m where K varies between 0.002 and 0.0035 for gradients up to 220. However, for a very high dam, it is necessary to define values that do not exceed that gradient. An example

is given by Campos Novos and Barra Grande dams, where the formula was changed to $T = 0{,}005$ H for $H \geq 100$ m.

- Experience has shown that in a narrow valley where the ratio A/H^2 is smaller than 4 and the compressibility of the rockfill is low, the compression efforts that develop at 30–40% *H* can generate failure of the central compressive joints with deformation of the reinforcement that may spread laterally as observed in Campos Novos, Barra Grande and Mohale dams.
- It is desirable in high dams to compact the rockfill with a smaller difference in layer thicknesses in zones 3B, *T* and 3C, with the aim of reducing the differences in compressibility. Well-graded materials (CNU > 15) should be compacted in relativity thin layers (0.80; 1.00; 1,20 m) in zones 3B, *T* and 3C respectively using vibratory rollers that apply a pressure of 5 t/m and with generous wetting.
- In high dams with uniform rockfill (CNU < 15) the previous recommendations are applicable. However, it is important in the central compression joints to follow these procedures: i) Placing the mortar pad to support the waterstops below the theoretical line of the slab thickness, so that the thickness of the slab is preserved. The wing of the cooper waterstop should be reduced or eliminated. ii) Placing vertical elements of wood or compressive elements (neoprene, EPDM) that allow displacements between the joints and reduce the compression efforts. iii) Reducing or eliminating the upper V notch in the central compression joints. iv) Using anti-spalling reinforcements.
- Numerical analysis (FEM) has been used in some dam designs, such as Cethana, Alto Anchicayá, Foz do Areia, etc., with good results for parametric comparisons. However, the final design decision has always been dictated by experience and precedence established in previous structures and using the empirical criteria as recommended in chapter 2.

According to Cooke (1999), empirical design is based on precedents, which are defined as something that can be used as an example for similar cases. Another form of precedent is the engineering precedent: knowledge of design, of dams under construction, in design and in planning. It is useful to know and to take into consideration what other professionals are doing today.

Chapter 4

The mechanics of rockfill

4.1 INTRODUCTION

The rockfill mass is the dam's structural body, which guarantees the stability of the slopes and controls of the flow coming from the concrete slab and from the foundation as well as the concrete slab support. It is a material with such a particular behavior that, according to Maranha das Neves (2002), its mechanics should be considered separately and not regarded as an extension of the granular soil mechanics of materials such as gravel and sand.

According to Maranha das Neves (2002), rockfills differ from gravel and sands (both particulate materials) because rockfills display fracturing and crushing at very low pressures.

Phenomena occurring in the block-to-block contacts are a determining factor in the mechanical behavior of these materials. And, although in the rockfill embankments analysis we commonly use exclusively the mechanics of the continuum, only a micromechanic approach helps to explain such behavior.

Phenomena like collapse and creep are very important. In a rockfill, the change in the level of stresses results not only in a change of the specific volume but also in changes in the solid matter, which becomes different due to fracturing and crushing.

A proof of these statements is shown in Figures 4.1 and 4.2. Figure 4.1 refers to the particle size changes measured between the top and the bottom of the compacted rockfill layers in Jaguara (quartzite) and in Foz do Areia (basalt) (De Mello, 1977). Such changes are due not only to the segregation that occurs during dumping, but also as a result of breakage of the particles under the pressure of the compaction roller. Figure 4.2 shows the specific deformation measured in deformability tests with the addition of water in limestone samples from the Angostura Dam (Marsal, 1971).

Regarding creep, the well-known Figure 4.3 (Sowers, Williams & Wallace, 1965) adapted by Maranha das Neves (2002) shows that some rockfills continue deforming with decreasing rates of deformation for years after the end of construction.

Additional values were registered for well-graded rockfills compacted by vibratory rollers. Well-graded rockfills show a reduction in settlement when well compacted.

After one year the crest of a dam's post-construction settlement measured as a percentage of height varied from 0.1% to 0.4%. After 10 years they reached 0.4% to 0.8% and in the few records available, they reached 1.5% after 30 years. Construction compaction seems to play an important role in long-term settlement. Table 4.1 presents the average rate of crest settlement over the years.

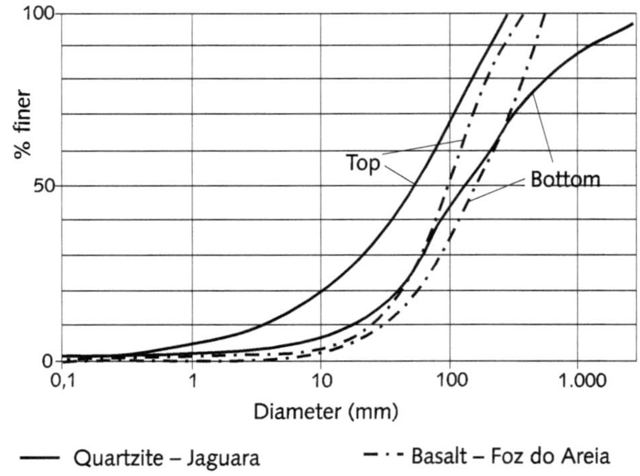

Figure 4.1 Particle size distribution in a compacted layer (Narvaez, 1980).

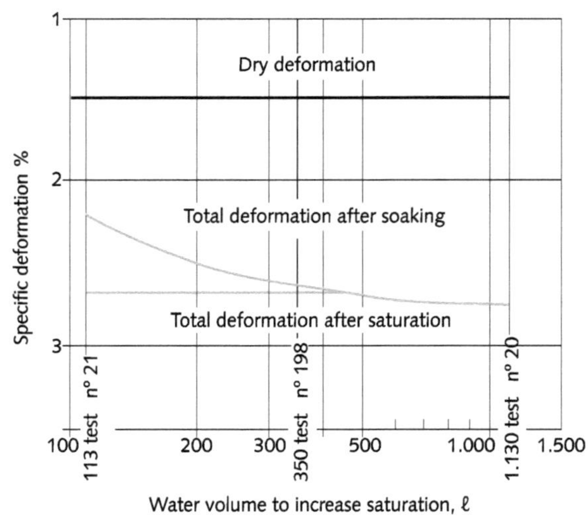

Test nº	Water addition (ℓ/m³)	Density (kg/m³)	Applied pressure (kg/cm²)
21	107	1.636	5,6
198	330	1.634	5,7
20	1.074	1.642	5,7

Figure 4.2 Addition of water – deformation test on Angostura Dam limestone (Marsal, 1971).

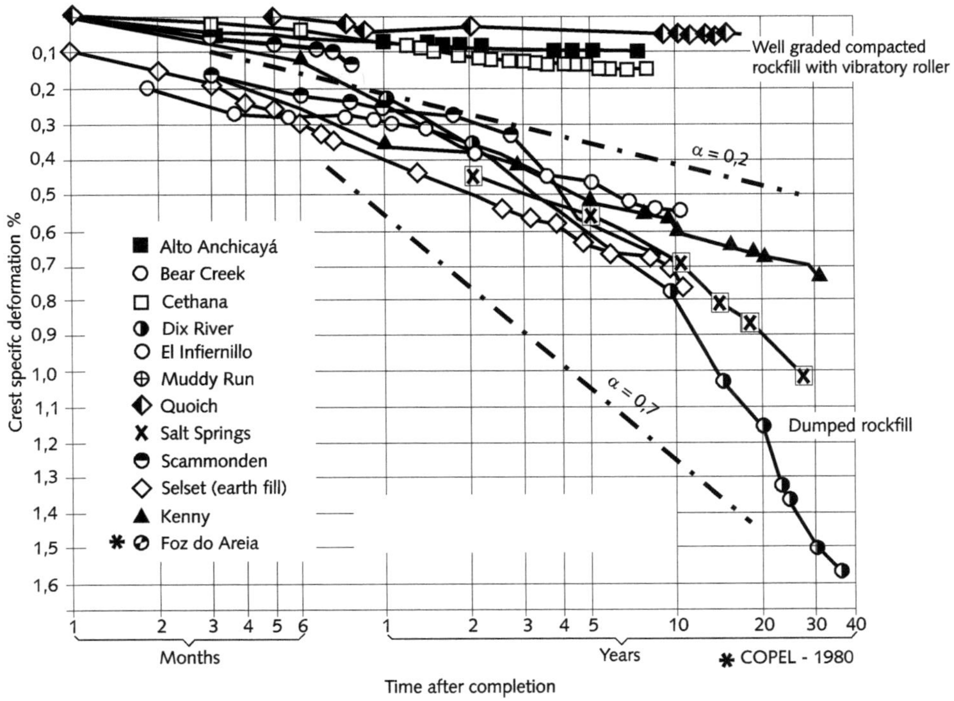

Figure 4.3 Long term vertical displacements in some dams (Maranha das Neves, 2002, adapted from Sowers, Williams & Wallace, 1965).

Table 4.1 Settlement rates in CFRDs.

Construction procedure	*Average rate of crest settlement for a 100 m high dam (mm/year)*		
	After 5 years	*After 10 years*	*After 30 years*
Compacted rockfill	3.5	1.5	0.6
Dumped rockfill	45.0	30.0	10.0

Penman and Rocha Filho (2000) show how the displacements measured on the concrete face of Xingó Dam changed in a six-year period due to the reorganization of the particles, see Figure 4.4. The displacements practically doubled in value.These displacements were also due to water leakage through cracks that opened on the concrete face.

It is worth mentioning that the shape of the valley interferes with the rockfill mass stress distribution, leading to an arching effect in narrow valleys which relaxes progressively. In such cases, the settlement resulting from creep can last for a longer period (Alto Anchicayá and Cethana; Cooke & Sherard, 1987).

Instrum.	Deflections (m)	
Elev. (m)	11/94	05/00
(EL1) 48	0,08	0,23
(EL2) 64	0,24	0,38
(EL3) 80	0,29	0,46
(EL4) 88	0,26	0,47
(EL5) 96	0,24	0,48
(EL6) 104	0,23	0,49
(EL7) 112	0,22	0,48
(EL8) 120	0,20	0,45
(EL9) 127	0,18	0,45
(EL10) 138	0,17	0,51

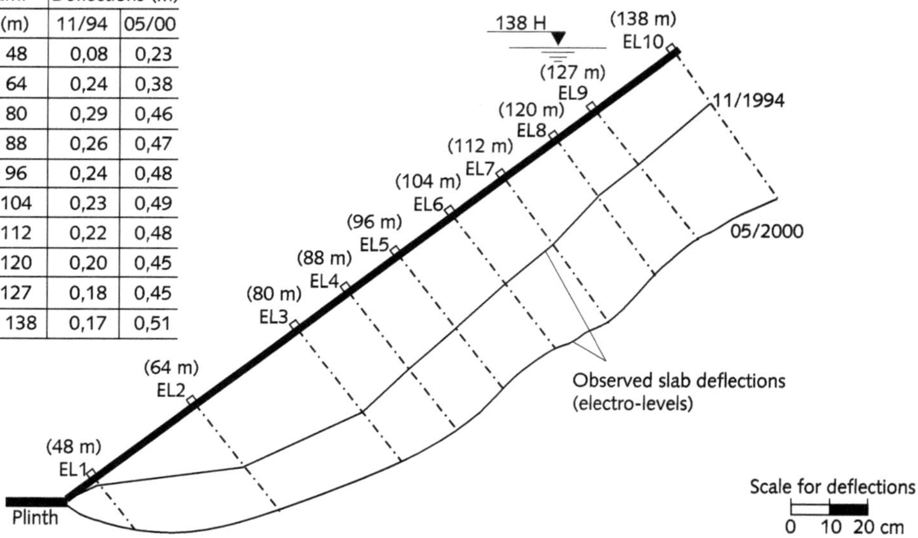

Figure 4.4 Deflections observed at Xingó Dam slab. Short- and long-term behavior (Penman & Rocha Filho, 2000).

4.2 ROCKFILL EMBANKMENTS EVOLUTION

Without knowing the past one cannot understand the present, and without ruptures in knowledge itself there is no evolution and, therefore, no future.

Many authors have written about the evolution of CFRDs. Bulletin 70 (ICOLD, 1989), and Penman and Rocha Filho (2000) show that in one way or another the history of CFRDs has developed in parallel to J. Barry Cooke's life and work.

This chapter focuses on the changes that have occured in the treatment and construction of rockfills with no regard to the chronological order in which the dams were built.

When one reads these authors, two factors seem to have been central to the evolution of CFRDs:

- the change from dumped rockfills to compacted rockfills, which has been facilitated by advancements in earth-moving equipment;
- the need to build progressively higher dams, which would result in a major disaster if any of them failed.

The first factor is related to the control of settlement or deformations. The second has led to a slope flattering, maybe due to laboratory studies and analyses developed to support rockfill dams with a clay core in which stability analysis – either conservative or not – suggested flatter slopes.

A review shows that rockfills have been built using different techniques throughout the years. In the early dams, rockfills were made of rock block piles with hand-placed rocks on the upstream face to receive the timber planking and form the waterproof element. Steep slopes were adopted, such as 0.75(H):1.0(V) in the first 15 m of the Upper Bear River Dam, followed by another 8 m with a slope of 0.5(H):1.0(V). The Upper Bear River Dam was finished in 1900 (Steele & Cooke, 1960).

After 29 years in operation, the downstream slope was flattened to 1.3(H):1.0(V) with dumped rockfill. Fifty three years later, the decayed timber planking was replaced by a continuous layer of gunite reinforced by a steel mesh. A 13 cm thick concrete layer was placed at the bottom, thinning down to 8 cm at the top. According to Penman and Rocha Filho (2000), it was the beginning of the concept that evolved into concrete face rockfill dams.

Among some curious facts, it is worth mentioning the very steep slopes and the timber facing of the Meadow Lake Dam. In 1930, a fire broke out in the forest and burnt down the timber facing, which was then replaced by a continuous sheet of lightly reinforced gunite, 10 cm thick at the bottom and 5 cm at the crest (Penman & Rocha Filho, 2000).

The piled rockfill building technique changed to dumped rockfill, as in the Salt Springs Dam (100 m high), built from 1928 to 1931. Salt Springs Dam held the title of highest dam in the world up until the completion of the Paradela Dam in Portugal in 1957 (108 m high). Large granite blocks with unconfined compression strength ranging from 100 MN/m^2 to 130 MN/m^2 were dumped on ramps from heights of 20 m to 50 m, because it was believed that the impact of 10 ton to 25 ton blocks would produce considerable compaction and promote interlocking within the rockfill mass.

> "It was also common to use water jets during placement with the intention of washing out any fines from between the larger pieces of rock to ensure good rock-to-rock contact and therefore minimize subsequent settlement."
>
> (Penman & Rocha Filho, 2000).

At Salt Springs Dam there was no instrumentation to measure settlement other than observation of surface marks. Settlement and horizontal movements were carefully measured by topographical surveying. A central point on the dam crest settled 58 cm and had a horizontal displacement of 32 cm.

Another 30 cm was added to the settlement as an estimate of what would have happened during the first two weeks of impounding, which lead to a total settlement of 91 cm. This settlement was considered low in comparison to the benefit of compaction in high lifts.

In order to get a smooth face on the rockfill prepared to receive the concrete, rock blocks were arranged in a layer 4.5 m thick with their flat side parallel to the upstream surface plane. The displacements of the face were measured at observational points. A displacement of 98.4 cm at two fifths of the height was measured after the first partial filling, followed by the reservoir drawdown.

In the next filling, 11.6 m above the previous one, another 33.5 cm of displacement was registered. Seventeen years later the displacement had reached 163 cm after the water level was raised another 3 m.

The new displacements that took place after the new filling occurred close to the crest. This fact was interpreted as the result of block breakage in the lower part of the dam due to cycles of reservoir loading and unloading. At the upper end the loads were smaller and because this portion was also exposed to a smaller impact during construction the rockfill settled at a greater rate during the filling process.

The measurements at the observation points displayed not only movements normal to the face but also movements within the face plane, indicating zones of compression and tension on the concrete face slabs. Open joints and crashed joints led to leakage. The joints were repaired during the annual drawdowns, which were part of the reservoir operation. After a 27 years period the crest settlement was 110 cm. At that time the settlement was thought to be due to weak rock and fines that prevented rock-to-rock contact. The data presented has been obtained from Penman and Rocha Filho (2000).

At Salt Springs Dam the upstream slope varied from 1.1(H):1.0(V) to 1.4(H):1.0(V), and it was 1.3(H):1.0(V) at the Paradela for both upstream and downstream slopes. Both CFRDs were built with granite blocks.

The question of whether to use water in the deformability of the dumped rockfill is examined by Penman and Rocha Filho (2000). The construction of the Cogswell Dam (85 m high) began in March 1932. Due to an acute water shortage, no sluicing was used initially. Construction had been 80% completed by the end of 1933, when there was an exceptional rainfall in the 24 hours from December 31, 1933 to January 1, 1934. The rainfall amounted to 382 mm, causing a crest settlement of 1.8 m.

The settlement continued due to creep down to 4.1 m. After the rainfall, there was plenty of water to keep the rockfill intensively watered in order to reduce the post-construction settlements, which may have reached 5.3 m.

The Lower Bear River Dam (77 m high) was built between 1951 and 1952, a few miles from Salt Springs. The rockfill was dumped from great heights (up to 65 m). Three water pumps able to push out over 320 liters/sec under a pressure of 820 kN/m^2 were used for sluicing the rockfill during construction. The water volume was three times the rockfill volume. The surface settlement measured during construction was small and maximum depression on the face after the first filling was 62.5 cm at one third of the height. The engineers were convinced of the benefits of water in removing the fines between rock blocks and keeping a good rock-to-rock contact.

Both the Salt Springs and the Paradela dumped rockfill dams presented serious leakage and maintenance problems. The new Exchequer Dam in California (150 m high) was built downstream of a pre-existing gravity concrete dam. Part of the rockfill was compacted in layers of 1.2 m to 3 m, and the downstream zone was built with dumped rockfill in layers of 18 m. It also had a very high level of leakage causing major water loss from an economic point of view requiring subsequent repairs (Cooke, 1969). The leakage occurred because the horizontal joints were no longer working properly due to inefficient compaction.

Cooke even says that the construction of dams above 92 m high was interrupted between the years of 1960 and 1965. He does not mention why the height of 92 m was significant, but on many occasions he states that this interruption could be accounted for by the fear that large deformations of dumped rockfills could lead to failures.

The move towards a third phase (i.e., compacted rockfills) began in the 1950s and 1960s thanks to the contribution of Terzaghi (1960b) on the influence of water and the fines in the rockfill mass and to the good performance of the Nissaström Dam in Sweden and the Quoich Dam in Scotland (320 m long, 38 m high). The development of equipment for transportation and compaction of the rockfill also contributed to the transition.

The Nissaström Dam (15 m high) was completed in January 1950. The rockfill was compacted in layers 60 cm thick by a steam roller of 98 kN, followed by a vibrating plate of 19 kN static weight. The layers were compressed by 18% of their original height (Hellström, 1955).

Water was added, in five times the quantity of the rockfill volume, in order to facilitate the compaction, yet not to remove the fines. Settlement measured on the crest was very small.

The second compacted rockfill dam was the Quoich, Scotland. This one followed the same procedures of the Swedish dam. According to Roberts (1958), the Quoich Dam was completed in 1954. After 15 years in operation, settlement on the crest was only 0.05% of its height, in other words, 1.9 cm for a 38 m dam. The concrete face was placed in squares of 6 m^2, 0.30 + 0.002 H thick, H being the height in meters. The squares were jointed together with 3 mm thick copper waterstops, providing for a flexible face that did not crack.

In summary, the very steep sloped, piled rockfills were followed by rockfills dumped from great heights, for which sound rock was required and plenty of water was used to remove the fines.

The first dams, the 100 m-high (Salt Springs and Paradela) and the 150 m-high (New Exchequer) dams, suffered large settlement and compression on the concrete face and produced leakages that led to high water loss and high maintenance costs. The rockfill masses, however, kept stable.

The Nissaström and the Quoich, both medium height dams, represent the beginning of compacted rockfill dams.

The use of water in large volumes was required in order to facilitate compaction but not to remove the fines.

It is worth mentioning Galiolli, an Italian engineer who in the 1950s called for the use of "pure rockfill", as he referred to the sound rockfill without fines required on the design of a dam in Patagonia.

4.3 THE COMPACTED ROCKFILL

Terzaghi (1960a) when reviewing the works of Steele and Cooke (1960) emphasized the use of water to remove the fines and allow for a good contact between the rock blocks, and presented some fields test results made on rockfills.

He observed that during the placement of a new rockfill load segregation took place causing the average particle size and degree of uniformity to increase as one got farther down the slope. The finest materials settled on the upper part of the slope, and the larger blocks rolled down to its base (Fig. 4.5).

- Zone A: small rock fragments acted on by water.
- Zone B: medium rock fragments with smaller particles in the voids.
- Zone C: large rock blocks, more compressible.

It was also observed that the water action was limited to the upper 10 m of the slope.

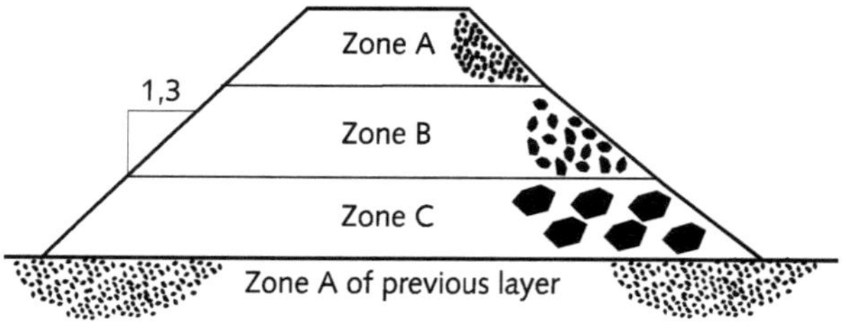

Figure 4.5 Dumped rockfill (Terzaghi, 1960b).

In order to analyze the penetration of fines into the rockfill, Terzaghi carried out a field trial test with a rockfill consisting of blocks up to 0.75 m in size. A mixture of sand and gravel was dumped on the rockfill and sluiced with a jet of water at close range. A one rock thick layer on the surface soon clogged up with the gravel and no significant amount of fines could be jetted into the rockfill below this layer.

In a second test, sand alone was used, but it was only washed down to a depth of two blocks of rock by a jet of water, and its kinetic energy had mostly dissipated in the first row of rock. Below the surface, the water flowed through the voids with the force of gravity, unable to perform any greater mechanical action than a heavy rainfall.

Terzaghi (1960) concluded that the beneficial effects of the sluicing of water in promoting settlement was a consequence of the loss in the strength of the rock with wetting and not due to removal of the fines among the rock blocks. He presented data on the compressive strength of rocks (Hirschwald, 1912; McHenry, 1945) showing the reduction of the rock strength with wetting (Table 4.2).

Penman and Rocha Filho (2000) present data of direct shear tests on polished quartz. For normal pressures between 17 kN/m^2 and 850 kN/m^2 the friction angle was 33°. When the surface was dry at 105°C for 28 hours and the tests were repeated, the friction angle dropped to 11°.

Similar results were shown by Terzaghi (1960c), who demonstrated that wetting did not reduce the friction angle of unpolished rock-to-rock surface and that the water did not act as a lubricant.

Direct shear tests in a stress-controlled shear box were performed by Tschebotarioff and Welch (1948) on a polished surface of quartz against quartz grains bedded in plaster with and without water. The friction angle measured in both conditions was of 25°. Tests results were:

- non-submersed tests, $\varphi = 25°$;
- non-submersed tests, but with humid surfaces, $\varphi = 25°$;
- oven dry surfaces without fines revealed an angle of friction of only 7°.

To investigate friction at high pressures, Penman (1953) used three quartz fragments that had been broken into pieces over a polished quartz surface. The applied load was initially 6.08 N per contact and the measured φ on this test was 29°. When increasing the normal load to 215 N, the measured φ was 19°. Crushing could be

Table 4.2 Compressive strength of wet and dry rocks (Terzaghi, 1960c).

Rock	Dry (MN/m²)	Wet (MN/m²)
Sound Austrian granite	145	128
Granites from Sweden and Germany	240 (average)	197 a 237
Crystalline limestone	97	87
Schist from Tennessee	96	45

heard with all loads greater than 148 N per contact. Once the test was over the three contact points were found to be badly damaged and the lower polished surface had been strongly scratched.

This group of tests showed that, first, water is not a lubricant; on the contrary, it has an anti-lubricant effect if we take into consideration that the test was performed on rock dried over 100°C. Second, the rock strength diminishes once the load is high enough to affect and crush the contact between the rock blocks.

Many researchers have studied the problem of the point crushing at sharp edges of rock blocks. Clements (1981) used prismatic samples of sandstone from the Scammonden Dam site (England) with different angles of sharp modelled shapes as shown in Figure 4.6. As we can see, the deformation increases with the reduction of the central angle (β). Figure 4.7 shows the evolution of creep displacements over time, after wetting the rock contact.

From Terzaghi's experiments (1960c) and other tests on dry, humid, and saturated rock some conclusions can be drawn.

- Water does not act as a lubricant on rock-to-rock contacts, and humid and saturated rockfills have the same friction angle.
- Water, however, weakens rock and deformation increases due to the breakage and crushing of the rock particles.
- The strength of rock blocks is affected by the stress level and progressive breakage. The value of φ decreases as soon as there is enough stress applied to initiate the breakage process.
- Rockfills with angular blocks are far more compressible than rockfills with rounded blocks.

As an overall conclusion, it can be said that compacted rockfills in layers ranging from 0.8 m to 2.0 m have characteristics of their own, which are very different from piled or dumped rockfills. The first are less compressible, have higher shear strength, and allow for the construction of both CFRD and CCRD to heights of 200 m high or more. Segregation persists as an inherent process as shown in Figures 4.8 and 4.9.

Cooke and Sherard (1987) suggest that there are benefits from the segregation in terms of flow control because the water flows through the lower part of each layer, avoiding a possible build up of pore pressure in the lower zone of the dam.

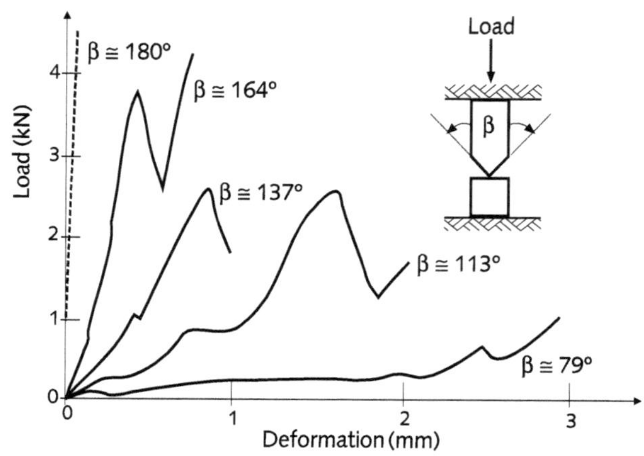

Figure 4.6 Load deflection curves at contact point – sandstone (Penman & Rocha Filho, 2000).

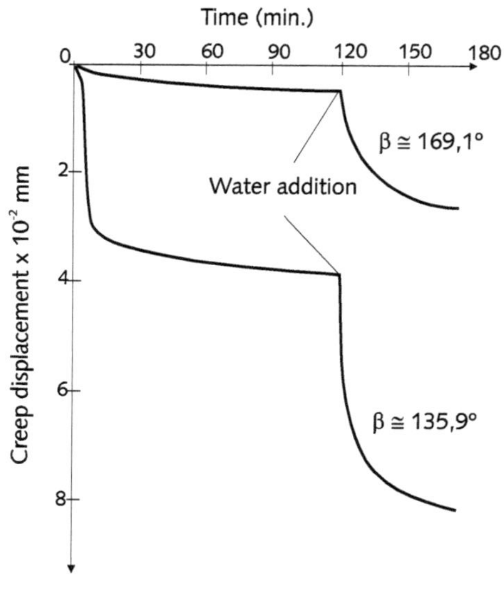

Figure 4.7 Creep displacement with time and wetting (Penman & Rocha Filho, 2000).

However, the inherent anisotropy in the rockfill is detrimental in the case of a flow during construction, if say a flood occurs before the conclusion of the concrete face of the dam (Pinto, 1999; Cruz, 2005). More details on this subject are given in chapter 6.

Still concerning fines, the observation of Cooke and Sherard (1987) regarding the conditions of trafficability on the surface of the rockfill during compaction is very important:

> "A stable construction surface under travel of heavy trucks demonstrates that the wheel loads are being carried by a rockfill skeleton. An unstable construction surface, with springing, rutting and difficult truck travel, shows that the volume

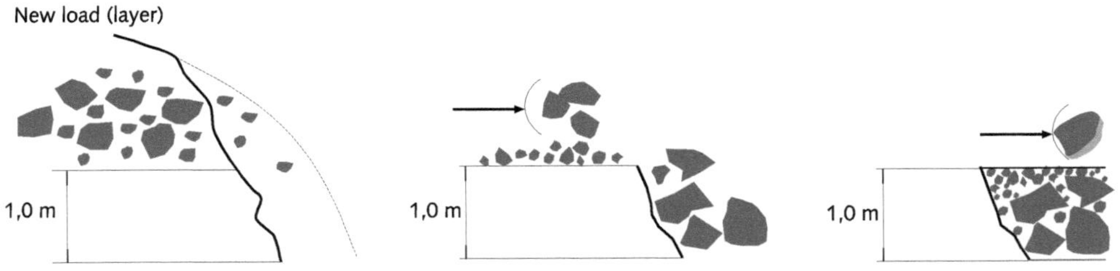

Figure 4.8 Stratification and spreading of compacted rockfill (Cooke, 1984).

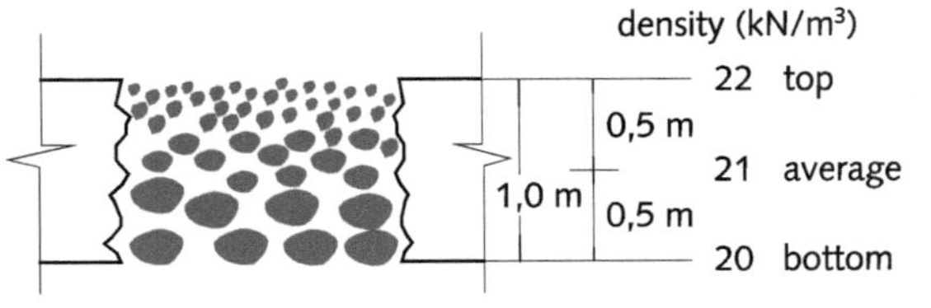

Figure 4.9 Density variations within a layer of compacted rockfill (Cooke, 1984).

of soil-like fines is sufficient to make the rockfill relatively impervious. Where the surface is unstable, the fines dominate the behavior and the resulting embankment may not have the properties desired for a rockfill zone."

4.4 ROCKFILLS GEOMECHANIC PROPERTIES

4.4.1 Intervenient factors

Materón (1983) presents a list of factors that affect the geomechanic properties of rockfills (see Table 4.3).

However, the inter relation of these factors in nature is complex and of difficult to interpret. There are no absolute rules that can be used to increase the strength or to reduce the compressibility of a rockfill, once the natural characteristics are influenced by the simultaneous action of different factors (mineralogy, moisture, rock density, etc.).

The knowledge we have at the present time allows us to make some assumptions.

- Well-graded rockfills, subject to compaction with the same equipment, result in denser rockfills, which are less deformable and therefore less likely to suffer internal fracturing than uniform rockfills of the same rock type.
- The shear strength of rockfills can be expressed by an exponential law:

$$\tau = A\,(\sigma_n)^b \tag{4.1}$$

where A and b are constants of the material and σ_n the normal stress.

Table 4.3 Factors that affect strength and compressibility.

Factor	*Effects*
Mineralogy	Affects the friction angle φ
Grain size distribution	Uniform grain sizes are more compressible. It is useful to add finer granular material to reduce the compressibility
Void ratio	The smaller the void ratio, the lower the compressibility
Particle shape	Shape is related to the amount of breakage. Angular particles break more. The larger the breakage, the larger the compressibility; there are some discrepancies among authors
Moisture conditions	In general, adding water results in more compression
Compressive strength of particles	The higher the strength, the lower the breakage
Size and texture	Larger blocks may result in larger breakage, lower resistance and higher compressibility
Time	A Long periode of time increases compressibility with creep
Factors affecting test results	
Loading rate	Apparently does not affect the strength and compressibility if drainage is free
Load sequence and intensity of load	Depends on the type of test – plane strain or triaxial

Alternatively, the friction angle can be computed by the following equation:

$$\varphi = \varphi_0 - \Delta\varphi \log(\sigma/Pa) \tag{4.2}$$

where:
φ – friction angle
φ_0 – friction angle for Pa
$\Delta\varphi$ – variation of φ for each material
σ – confining pressure
Pa – atmospheric pressure taken as reference

4.4.2 Molding problems

As stated in previous chapters, rockfill contains large blocks, which can be 50 cm, 1.0 m and 2.0 m in size. For this reason they have not been tested in their original particle size – neither in the field nor in laboratory. Regarding compressibility, it is possible to measure the displacements at internal points within the CFRD and relate these values to the acting pressures estimated by mathematical models.

However, regarding shear strength, the problem is less easily addressed because there are no well-documented data on failures of rockfills, which would allow for a back analysis that would lead to an equation for shear strength.

The failures that have occurred were due to flood and overtopping, and in such cases back analyses are more complex due to the pressure of seepage forces that were the main cause of the failures.

Thus, when modeling rockfill samples in the laboratory one has to consider the choice of particle size and the choice of relative density. For the particle size, two techniques are used.

1 Cutting the particle size to a maximum ϕ, and replacing the larger particles by an equivalent weight of smaller particles.
2 Adopting a grain size distribuition parallel to the one of the field rockfill starting at a laboratory $\phi_{max.}$.

The $\phi_{max.}$ is limited to a fraction of the sample diameter of the triaxial test, or a fraction of the side of a cubic or rectangular sample in direct shear tests. In general, $\phi_{max.}$ is ¼ to ⅙ of either the diameter or the sample side.

In the first case, the particle size may become uniform, which directly affects the interlocking between the particles; and in the second case, the finest fraction may result in sand size and the occurrence of loose particles in the voids, building up pore pressures during test.

Another aspect is modeling the relative density. Different particle sizes lead to different relative densities and if one tries to reproduce the field density in the laboratory, one may achieve a much higher relative density compared to the one achieved in the field.

These limitations, which exaggerate the laboratory parameters by comparison with the real conditions of the dams, have been discussed and pursued in the works of Marsal (1973), Seed and Lee (1967), and others in the 1950s and 1960s, and they are still under discussion today.

4.5 SHEAR STRENGTH

A complete and detailed review of the shear strength of rockfill is not the purpose of this book. References on the subject can be found in Marsal (1973), Seed and Lee (1967), Marachi et al. (1969), Midea (1973), Cruz and Nieble (1970), Signer (1973), and Materón (1983) among many others.

What we sought in these references were good test results on "rockfill" samples which, given all the already discussed limitations in molding, could represent at least an estimate of the field shear strength at the dam.

Table 4.4 summarizes triaxial test results from different sources and researchers including the parameters A and b; Figure 4.10 shows the shear strength envelopes of Marsal's tests (1973).

Figure 4.11 gives the particle size distribution of the rockfills and Table 4.5 the main characteristics of the same samples.

The strength envelopes are curved, reflecting the decrease of shear strength along with an increase of octaedric stress or/and normal stress during direct shear tests. The Mohr-Coulomb envelope can be expressed by De Mello's (1977) equation:

$$\tau = A\sigma^b \tag{4.3}$$

where τ is the shear strength and σ the normal stress in the shear plane. A and b are the parameters of each material and are defined by trial and error from test results.

Table 4.4 Shear strength of rockfills.

Material	*Test*	$\phi_{max.}$(mm)	*A kg/cm²*	*b*	φ_l	*References*
Basalt – San Francisco	TR	200	1.68	0.79	59	Marsal, 1971
Diorite – El Infiernillo	TR	200	1.00	0.90	45	Marsal, 1973
Gneiss-granite mica	TR	200	0.87	0.90	41	Marsal, 1973
Sand/gravel – Pinzandarán	TR	200	1.57	0.82	57	Marsal, 1967
Silicified conglomerate – El Infiernillo	TR	200	1.53	0.79	57	Marsal, 1973
Slate – El Granero	TR	200	1.44	0.77	55	Marsal, 1973
Gneiss-granite-schist – Mica Dam	TR	200	0.80	0.94	38	Marsal, 1973
Basalt – San Francisco	TR	90	1.69	0.78	59	Marsal, 1973
Slate – El Granero	TR	200	1.40	0.82	54	Marsal, 1973
Schist-gneiss-granite – Mica Dam	TR	200	1.15	0.80	49	Marsal, 1973
Basalts			1.54	0.82	57	De Mello, 1977
Basalt	TR		2.25	0.75	65	M. Neves, 2002
Graywacke – Bakún	TR		2.13	0.75	64	Intertechne
Graywacke 30%, Shale 70% – Bakún			1.41	0.77	55	Intertechne
Oroville – RD = 85%			1.34	0.86	53	Marachi, 1969
Sandstone – RD = 85%			0.85	0.96	40	Marachi, 1969
Basalt Marimbondo Dam						
1. quarry	TR		2.18	0.79	64	
2. dam (after 25 years)	TR		1.75	0.85	62	Maia, 2001
3. weathered (32 h – soxhlet)	TR		1.60	0.88	61	
Basalt – Campos Novos	DS	51	1.38	0.892	54	Basso e Cruz, 2006
Gneiss-schist-micaschist	DS	25.0 75.0	0.83 0.90	11		Fleury et al., 2004
Rockfill with 25% of cemented material – Tankang Hp	TR	-	3.0	0.60	72°	Peng Yii (in Zeping, 2006)
Oroville Dam	TR	-	1.12	0.82	48	Marachi et al., 1969

φ_l – friction angle for $\sigma = 1$ kg/cm²;
TR – triaxial test;
DS – direct shear test.

The parameter A is equal to $tg\varphi$ for normal stress equal to an unit and b reflects the changes of φ with σ.

A is a dimensional parameter. To avoid confusion, Indraratna et al. (cited in Maranha das Neves, 1993) suggested that τ and σ be divided by the unconfined compression strength σ_c of the rock fragments forming the new parameters A' and b', both non-dimensional:

$$\frac{\tau}{\sigma_c} = A'\left(\frac{\sigma}{\sigma_c}\right)^{b'} \tag{4.4}$$

As σ_c is not always known, here we have kept the original values of A and b.

In the 1960s and 1970s Midea (1973) ran a large number of direct shear tests and triaxial tests on samples of basalt and basaltic breccia from the rockfills of Ilha Solteira, Salto Ozório, Capivara, and Itaipu dams in boxes of 20 cm × 20 cm × 20 cm and 100 cm × 100 cm × 40 cm. In the triaxial tests the sample was 10 cm in diameter.

In these tests the influence of weathering cycles of water-stove or of glycol ethylene attached to the shear strength of the basalt was tested. In the case of the gneiss

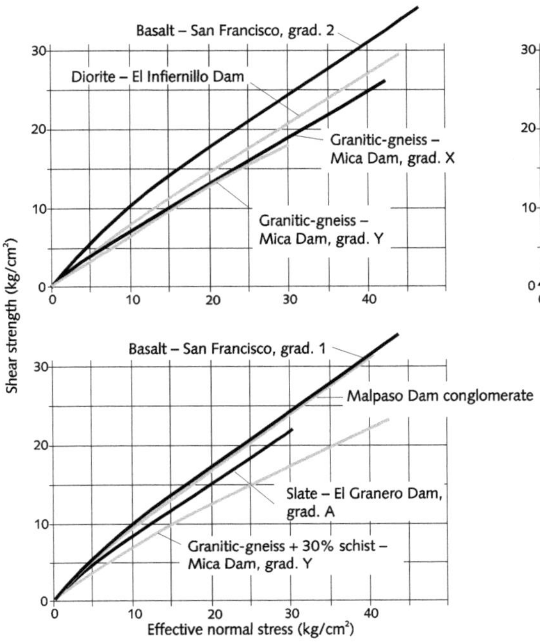

Figure 4.10 Mohr envelopes for several granular materials (Marsal, 1973).

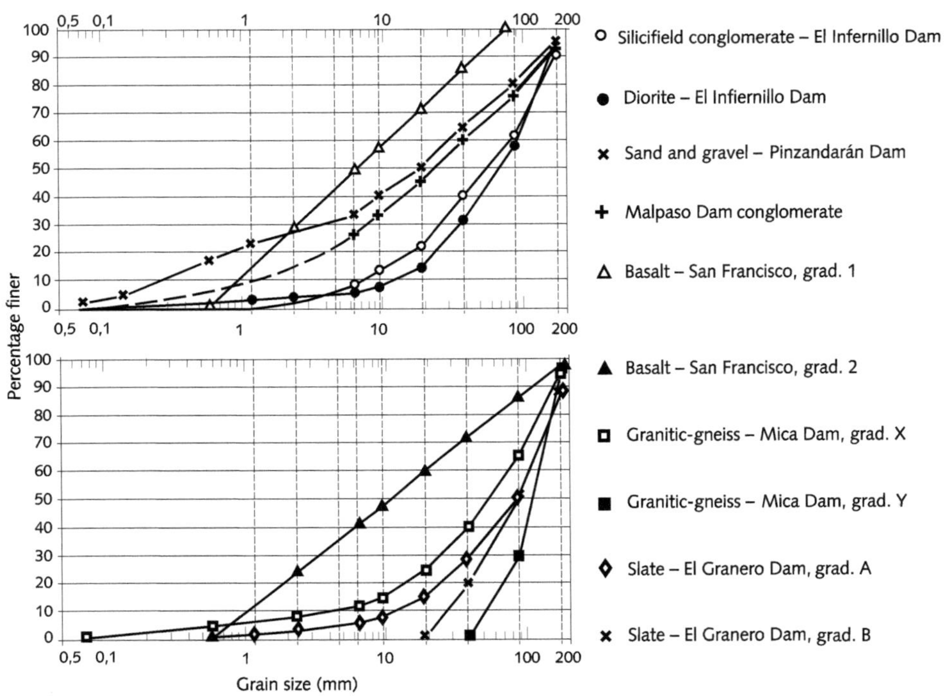

Figure 4.11 Particle size distribution of materials tested in Table 4.4 (Marsal, 1973).

Table 4.5 General characteristics of materials tested in Table 4.4 (Marsal, 1973).

Material	*Origin*	*Particle shape*	*d^e (mm)*	*Cu*	*δ*	*Void ratio*			*Pa (kg/cm²)*
						Initial	*Densest*	*Loosest*	
Silicifield conglomerate – El Infernillo Dam	Quarry blasted	Angular	5	10	2.73	0.45	0.40	0.55	
Diorite – El Infiernillo Dam	Quarry blasted	Angular	20	5	2.69	0.56	0.48	0.63	700
Sand and gravel – Pinzandarán Dam	Alluvium	Rounded	0.2	100	2.77	0.34	0.29	0.43	1,200
Malpaso Dam conglomerate	Quarry blasted	Subangular	0.8	64	2.70	0.38	0.31	0.51	500
Basalt – San Francisco, grad. 1	Blasted and	Angular	1	11	2.78	0.35	0.33	0.55	950
Basalt – San Francisco, grad. 2	crushed	Angular	1.1	18	2.78	0.37	0.29	0.46	
Granitic-gneiss – Mica Dam, grad. X	Crushed	Subangular	6	14	2.62	0.32	0.31	0.50	600
Granitic-gneiss – Mica Dam, grad. Y	Crushed	Subangular	53	2.5	2.62	0.62	0.58	0.77	
Granitic-gneiss + 30% schist – Mica Dam, grad. X			4	19	2.64	0.32	0.29	0.51	
Granitic-gneiss + 30% schist – Mica Dam, grad. Y			53	2.5	2.64	0.63	0.60	0.79	
El Granero Dam grad. A Dense slate Loose slate	Quarry blasted	Angular	11	10	2.68	0.49 0.70	0.45	0.70	590
El Granero Dam grad. B Dense slate Loose slate			27	4.3		0.59 0.69	0.64	0.84	

Cu – Coefficient of uniformity;
δ – Specific gravity;
d^e – Effective size;
Pa – Crushing strength.

from the Paraibuna Dam, the tests were run with different percentages of fines on the surface of the compacted layers in the field (Midea, 1973). Table 4.6 summarizes the main test results.

More recent direct shear tests were run at Furnas laboratory (Goiânia, Brazil) on rockfill samples from the Itapebi and the Campos Novos dams.

For the gneiss of Itapebi the direct shear strength tests were run on samples compacted to different densities. Figure 4.12 shows the test results. The value of φ corresponds to the average.

Considering the data from Tables 4.4 and 4.6 regarding basalt rockfills (excluding those that were weathered) we can see that parameter A varies from 1.13 to 2.25 kg/cm² with an average of 1.58 and that parameter b varies from 0.75 to 0.87, with an average of 0.80. These average parameters are very close to those proposed by De Mello (1977), $A = 1.54$ kg/cm² and $b = 0.82$.

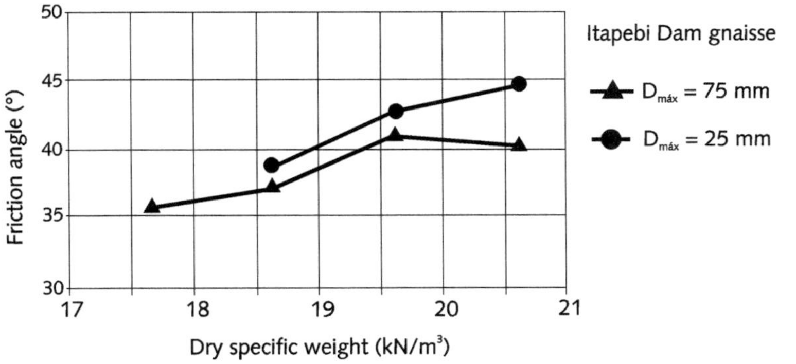

Figure 4.12 Variation of friction angle with the dry specific weight (Fleury et al., 2004).

Table 4.6 Shear strength of wheathered rockfills.

Material	*Test*	$\phi_{max.}$	*A kg/cm²*	*b*	φ_l	*Reference*
Compacted sound basalt (A)	DS	80	1.34	0.87	53°	Cruz e Nieble, 1970
in natura	DS	50				
	TR	25				
Vesicular amigdaloid basalt	DS	80	1.13	0.85	52°	Cruz e Nieble, 1970
	TR	40				
		25				
Micro-vesicular basalt in natura	DS	80	1.45	0.78	55°	Cruz e Nieble, 1970
	TR	40				
		25				
Vesicular and amigdaloid basalt	DS	80	0.96	0.94	44°	Cruz e Nieble, 1970
(B) wheathered	TR	40				
		25				
Vesicular and amigdaloid basalt	DS	80	0.93	0.90	43°	Cruz e Nieble, 1970
(C) wheathered	TR	40				
		25				
Gneiss with 3% to 9% of fines	DS	50	1.18	0.86	50°	Midea, 1973
Gneiss with 26% to 48% of fines	DS	5	1.08	0.85	47°	Midea, 1973
Gneiss with 100% of fines	DS		0.81	0.95	40°	Midea, 1973

φ_l – friction angle for σ = 1 kg/cm²;
DS – Direct shear tests in box: 20 × 20 × 20 cm e/and 100 × 100 × 40 cm;
TR – Triaxial tests on samples of 10 cm.

Graywacke from the Bakún Dam (A = 2.13 kg/cm² and b = 0.75) is a very strong rockfill. When mixed with shale, the strength drops to A = 1.41 kg/cm² and b = 0.77 (Intertechne).

Granitic-gneiss rockfill and granitic-gneiss with schist from the Mica Dam have a lower shear strength. Average parameters are A = 0.94 kg/cm² and b = 0.88. These are close to the Paraibuna's gneiss rockfill with 30% of fines (A = 1.13 kg/cm² and b = 0.85). Slate El Granero rockfill has A = 1.42 kg/cm² and b = 0.78 on average. The gravel of Oroville and Pinzandarán have high values of A = 1.74 kg/cm² and 1.53 kg/cm², and b values of 0.86 and 0.82, respectively.

Figure 4.13 shows the triaxial equipment built in China to perform large tests on rockfills.

Figure 4.14 shows particle size distribution of gravels and transitions used in some Brazilian dams and Table 4.7 contains data on the shear strength of these materials.

An interesting result from the direct shear tests on the Capivara dam basaltic rockfill is the high influence of the pre-consolidation ($\sigma_{Nmax.}$) pressure applied before the test run at a lower normal pressure σ_{test} as can be seen in Figure 4.15. This fact may explain the curvature of the strength envelope of rockfill because the compaction of the samples or of the rockfill at the dam generates a kind of rockfill pre-consolidation.

Miliroje Barbarez from Peru uses the $\Delta\varphi$ equation to reproduce the strength envelope of four sand and gravel materials. If the strength envelope is referred to as $\tau = A\sigma^b$, the parameters are those shown in Table 4.8.

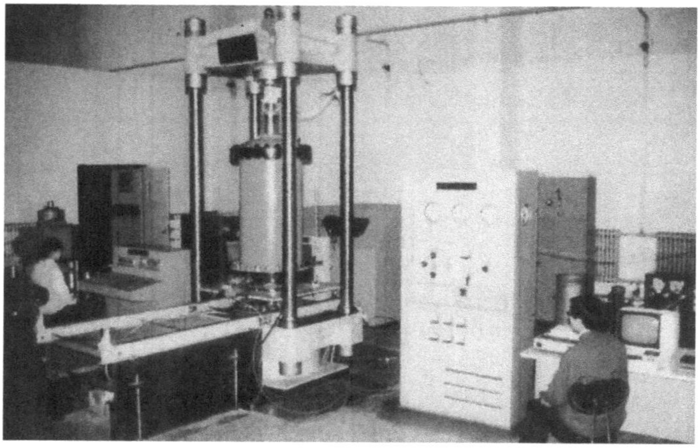

Figure 4.13 Triaxial equipment built in China to perform tests on rockfills (courtesy of Xu Zeping, 2006).

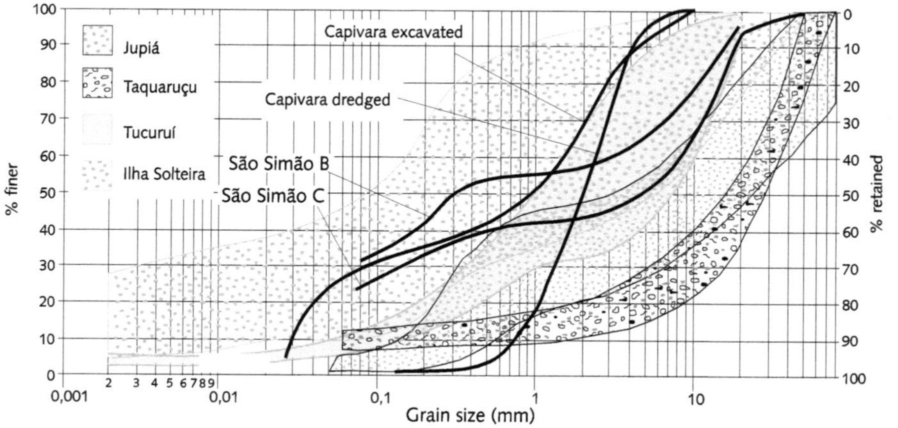

Figure 4.14 Grain size distribution of gravels (Cruz, Quadros & Corrêa, 1983).

Table 4.7 Gravel strength parameters (Cruz, Quadros & Corrêa, 1983).

Dam	Shear strength parameters				$\overline{B} = \frac{u}{\tau_v}$
	c' (kg/cm²)	φ'	c_{sat} (kg/cm²)	φ'_{sat}	
Terrace gravel Ilha Solteira, São Simão	0.34–0.67	31°–33°	0–0.39	29°–39°	0–16%
Terrace gravel – Tucuruí	0	38°–43°	0	37°–43°	2–13%
Clean gravel – Capivara	0	32°–35°			

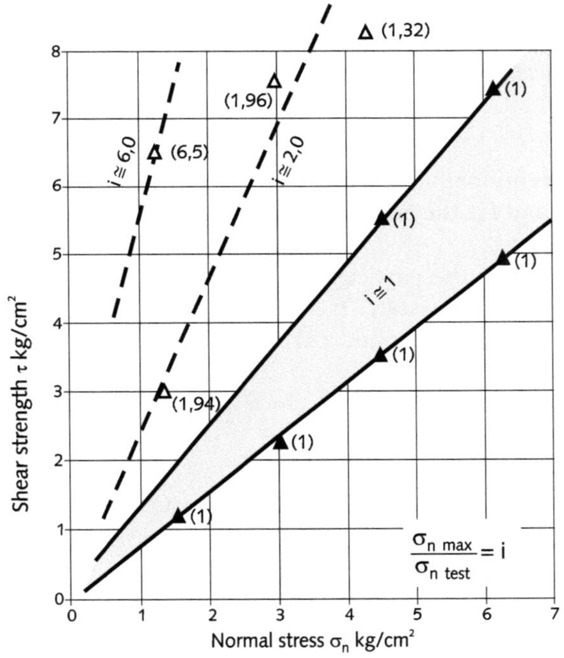

Figure 4.15 Direct shear tests on basaltic rockfill from Capivara Dam (Cruz, 1996).

Table 4.8 Strength parameters.

Oroville gravel (Marachi et al., 1969)	A = 1.12 kg/cm² b = 0.82
Calcareous sand (Billan)	A = 0.88 kg/cm² b = 0.90
Very dense sand (Bolton & Lau)	A = 0.92 kg/cm² b = 0.92
Compacted alluvium (estimated by Barbarez, 2007)	A = 0.82 kg/cm² b = 0.92

4.6 COMPRESSIBILITY

The deformability of compacted rockfill is highly influenced by two factors: 1 – the breakage and 2 – the crushing of the rock blocks.

A construction process in layers with a thickness of one maximum block diameter (0.8 to 1.0 m) or even two maximum diameters (1.6 to 2.0 m) with four passes of a 10 ton vibratory roller plus watering is enough to promote the interlocking and accommodation of the rock blocks and fragments into a dense state (low void ratio) that brings the rockfill to a condition of low compressibility. In higher dams the thickness of the layers should not exceed 1.20 m.

A skeleton of solid particles in which the larger blocks are in contact with each other is formed, leaving voids to be filled by smaller particles, some of which might even remain loose and not be subjected to any stress.

To account for this condition Marsal (1973) has even proposed a new void ratio, defined as the structural void ratio and equal to:

$$e_s = (e + i)/(1 - i) \tag{4.5}$$

where e is the conventional void ratio, equal to V_v/V_s and $i = \Delta V_s/V_s \cdot \Delta V_s$, the volume of loose particles, and i is the percentage of loose particles in relation to the total volume of solid particles.

ΔV_s is a function of the particle size distribution, the fabric, the particle shape, the void ratio and the stress state. It is easy to recognize that angular particles will produce larger ΔV_s, and that round particles would have i close to zero at relative densities near 100%.

Let us consider a rockfill with $e_{max.} = 0.42$ (loose) and $e_{min.} = 0.28$ (dense) and measured e of 0.30:

$$D_r = (0{,}42 - 0{,}30)(0{,}42 - 0{,}28) = 0{,}857 \text{ ou/or } 85.7\% \tag{4.6}$$

If, however, the volume of loose particles were 3%, e_s would be $(0.30 + 0.03)/(1 - 0.03) = 0.34$ and D_r would be only 57%.

This computation may be not correct because when $e_{max.}$ and $e_{min.}$ were determined, loose particles also existed and ΔV_s was not taken into consideration.

Even the maximum and minimum densities of a rockfill cannot be obtained at the laboratory, thus field tests are used. Figure 4.16 shows compaction test results related to layer thickness, number of passes and type of roller.

Figure 4.16 provides two important conclusions.

- The compaction of the layer decreases as initial layer thickness increases, but it should be pointed out that in thinner layers only smaller blocks can be accommodated. The smaller the blocks the larger the number of contacts, which leads to a denser structural skeleton.
- Above 1.0 m thick, the compaction of the layer only increases with the number of passes, as long as the rockfill particle size distribution is the same.

From this examination we can infer that with vibratory rollers it is possible to produce a dense rockfill (with void ratios of 0.35 to 0.25) in which particle adjustments and accommodation have produced a less deformable skeleton.

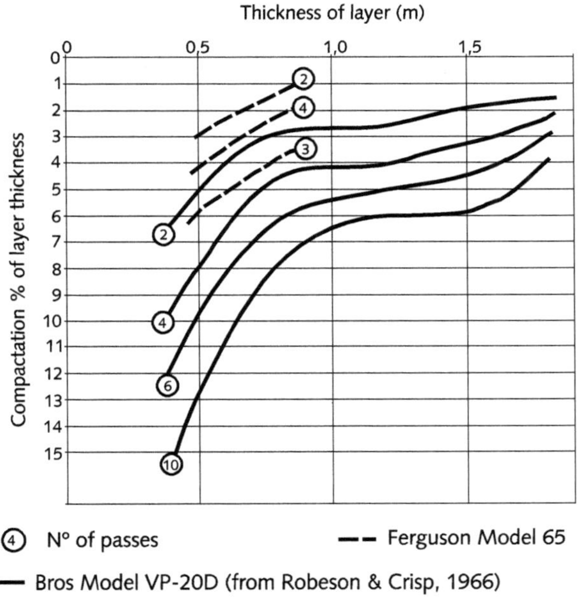

Figure 4.16 Compaction field tests (Penman, 1971).

For practical and economical reasons, four to six passes of a vibratory roller weighing 10 ton (5 ton/meter) on layers between 0.8 m to 1.0 m thick (zone 3B) and 1.6 m to 2.0 m thick (zone 3C) have given satisfactory results for CFRDs. Although layers up to 2.0 m thick have been used in zone 3C of some dams, the actual tendency is to limit the layers to 1.2 m thickness in the high dams.

The blocks in contact with each other are subjected to intergranular forces, as suggested by Marsal (1973); see Figure 4.17. He suggests that the contact forces are normal to the contact plane, but if we consider any other plane (θ in the figure) that crosses those forces, they may move in different directions. Marsal, investigating sands, gravels, and rockfills, estimated that at an average pressure of 1 kg/cm^2, the contact forces would be:

- for a medium sand – $P = 1$ g;
- gravel – $P = 1$ kg;
- rockfill ($\phi = 0.70$ m) – $P = 1$ t.

According to Marsal (1973) the contact forces P between individual particle had approximately a normal distribution. The average force P_{ave} can be expressed by:

$$P_{ave} = kD^b \qquad (4.7)$$

According to Marsal (1969) the forces that cause the contact crushing between particles, follow a similar law:

$$P_a = \eta D^\beta \qquad (4.8)$$

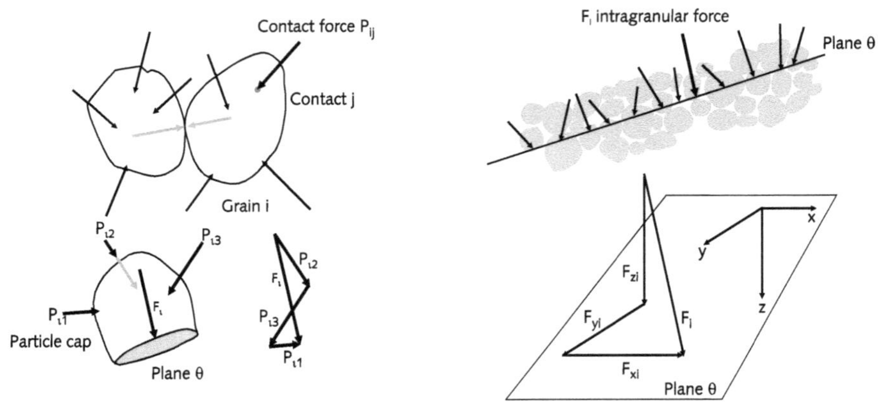

Figure 4.17 Contact and intergranular forces (Marsal, 1973).

where k, b, η and β are constants and β is smaller than b.

For two particle diameters D_1 and D_2 it is possible to build the distributions of P_{ave} and P_a as shown in Figure 4.18 ($D_2 > D_1$).

As the superposition of areas of P_{ave} and P_a is larger for D_2, it demonstrates that breakage or crushing will be higher at larger diameters.

If the internal pressure increases, the contact forces increase as well, and due to plastification, the contact areas are also boosted. Whenever the contact forces exceed the breakage strength, the particles break and a new skeleton or rockfill structure is formed. Grain breakage occurs more frequently in rockfill due to the high contact forces that are present even at moderate average pressures. The natural fissures already existing in the rock and the lower strength of weathered rock within the mass also contributes to increased breakage. When dealing with sand or gravel, the rock breakage has already occurred and the weathered particles have been washed away, so the smaller the grains or particles, the stronger they remain.

Of course it is necessary to distinguish riverbed sands and gravels from the sands and gravels of old terraces. These are already subject to a process of weathering except for the portion close to the river banks that contains hard particles of sand and gravel.

Within this context it is proper to include Maranha das Neves (2002) considerations, which are focused on the separation of the deformability of the rockfill from the creep deformability due to crushing:

> "This behavior was verified in rockfill laboratory tests. We cannot forget that the breakage of a rock fragment itself gives rise to stronger particles than the original material from where they came.
>
> It happens that the compacted rockfills have an appreciable rigidity. This behavior is improved by the increase in the creep yielding pressure due to compaction (pre-compression). The effect is not only a reflex of crushing at the contact points (De Mello, 1986) but a result of the breakage and the reduction of the void ratio. These yielding stresses (pressures) have been determined in the

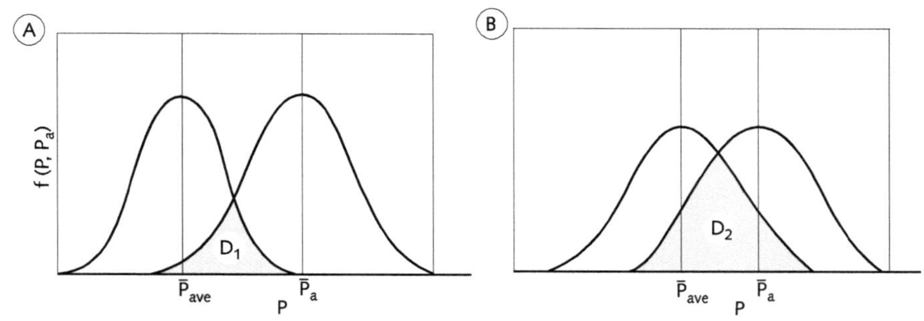

Figure 4.18 Distribution frequency of P_{ave} and P_a. A) Nominal diameter equal to D_1; B) Nominal diameter equal to D_2 (Maranha das Neves, 2002).

laboratory (Veiga Pinto, 1983 – among others) and on site (De Mello, 1986, Biarez et al., 1994).

It is interesting to observe that even when compared to the normal compression line (NCL) rockfills have a higher rigidity i.e., lower values of λ (an intrinsic parameter that reflects the deformability for average effective pressures above the average effective yielding pressures). In comparison with a large range of other granular materials, the λ values of rockfill fall in the range of 0.1–0.4 (Novello & Johnston, 1989; McDowel & Bolton, 1998). This situation corresponds to a phase in which the reduction of the void ratio would only be possible with the breakage of the particles.

The possibility of reducing void ratio by the rearrangement of the particles was entirely used at stress levels below the yielding elastic stress. This means that in sands, the NCL coincides with what those authors call the elastic yielding, and means that the deformability is entirely dependent upon breakage. As mentioned before, the yielding pressure in sands is quite high, of the order of 10 MPa.

So, from the micro-mechanics point of view, rockfills behave quite differently when compared to sands within the range of pressures to which they are usually subjected.

Thus, in rockfill the breakage (including crushing of the contacts) and the rearrangement of the particles coexist even under low pressures. It is not surprising that the yielding pressures as well as the λ values are lower. And it can also be concluded that creep in rockfills differs from creep in sands because it is more related to the crushing of the contacts than to the breakage of particles."

Rockfill compressibility can be measured in the laboratory in large compression cells (up to 1.0 m in diameter) or measured in the field using settlement cells installed at different points within the rockfill embankment.

Laboratory test results are shown in Figure 4.19 (see Maranha das Neves, 2002) and field measurements in Figure 4.20. Both curves have similarites, and both of them show a breaking point equivalent to a "pre-consolidation pressure" of soils, being the breaking point between the normal pressure (log scale) and the specific linear compression, defined by a λ parameter. Values of λ are reproduced in Table 4.9.

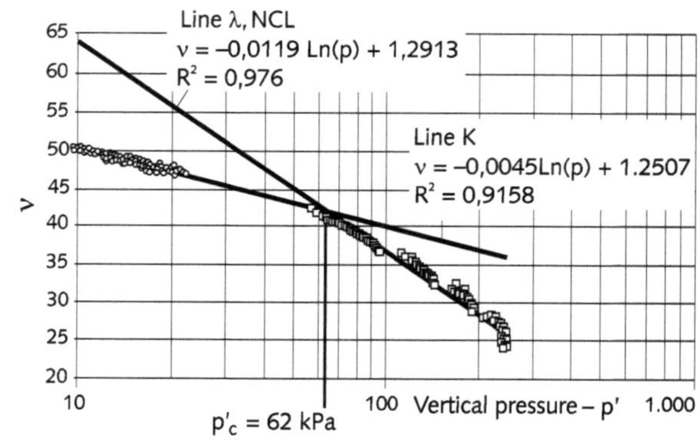

Figure 4.19 **Determination of *N*, λ and *k* parameters for dry rockfill (Mateus da Silva, 1996, cited in Maranha das Neves, 2002).**

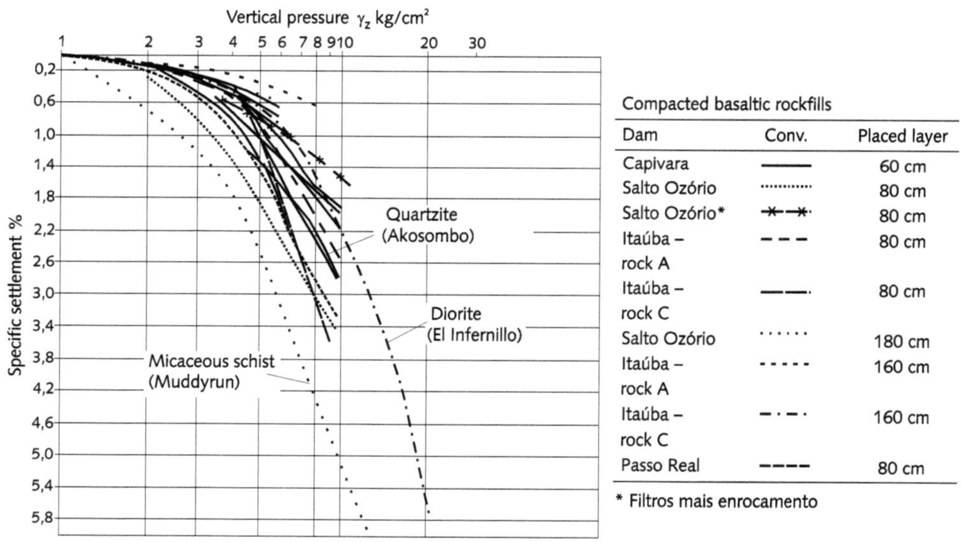

Figure 4.20 **Compacted rockfill compressibility (Signer, 1973).**

The λ parameter for rockfill measured in laboratory tests is in the range of 0.01 to 0.10, well below the λ values of other materials (0.1 to 0.4) according to Novello and Johnston (1989, cited in Maranha das Neves, 2002).

Figure 4.21 shows the equipment used for compression tests in laboratory.

Table 4.10 reproduces field data from rockfill dams. The third line for each material is the compression modulus *E*. The values of *E* range from ~5000 kg/cm² (500 MPa) at low pressures to 200 kg/cm² at high vertical pressures.

The average values of *E* measured recently in new dams show the same range of values. In high dams the average pressure is already of the order of 6 to 10 kg/cm² or

Table 4.9 λ values for different rockfills.

Rockfill	*Test*	λ *(MPa)*	*Reference*
Dry sound Graywacke (Beliche Dam)	Consolidation	0.020	Veiga Pinto (1983)
Saturated sound Graywacke (Beliche Dam)	Consolidation	0.080	Veiga Pinto (1983)
Weathered dry schist and graywacke (Beliche Dam)	Consolidation	0.055	Veiga Pinto (1983)
Weathered saturated schist and graywacke (Beliche Dam)	Consolidation	0.085	Veiga Pinto (1983)
Dry Schist (Pancrudo)	Consolidation	0.035	Oldcop (2000)
Saturated schist (Pancrudo)	Consolidation	0.092	Oldcop (2000)
Bentinck Colliery spoil (C_r = 75%; w = 8%)	Consolidation	0.045	Charles & Skinner (2001)
Bentinck Colliery spoil (C_r = 83%; w = 10%)	Consolidation	0.095	Charles & Skinner (2001)
Dry schist (Odeleite Dam)	Triaxial (hydrostatic compression)	0.013	Mateus da Silva (1996)
Saturated schist (Odeleite Dam)	Triaxial (hydrostatic compression)	0.038	Mateus da Silva (1996)
Dry schist (Odeleite Dam)	Consolidation	0.010	Mateus da Silva (1996)
Saturated schist (Odeleite Dam)	Consolidation	0.027	Mateus da Silva (1996)
Riolite slightly weathered (Talbingo Dam)	Consolidation	0.008	Parkin e Adikari (1981)

Figure 4.21 Equipment for large compression tests in laboratory, 0.5 m diameter – Furnas Centrais Elétricas S.A. (Fleury et al., 2004).

Table 4.10 Field compression data for rockfills.

			Vertical pressure (kg/cm²)					
Dam	*Material*	*Parameter**	*1*	*2*	*4*	*6*	*10*	*15*
Capivara	Basaltic rockfill, layer	1	0	0.12	0.50	1.20		
H = 60 m	0.6 m, vibratory	2	0	0.06	0.125	0.20		
(Brazil)	roller	3		1667	800	500		
Salto Osório	Compacted basal-	1	0	0.10	0.70	1.4–1.8		
H = 65 m	tic rockfill, layer	2		0.05	0.175	0.233–0.300		
(Brazil)	0.8 m, vibratory	3		2000	571	428–333		
	roller							
Salto Osório	Compacted basal-	1	0	0.10	0.50	1.20		
H = 65 m	tic rockfill, layer	2		0.05	0.125	0.08		
(Brazil)	1.6 m, vibratory	3		2000	800	500		
	roller							
Itaúba	Compacted basal-	1		0.04	0.28	0.60	1.50	285
H = 92 m	tic rockfill	2		0.02	0.07	0.10	0.15	0.19
(Brazil)		3		5000	1428	1000	666	526
Itaúba	Compacted rock-	1		0.04	0.20	0.35	0.68	1.20
H = 92 m	fill transition	2		0.02	0.05	0.06	0.068	0.08
(Brazil)		3		5000	2000	1714	1470	1250
Emborcação	Gneiss, layer 0.6 m	1	0.20	0.38	0.60			
H = 158 m		2	0.20	0.19	0.15			
(Brazil)		3	500	666	666			
Pedra do	Compacted	1	0.15	0.22	0.50	1.00	1.85	3.90
Cavalo	rock-fill	2	0.15	0.11	0.125	0.166	0.185	0.260
H = 140 m		3	666	909	800	600	540	384
(Brazil)								
Infiernillo	Compacted dio-	1		0.15	0.40	0.75	2.00	3.20
H = 148 m	ritic rockfill, layer	2		0.075	0.20	0.125	0.20	0.246
México	1.0 m D-8	3		1333	1000	800	500	405
(Brazil)								
Muddy Run	Micaceous schist	1		0.80	1.60	2.80	5.00	6.40
H = 75 m	rockfill, layer	2		0.40	0.40	0.466	0.50	0.426
(USA)	0.3–0.9 m,	3		1000	250	214	200	234
(Brazil)	vibratory roller							
Akosombo	Compacted quartzite	1		0.20	1.00	1.60	2.80	
H = 111 m	rockfill, layer	2		0.10	0.250	0.266	0.28	
Gana	0.90 m–4.0 t	3		1000	400	375	357	
(Ghana)								

*1 – Δh/h %; 2 – (cm/m)/(kg/cm²); 3 – E_v (kg/cm²).

even more, so the average modulus *E* very seldom exceed 1000 kg/cm² (100 MPa). Usually the value of *E* is in the order of 40 to 60 MPa, depending upon the layer thickness and placement technique as well as upon the compression effort.

It is important to emphasize the non linearity of the compressive modulus against pressure. It is also interesting to point out the sudden turning point of the modulus of *E* around a vertical pressure of 0.40 MPa.

Figure 4.22 illustrates rockfill breakage with compaction at Shiroro Dam.

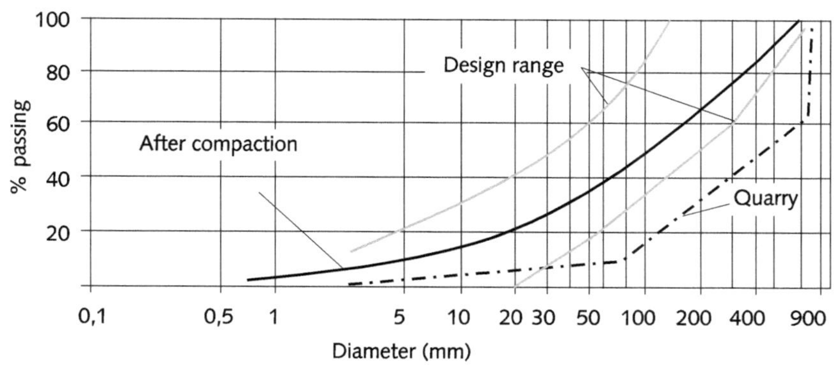

Figure 4.22 Shiroro Dam rockfill before and after compaction.

4.7 COLLAPSE

Figures 4.23 and 4.24 show that collapse has occured when water was added to the rockfill in the oedometer test at a constant vertical pressure.

Most commonly collapse occurs as a result of the crushing of rock edges, which lose strength when soaked. The phenomenon is analogous to a collapse in soils. Collapse is basically the disruption of a stable fabric due to the presence of water.

In soil mechanics the assumption is that collapse happens due to the destruction of the negative water pore pressures. In rockfill, it is due to the rock breakage or to the crushing of the block edges.

An interesting test is shown in Figure 4.25. Particles of a schist ($\sigma_c = 20{,}5$ MPa) meant to be used as rockfill in a dam were submitted to an increase of water content from 0.45% to 3.20%. A collapse of 1.34% was measured.

It suggests that collapse occurs independently of the rockfill submersion. As long as the water reaches the contact points, it is sufficient to reduce the strength and produce collapse. That means that rain-water might be enough to provoke a collapse withing the rockfill even if the rockfill is not saturated.

4.8 CREEP

The Xingó Dam rockfill creep phenomenon is shown in Figure 4.4. In less than a six year period the displacements in the concrete face doubled.

These displacements might have occurred due to water flow though the slab fissures.

Rockfill creep is a process of progressive block accommodation due to edge crushing and block breakage, which changes the original grain size distribution and the rockfill fabric over time.

In general, creep has two aspects, volumetric creep associated with hardening phenomena, and distortional creep associated with weakening processes.

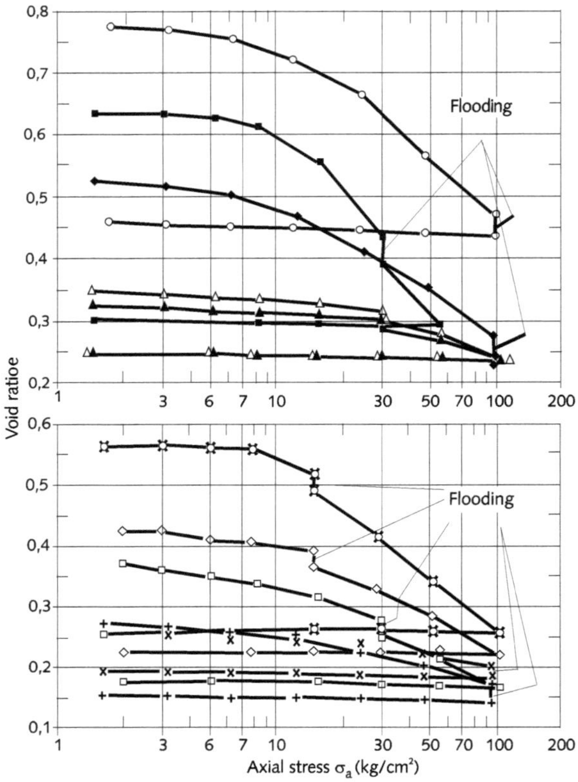

Material	Sample/symbol
Sand and gravel from Pinzandarán	1 ×
Infiernillo Dam quarry n° 1	2 ○
Infiernillo Dam quarry n° 5	3 ●
Malpaso Dam	4 +
San rancisco grad. 1	5 △
San Francisco grad. 2	6 ▲
Mica Dam grad. X	7 □
Mica Dam grad. Y	8 ■
El Granero Dam grad. A	9 ◇
El Granero Dam grad. B	10 ⌘

Figure 4.23 Collapses in rockfills (Marsal, 1973).

According to Maranha das Neves (2002):

"Deformation behavior shows three phases. The first phase corresponds to a non-linear deformation rate, in which the deformation rate decreases over time. It is mainly volumetric deformation, with increasing rigidity. When the deformation rate becomes constant, we reach the second phase. At this phase both volumetric and distortional deformations occur, the last being dominant. The third phase would only occur if the deviatory stresses were much higher then those within the rockfill mass and would generate deformations at increasing rates with a failure process dominating."

The phase that concerns concrete face rockfill dams is the second one because it will occur independent from all the actions taken during construction such as wetting, reduction of layer thickness, and increasing the number of roller passes. It is inherent to rockfills that they require some time to eventually create a stable skeleton.

See chapter 10 for more data on settlement, horizontal displacements and slab movements in CFRDs.

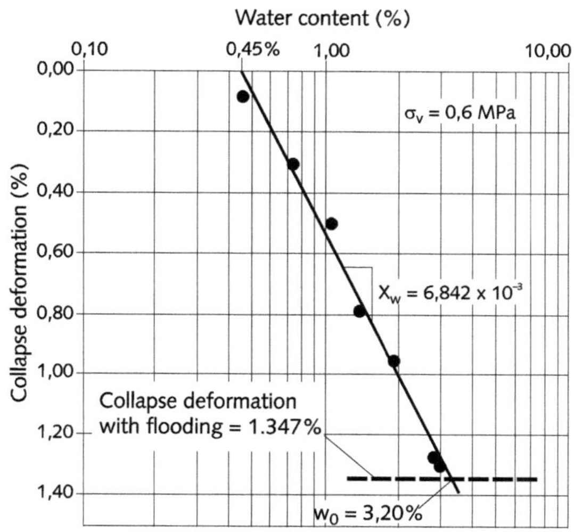

Figure 4.24 Collapse due to the increase of the water content (Oldcop, 2000).

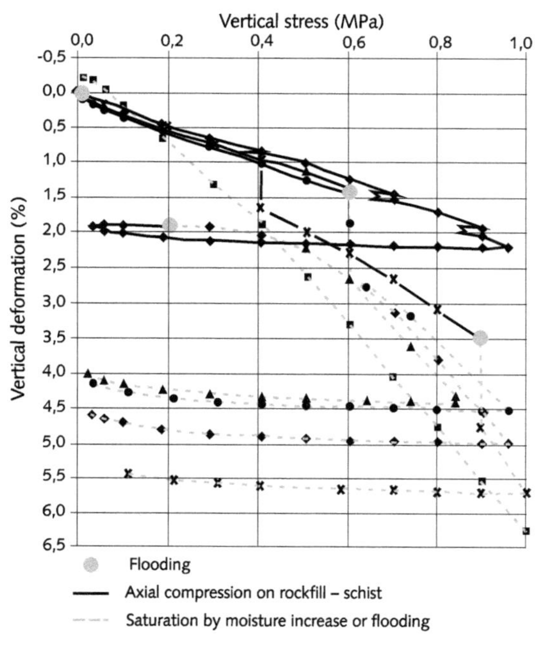

Figure 4.25 Collapse tests with flooding (Oldcop, 2000).

4.9 ROCKFILLS AS CONSTRUCTION MATERIALS

Chapter 3 presented cross sections of dams with the materials used and the corresponding construction specifications for each zone. These dams were built with a wide variety of rockfills and gravels.

As will be discussed in chapters 5 and 6, there have been no cases of concrete face rockfill dam failure, even in seismic areas, and there is unlikely to be one as long as the dams are not overtopped or subjected to excessive flow.

So, the basic requirements for design are the deformation control in all zones of the dam, and the flow control during the construction period and before the concrete face is completed.

The theoretical aspects of the rockfill mechanics have already been discussed. The observation data and the performance of these dams are considered in chapter 10.

4.9.1 Some of rockfills used in ECRDs and CFRDs

Cooke and Sherard (1987) stated that *"any hard rock, with less than 20% of particles passing through sieve nr. 4 (4.8 mm) and 10% or less of fines passing through sieve 200 (0.074 mm) has the high strength and the low compressibility necessary for the rockfills of CFRDs"*. According to them, these limits are a better way to select a rockfill than the specifications that limit a percentage finer than 2.5 cm.

Watzko (2007) in his Master of Science dissertation on the use of rockfills in CFRDs and, in particular, looking at the case of the Machadinho Dam, says that:

> "During the 1970s, the word rockfill was defined as a material which the rock particles were 70% larger than ½″ (12.5 mm) with a maximum percentage of 30% (the ideal was 10%) fines passing sieve 4 (4.8 mm). Presently these percentages are far more flexible. There are particle size distributions where the maximum size is up to 1.5 m and the percentage of fines (<4.8 mm) is 35% to 40%, and up to 10% finer than sieve 200 (0.075 mm). One criterion that is important regards the permeability of the rockfill that should be higher than 10^{-3} cm/s."

At the Salvajina Dam, 50% of rock fragments were inferior to 1″ in size.

Penman and Rocha Filho (2000) noted that rockfills are materials that do not develop pore pressures during construction and that have a permeability equal or above 10^{-3} cm/s.

For illustration purposes, the rockfills used in the Machadinho CFRD are shown in Appendix 4.1.

More information on Machadinho Dam performance is presented in Chapter 10.

1. Rockfills are compressible materials, which are in the lower deformability range of construction materials.
2. The mechanics of deformations, however, follows particular rules or processes including breakage and crushing of rock particles at low average pressures. But in fact, high pressures occur at the few rock-to-rock contact points. This phenomenon can also occur in other materials, such as sands, but at much higher stress levels (~10 MPa).
3. Collapse occurs due to a drop in the strength of rock-to-rock contact. Whether such loss in strength is due to a loss in suction – as in soils – is open to research and discussion.
4. With time rockfills are subject to creep or slow deformation under a constant state of stress.

4.10 APPENDIX – MACHADINHO DAM

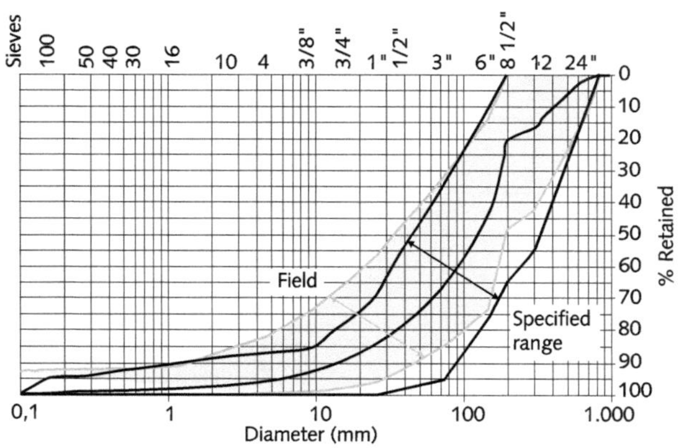

Figure 4.26 Machadinho Dam: grain size – fine rockfill – zone 3A (Watzko, 2007).

Table 4.11 Characteristics of the fine rockfill (Machadinho CFRD – 3A).

Characteristics of the fine rockfill		*Basaltic origin*
Litology	Uniform	Dense basalt
Grain sanity	Sound rock	Sound basaltic rock
Non-uniformity coefficient C_u	$5 < C_u < 30$	
Maximum diameter	$\phi_{max.} < 400$ mm	
f_{50}	10 mm $< Ø_{50} <$ 80 mm	
Permeability coefficient k	$k > 10^{-3}$ cm/s (usually $10^{-3} < k < 10$ cm/s)	
Deformability modulus	$E > 80$ MPa	

Figure 4.27 Machadinho Dam: A) Fine rockfill; B) Slope protection rockfill (Watzko, 2007).

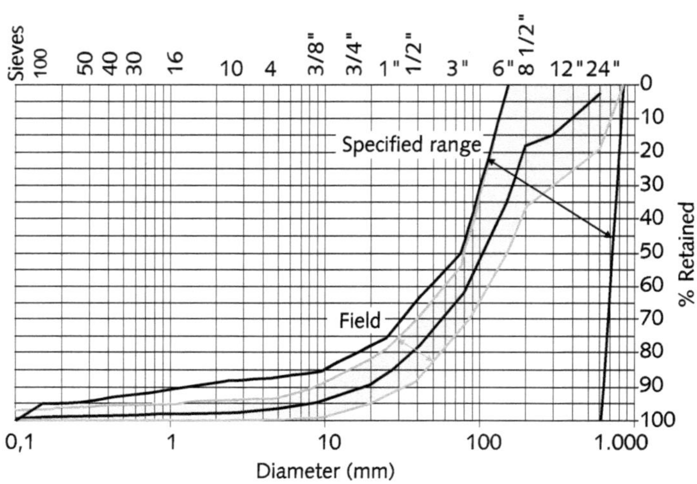

Figure 4.28 Machadinho Dam: grain size of the medium rockfill – 3B (Watzko, 2007).

Table 4.12 Characteristics of the medium rockfill – 3B (Machadinho CFRD).

Characteristics of the medium rockfill		*Basaltic origin*
Litology	Variable Basalt and Breccia	Dense basalt 70%; Breccia 30%
Grain sanity	Sound rock	Sound basaltic rock
Non-uniformity coefficient C_u	$5 < C_u < 20$	
Maximum diameter	$\phi_{max.} < 800$ mm	
f_{50}	30 mm $< \varnothing_{50} <$ 100 mm	
Permeability coefficient k	$k > 10^{-3}$ cm/s (usually $10^{-3} < k < 10$ cm/s)	
Deformability modulus	$50 < E < 90$ MPa	

Figure 4.29 Machadinho CFRD: medium rockfill (Watzko, 2007).

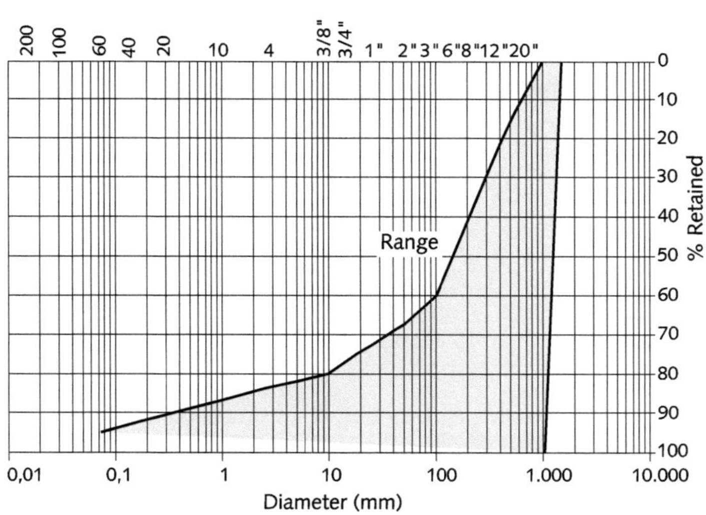

Figure 4.30 Machadinho Dam: grain size of the large and hard rockfill – 3C (Watzko, 2007).

Table 4.13 Characteristics of the large and hard rockfill – 3C (Machadinho CFRD).

Characteristics of the large and hard rockfill		*Basaltic origin*
Litology	Variable Basalt and Breccia	Dense basalt 70%; Breccia 30%
Grain sanity	Sound rock	Sound basaltic rock
Non-uniformity coefficient C_u	$C_u < 10$	
Maximum diameter	$\phi_{max.} < 1600$ mm	
f_{50}	100 mm $< \varnothing_{50} <$ 600 mm	
Permeability coefficient k	$k > 10$ cm/s	
Deformability modulus	$50 < E < 90$ MPa	

Figure 4.31 Machadinho Dam: large and hard rockfill (Watzko, 2007).

Table 4.14 Characteristics of the large and weak rockfill – 3C (Machadinho CFRD).

Characteristics of the large and weak rockfill		*Basaltic origin*
Litology	Variable Basalt and Breccia	Dense basalt 70%; Breccia 30%
Grain sanity	Sound to weathered rock	Sound to weathered basaltic rock
Non-uniformity coefficient C_u	C_u<10	
Maximum diameter	$\phi_{max.}$ < 1600 mm	
f_{50}	100 mm < $\emptyset_{50}$< 600 mm	
Permeability coefficient k	$k > 10^{-3}$ cm/s	
Deformability modulus	$15 < E < 50$ MPa	

Figure 4.32 Machadinho Dam: large and weak rockfill (Watzko, 2007).

Chapter 5

Stability

5.1 STATIC STABILITY

It may look surprising for designers of earth dams and earth-rockfill dams that in a project of a 200 m high CFRD there are no references made to stability analyses, even though this aspect is considered of major importance in the design of any other dam.

Cooke and Sherard (1987) strongly state on this subject that:

> "...rockfills cannot fail along plane or circular surfaces, whenever dumped or compacted, if the external slopes are 1.3(H):1.0(V) or 1.4(H):1.0(V), which are the usual slopes in CFRDs, because the friction angle of the rockfills are at least 45°, and this is already a guarantee of stability.
>
> Rockfills are materials of high shear strength and are dry, which means they do not have water in the voids to generate pore pressure as is the case with compacted soils. If the foundation is in rock, there is no risk of failure through the foundation."

In such a case, failure would have to develop in surfaces parallel to the slope or in deep circular surfaces throughout the rockfill, and this has not been recorded in any of the over 300 dams already built.

A case of failure in a dumped "dirty" rockfill was observed by Cruz (1996) in a mine spoil in the vicinity of Poços de Caldas, Brazil. The rockfill was being dumped in a valley as can be seen in Figure 5.1. As the bottom of the valley deepened progressively, the height of the rockfill increased while the placement was advancing. At a certain point cracks began to appear on the surface and the truck drivers refused to continue the work because they were afraid of a possible failure. The slope inclination of the dumped rockfill was much steeper than 1.3(H):1.0(V). The solution was to stop the depositing and divide the height of the fill by building a berm.

In CFRDs the upstream slope (which usually has the same inclination as the downstream one) is submitted to the stabilizing pressure of water after the filling of the reservoir. Thus, it is always more stable than the downstream-unconfined slope.

In order to illustrate this issue, static stability analyses ignoring the water flow through rockfill slopes were developed based on the charts proposed by Charles and Soares (1984).

Stability analyses which consider the water flow through rockfill and overtopping are presented in chapter 6.

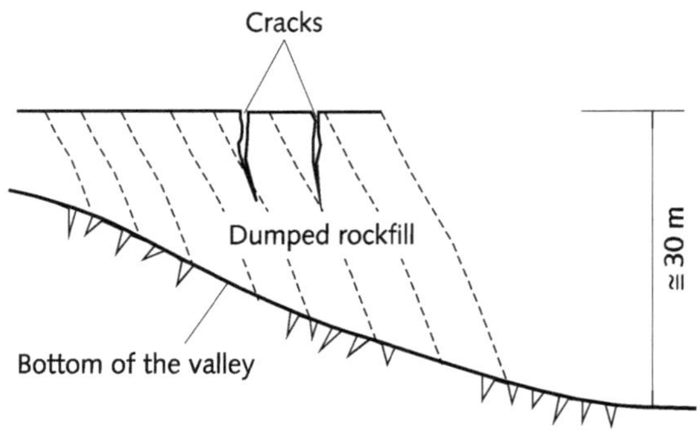

Figure 5.1 Signs of failure in the top of the dumped "dirty" rockfill.

Using the shear strength equations of rockfill presented in chapter 4, the safety factor of a sliding plane parallel to the slope is easily computed by the expression:

$$SF = (tg\varphi)/(tg\beta) \quad (5.1)$$

where φ is the friction angle of the rockfill for low normal stresses and β the slope inclination. In this case, the safety factor (SF) is independent of the dam height.

In the case of dams over 50 m high, and in cases where they reach 150 m to 200 m in height, the scenario changes. The critical slip surface develops deeper inside the rockfill mass because the shear strength of the rockfill decreases for higher stress levels.

Figure 5.2 reproduces the chart proposed by Charles and Soares (1984) allowing for a quick analysis to estimate the safety factor of a CFRD. The basis for the preparation of such charts is the slices method proposed by Fellenius and improved by Bishop.

Once both the value of b in the shear strength equation and the value of β – the slope inclination – are known, we get from Figure 5.2 the stability number or parameter Γ. The SF is computed by the expression:

$$SF = (\Gamma A)/(\gamma H)(1 - b) \quad (5.2)$$

where A is the strength parameter from equation $\tau = A\sigma^b$, γ the density and H the height of the dam.

In the expression for the SF all parameters should be expressed in the same measurement units: (ton and m), (kg and cm), (KN and m).

Figure 5.3 is a particular case of Figure 5.2 in which the exponent b is 0.75.

Figure 5.4 gives the coordinates of the critical slip surface, as a function of b and β. The critical slip surface is independent of A.

Figure 5.5A shows that for a constant value of b, the critical circles are deeper for flatter slopes. If the slope is constant, the critical circles get progressively shallower as values of b get higher. There is logic in such behavior. If b were equal to 1, the shear envelope would be linear and the critical slip surface would be a plane coincident with the slope.

Considering the expression for the SF:

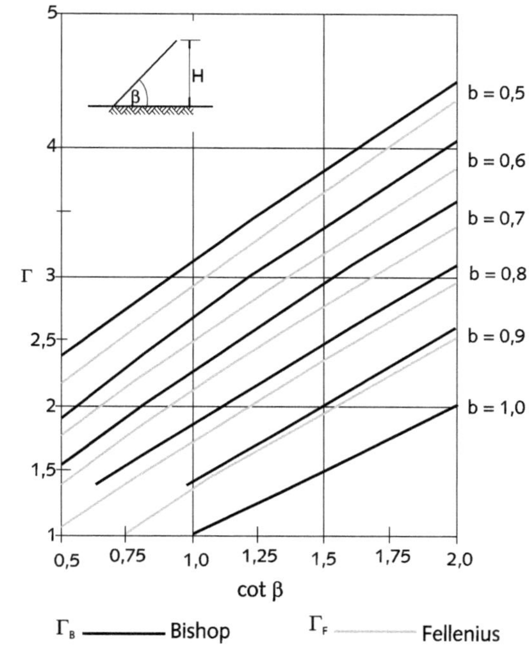

Figure 5.2 Stability numbers for circular arc analyses (Charles & Soares, 1984).

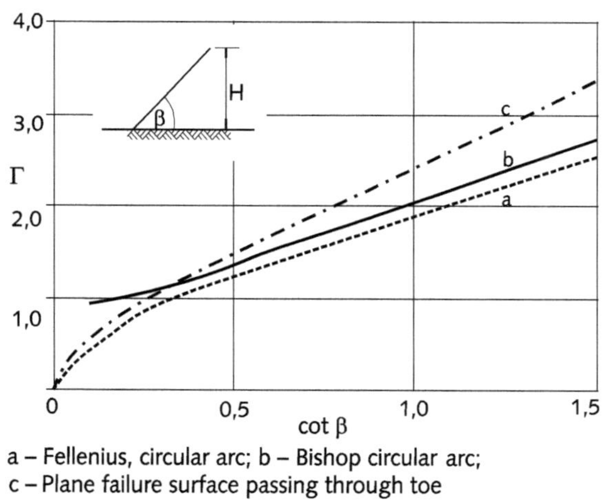

Figure 5.3 Stability analysis of a rockfill slope ($b = 0.75$) (Charles & Soares, 1984).

$$SF = (\Gamma A)/(\gamma H)^{(1-b)} \tag{5.3}$$

we see that: 1) SF decreases with increasing γ and H, and 2) SF increases with b. The parameter expresses the fall in the shear strength with increasing pressures. The lower b is, the larger the drop in strength.

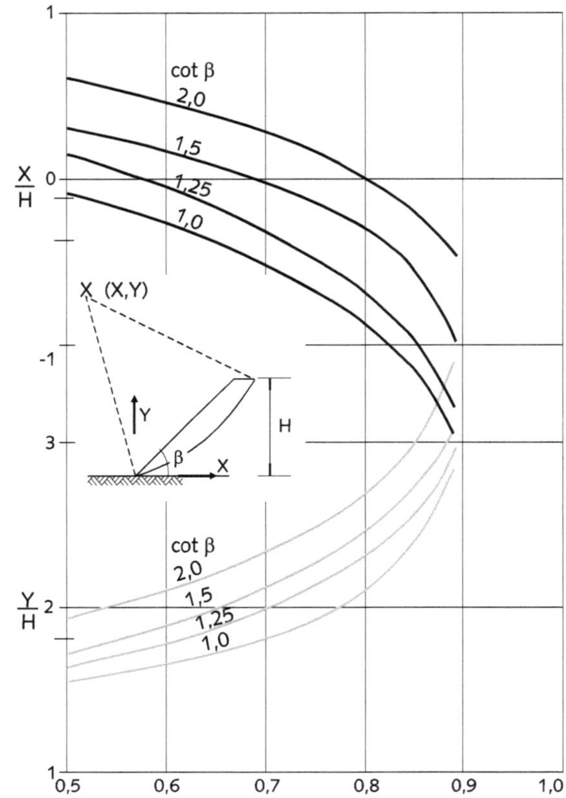

Figure 5.4 Location of circular slip surfaces centers in Bishop analysis.

For $A = 1.30$ (kg/cm^2), $\gamma = 2.2$ ton/m^3 and $H = 150$ m, and b = 0.7, 0.8 and 0.9 and a slope of 1.3(H):1.0(V), the Γ values are 2.55, 2.05 and 1.60 and the SF 1.16; 1.32 and 1.47 respectively.

The variation of SF with respect to b is practically linear (Figure 5.5B). The same happens in Figure 5.5A.

5.2 SAFETY FACTORS FOR TYPICAL ROCKFILLS EMBANKMENTS

Considering four average shear strength envelopes for rockfills of basalt, graywacke with mudstone, gneiss-granite and gravels (Table 5.1), and slopes of 1.1(H) to 1.6(H):1.0(V), it is possible to compute the safety factors shown in Tables 5.2 and 5.3. The results correspond to circular slip surfaces based on Charles and Soares, (1984) chart and for plane failure parallel to the slope respectively.

The shear strength envelopes were based on data obtained from chapter 4. The density is taken to be 2.0 t/m^3 (20 kN/m^3).

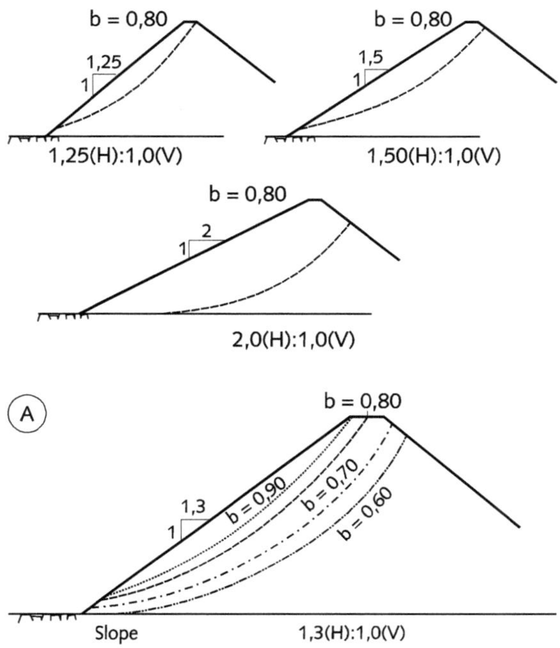

Figure 5.5A Critical slip surfaces.

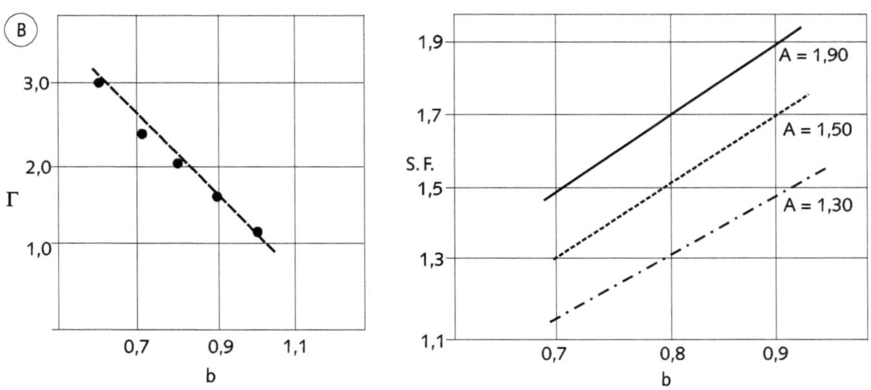

Figure 5.5B Safety factors versus *b* as a function of *A*.

Table 5.2 shows that slopes of 1.1(H):1.0(V) up to 1.6(H):1.0(V) are stable for dams ranging from 80 m to 200 m in height. For the usual slope of 1.3(H):1.0(V) the safety factors vary from 1.18 ($H = 200$ m, gravel) to 1.99 ($H = 80$ m, basalt).

Finally, let's consider the stability of two Brazilian CFRDs: Itapebi (120 m high) and Campos Novos (202 m high).

Table 5.1 Strength envelopes.

	Average values of shear strength			
Rockfill	*A kg/cm²*	*b*	φ_1	φ_2
Basalts	1.58	0.80	57.6	54.0
Graywacke and mudstone	1.41	0.87	54.6	52.2
Gneiss-granite	1.15	0.80	47.7	45.1
Gravel	1.05	0.85	46.3	43.4

φ_1' – Friction angle for $\sigma = 1$ kg/cm²
φ_2' – Friction angle for $\sigma = 2$ kg/cm²

Table 5.2 SF values for circular slip surfaces.

Rockfill	*H = 80 m*					*H = 150 m*					*H = 200 m*				
Slope	1.1	1.2	1.3	1.4	1.6	1.1	1.2	1.3	1.4	1.6	1.1	1.2	1.3	1.4	1.6
Basalt	1.77	1.85	1.99	2.15	2.30	1.56	1.67	1.75	1.89	2.03	1.48	1.54	1.64	1.79	1.91
Graywacke and mudstone	1.56	1.68	1.85	1.95	2.16	1.43	1.55	1.70	1.80	1.99	1.38	1.49	1.64	1.72	1.91
Gneiss-granite	1.21	1.31	1.44	1.52	1.69	1.12	1.20	1.33	1.40	1.55	1.07	1.16	1.27	1.34	1.49
Gravel	1.18	1.30	1.35	1.42	1.64	1.07	1.18	1.23	1.30	1.59	1.03	1.13	1.18	1.24	1.43

Table 5.3 SF values for plane failure parallel to slope.

Rockfill	*H = 80 m*					*H = 150 m*					*H = 200 m*				
Slope	1.1	1.2	1.3	1.4	1.6	1.1	1.2	1.3	1.4	1.6	1.1	1.2	1.3	1.4	1.6
Basalt	1.63	1.78	1.93	2.07	2.37	1.63	1.78	1.93	2.07	2.37	1.63	1.78	1.93	2.07	2.37
Graywacke and mudstone	1.46	1.59	1.72	1.85	2.12	1.46	1.59	1.72	1.85	2.12	1.46	1.59	1.72	1.85	2.12
Gneissgranite	1.13	1.24	1.34	1.45	1.65	1.13	1.24	1.34	1.45	1.65	1.13	1.24	1.34	1.45	1.65
Gravel	1.08	1.18	1.27	1.37	1.57	1.08	1.18	1.27	1.37	1.57	1.08	1.18	1.27	1.37	1.57

Itapebi:

- Upstream slope 1.25(H):1.0(V)
- Downstream slope 1.35(H):1.0(V)
- Rockfill $\gamma = 2.10$ t/m³
- Gneiss-schist-mica-schist $\tau = 0.90\ \sigma'$ (kg/cm²) (linear equation)
- Stability number: upstream $\Gamma = 1.20$; downstream $\Gamma = 1.40$

Safety Factors:

- Upstream slope SF = $(1.20 \times 0.90)/(2.10 \times 10^{-3} \times 120 \times 10^2)^0 = 1.08$
- Downstream slope SF = $(1.40 \times 0.90)/(2.10 \times 10^{-3} \times 120 \times 10^2)^0 = 1.26$

Campos Novos:

- Upstream slope 1.3(H):1.0(V)
- Downstream slope 1.4(H):1.0(V)

- Rockfill $\gamma = 2.25$ t/m^3
- Basalt $\tau = 1.38\ \sigma^{0.89}$ (kg/cm^2) (strength envelope)
- Stability number: upstream $\Gamma = 1.80$; downstream $\Gamma = 1.95$
  Safety Factors:
- Upstream slope SF = $(1.80 \times 1.38)/(2.25 \times 10^{-3} \times 202\ .\ 10^2)^{0.11} = 2.48/1.52 = 1.63$
- Downstream slope SF = $(1.95 \times 1.38)/(2.25 \times 10^{-3} \times 202 \times 10^2)^{0.11} = 2.69/1.52 = 1.77$

For the upstream slope of the Itapebi Dam, the shear strength equation $\tau = 0.90\ \sigma$ is a limit condition because the material tested came from the worse rockfill that was partly used in the downstream shell. If we considered data from Marsal (1973) for the gneiss-granite-schist of Mica Dam $\tau = 1.15\ \sigma^{0.80}$ or Midea's (1973) tests on gneisses from Paraibuna Dam $\tau = 1.18\ \sigma^{0.86}$, the SF of the Itapebi upstream slope would be 1.27 and 1.42 respectively. After the filling of the reservoir, the upstream stability increases significantly.

5.3 STABILITY IN SEISMIC AREAS

CFRDs have been reported as being resistant to the dynamic effects caused by seismic events. As the dam embankment is generally dry, the vibrations induced by an earthquake do not develop pore pressures that could affect the stability of the structure in a catastrophic way. Never the less, earthquakes can cause a densification of the rockfill with settlements and displacements of the slopes. In the case of strong earthquakes, the face slab can break leading to an increase in the flow in the downstream direction (Cooke & Sherard, 1987).

While this chapter was being written a strong earthquake took place in the province of Sichuan, China. This was one of the strongest seismic events ever to happen in that country, and it claimed a high number of victims.

The earthquake was of magnitude 8, and the epicenter was quite shallow – only 10 km deep – and it lasted for 1 minute. According to reports from Chengdu, the earthquake affected many dams in the area.

The Zipingpu CFRD, 160 m high, took the shake well, although some fractures seem to have occurred on the concrete face.

New criteria have been defined and applied to regions of known seismic activity using the parameters set out below according to the international terminology.

- **Maximum Possible Earthquake** (MPE) – The highest earthquake magnitude that could ever occur in a zone of seismicity.
- **Maximum Design Earthquake** (MDE) – The highest earthquake magnitude a dam can withstand. It may suffer severe damage to its structure but this can be repaired.
- **Operational Basic Earthquake** (OBE) – The earthquake that corresponds to a basic acceleration for which the damage can be repaired while the dam is still operational. Thus, this is an earthquake of smaller acceleration than the MDE.

Once the above parameters and the accepted risks have been deffined for an area, the geomechanical parameters are determined in triaxial tests and hyperbolic models.

One excellent example on how to compute these parameters is given by Romo (1991) for the 187 m high Aguamilpa Dam in Mexico.

The Mohr envelopes are curves that vary with the confining pressure, causing variations of the friction angle, as the confining pressure increases. An equation that represents such changes is:

$$\varphi = \varphi_0 - \Delta\varphi \log (\sigma / Pa) \qquad (5.4)$$

where φ is the friction angle; φ_0 is the friction angle for Pa; $\Delta\varphi$ is a material characteristic; σ is the confining pressure; Pa is the atmosphere pressure taken as reference.

Similar expressions were proposed by Leps (1970). Other geomechanical parameters such as density, specific weight, void ratio and Poisson modulus are determined in the laboratory or defined by correlation with similar materials that have already been tested.

5.3.1 Seismic safety factor

The seismic safety factor can be computed by the equation:

$$\mathrm{SF} = tg\ \varphi / tg\ (\beta + \delta) \qquad (5.5)$$

where φ is the average friction angle; β is the slope inclination angle; δ is the arctg α; α is the coefficient of seismic acceleration.

The coefficient of seismic acceleration is defined by the criteria mentioned earlier (MCE, MDE, OBE).

The stability equation is similar to the Bishop method and explains why rockfills with typical slopes such as 1.4(H):1.0(V) in areas of seismicity up to $\alpha \leq 0.3$ g do not fail. If the rockfill has average friction angles above 48°, $(\beta + \delta)$ must be less than φ.

In Brazil, where there is no seismic activity, dams built with slopes of 1.3(H):1.0(V) are stable as it is the case with Campos Novos (202 m) and Barra Grande (185 m) as demonstrated in Figure 5.2.

In countries with high seismic activity like Mexico, Argentina, Chile, Colombia, Ecuador and so on, it is a current practice to adopt a more refined analysis. The first step is to determine the state of stresses in the embankment before an earthquake. Such computation considers the construction history of the dam in its successive phases:

- elevation of the fill by superposition of compacted layers during construction;
- construction of the face slab;
- filling of the reservoir.

For the simulation, different finite elements programs are applied to consider the non-linear stress-strain behavior of the materials as expressed by a hyperbolic relation. Variable deformation modulus of the materials are calculated according to stress evolution. The deformations at the end of construction are obtained, as are the displacements due to the filling of the reservoir.

Using these parameters it is possible to define the settlements and the horizontal displacements at the end of construction and during the filling of the reservoir.

Figure 5.6 schematically shows the settlements and displacements calculated at the Santa Juana Dam in Chile (Troncoso, 1993) for the upstream slope before simulating an earthquake using the mathematical model.

Figures 5.7 and 5.8 show typical test results on the materials used in the Aguamilpa Dam (Romo, 1991). The drop in the friction angle with the confining pressure σ_3 is clearly noticeable.

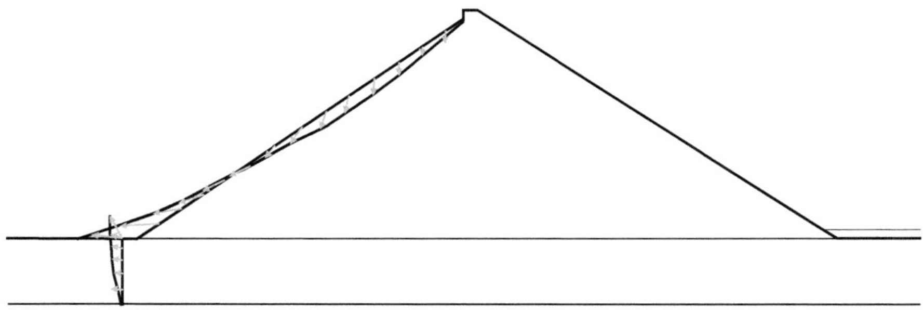

Figure 5.6 Schematic shape of settlements and displacements at Santa Juana (Troncoso, 1993).

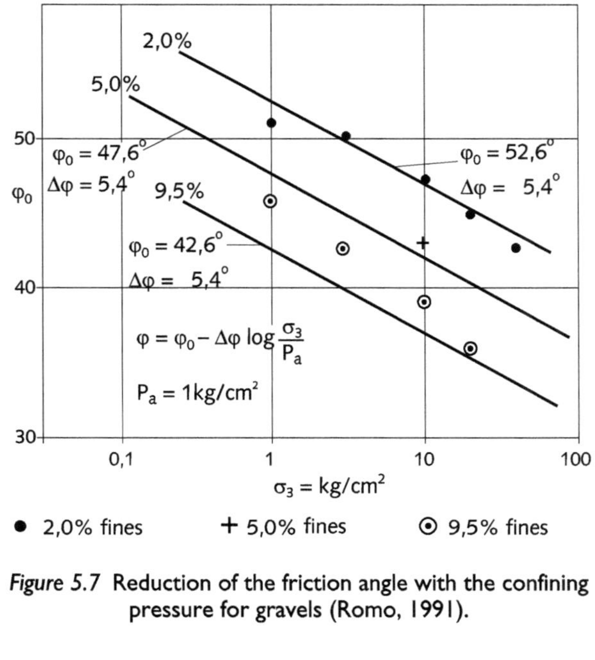

Figure 5.7 Reduction of the friction angle with the confining pressure for gravels (Romo, 1991).

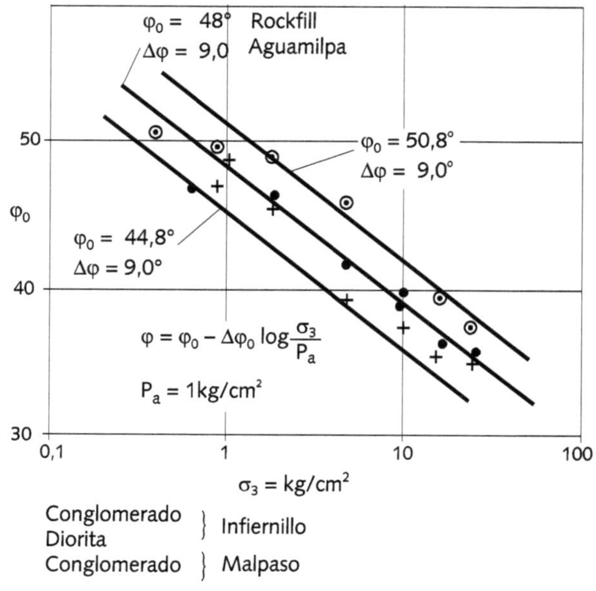

Figure 5.8 Similar reduction observed on rockfills (Romo, 1991).

5.4 DYNAMIC ANALYSIS

Whenever regional accelerations are high, a dynamic analysis is necessary in order to predict the behavior of a dam in an earthquake. During an earthquake unstable conditions may arise within the periods in which the pulses – due to the earthquake – exceed the safety seismic accelerations; i.e. when the acceleration coefficient α is larger than 0.3 g for the pseudostatic computation.

These accelerations lead to displacements whose magnitude depends on the active earthquake duration.

It is important to consider the propagation of the waves that have been generated all the way into the foundations of the structure, because the seismic waves are modified by the dynamic properties of the materials they travel through on the way.

To compute such effects inside the structure, finite element analysis programs are used. The dynamic parameters involved are the shear modulus G of the dam, the dumping ratio π, and the mass density γ for each element of the finite element mesh. Each element of the model has independent and variable dynamic properties, according to their position inside the dam.

The shear modulus and the dumping rates are determined for each material by using special programs. Correlations are estabilished through an interactive process with the purpose of relating the unit deformation to shear modulus and to the dumping - deformation percentage.

Figure 5.9 shows typical correlations for the Santa Juana Dam, Chile.

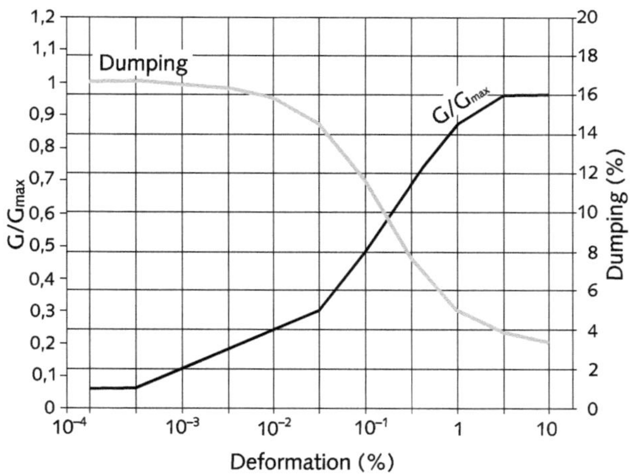

Figure 5.9 Shear modulus ratio and dumping effect as a function of deformation (Troncoso, 1993).

At Aguamilpa Dam the dynamic properties of the materials were defined by a finite element analysis that considered the non-linear effects in the fill when it is subjected to seismic loads and computed the G modulus and the dumping rates of the displacements within the dam.

5.5 SEISMIC DESIGN SELECTION

The seismic risk ratio of a dam site is determined by analyses of probabilities and determinants.

First, the probability that a given acceleration be surpassed during a certain period of time is determined. Then, after an analysis of the historical seismic activity of the area in which the dam is to be built, the seismic design is established. It is defined by the magnitude on the Richter scale, the distance from the epicenter, and the maximum probable acceleration. Empirical correlations by different authors are used to calculate the seismic accelerations as a percentage of g (for design periods from 50 to 100 years).

A supplementary analysis is performed in order to map out the seismic regional framework by surveying the registered seisms on regional projects. By using a deterministic approach, the historical seisms are defined in terms of magnitude, distance from the epicenter, maximum acceleration, characteristic acceleration, earthquake duration, and its dominant period.

Figure 5.10 shows a typical acceleration chart that would be analyzed in order to determine the seism characteristics to simulate in a dynamic stability analysis of the dam.

Once the maximum design earthquake (MDE) and the maximum possible earthquake (MPE) are defined, the next step is to analyze the effect of such a seism, which gets amplified as it gets closer to the crest of the dam due to the unconfined condition prevailing in that area.

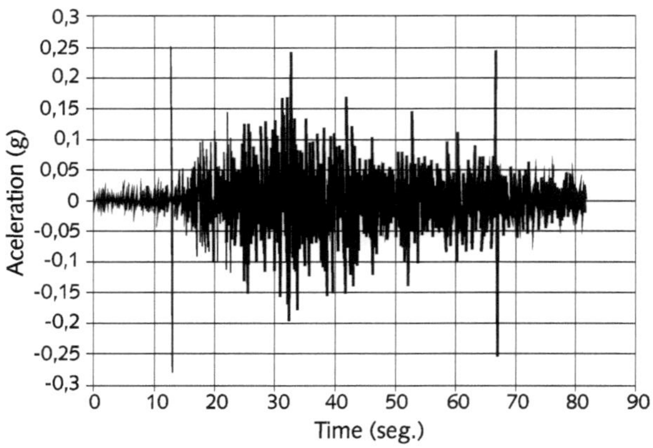

Figure 5.10 Acceleration chart.

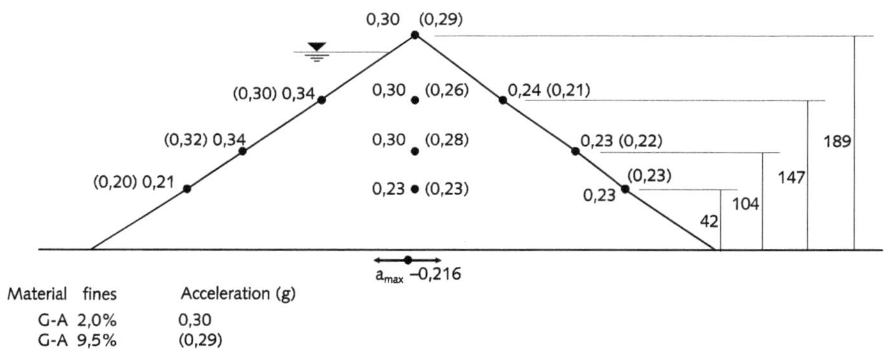

Figure 5.11 Acceleration distribution at Aguamilpa Dam.

Figure 5.11 shows the distribution of accelerations in the central zone of the Aguamilpa Dam, as well as on the upstream and downstream slopes.

5.6 SLOPE STABILITY

Once the accelerations are known, as shown in Figure 5.11, the safety factors for both upstream and downstream slopes are computed. Several slip circles can be analysed by using the Bishop method as illustrated in Figure 5.12.

The slip circles of the downstream slope are more critical because on the upstream slope the face slab compresses the upstream shell against the foundation through water pressure and this contributes to enhanced stability after the filling.

5.7 PERMANENT DEFORMATIONS

In the technical literature there are different methods used to estimate the permanent deformations of the slopes at the crest of a dam induced by earthquakes (Makdisi & Seed; Newark etc.)

Romo and Reséndiz (1980) presented a formula that allows for an equation to compute the freeboard loss L:

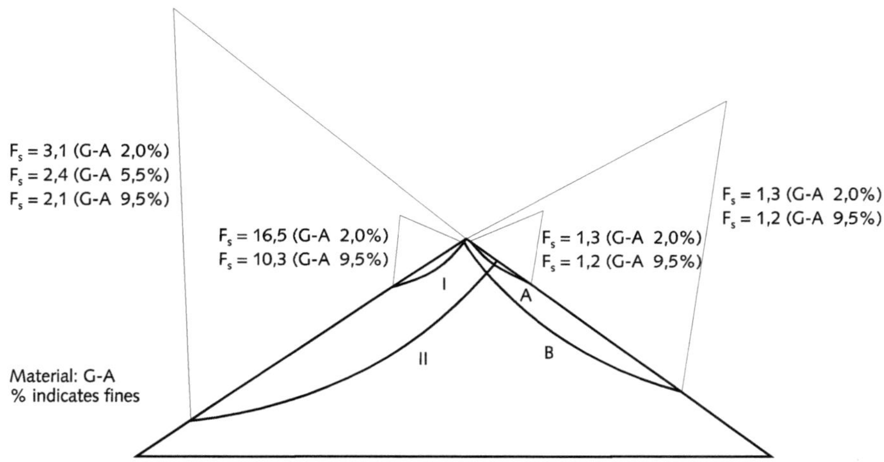

Figure 5.12 Slip surfaces at Aguamilpa Dam.

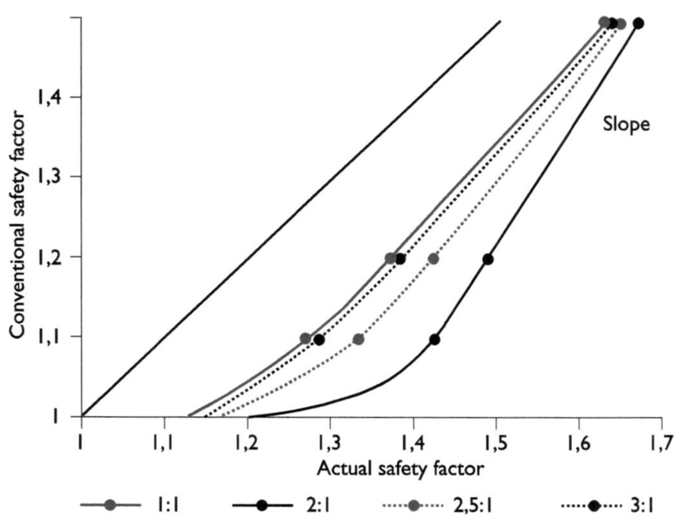

Figure 5.13 Comparison between actual and conventional safety factors.

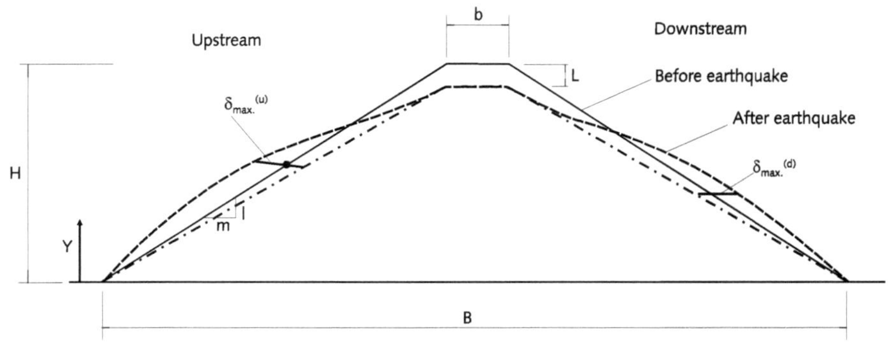

Figure 5.14 Displacements predicted by pseudostatic analyses.

$$L/H^2 = \frac{1}{(B+b)\times\left[\frac{\delta_{max.}{}^{(u)}}{H}+\frac{\delta_{max.}{}^{(d)}}{H}\right]} \tag{5.6}$$

where H is the dam height measured from the lowest point of the slip surface; B is the width of the dam at the point in which the slip surface reaches the slope; b is the crest width; u is the upstream slope d is the downstream slope.

The maximum displacements $\delta_{max.}/H$ are computed by the formula:

$$\frac{\delta_{max.}}{H} = \frac{1}{4,65\left[\frac{(F-1)Ei}{\sigma_f}\right]} - \frac{1}{1,34\left[\frac{(F-1)Ei}{\sigma_f}\right]^2} + \frac{1}{1,16\left[\frac{(F-1)Ei}{\sigma_f}\right]^3} \tag{5.7}$$

where Ei is the Young initial modulus; σ_f is the deviator stress; F is the safety factor for the dynamic conditions.

The F factor is defined by pseudo-static analysis. This factor is corrected with graphs as in Figure 5.13.

Figure 5.14 shows the parameters involved in the estimates of L. In areas of high seismicity, it is recommended to adopt a safe freeboard considering the potential loss after an earthquake.

Chapter 6

Seepage through rockfills

6.1 INTRODUCTION

Flow through in rockfills has got much attention from a fair number of researchers, but when compared to other subjects related to rockfills and dams, the references on throughflow are more limited and somewhat repetitive.

The excellent work *Flow Through Rockfill* by Leps (1973), published as part of the *Casagrande Volume*, brings 20 references.

In chapter 15 (*Flow Through and Over Rockfills)* by Thomas (1976), there are 21 bibliographical references and 21 quotations, some from Masters and Doctorate theses developed at the University of Melbourne, Australia.

The subject is revisited by Pinto (1999) in *Seepage through Concrete Face Rockfill Dams under Construction* (in Portuguese), in which he presents data from laboratory experiments. There are only three references, including Cooke and Sherard (1987) and the classical piece of work by Leps (1973) already mentioned.

Marulanda and Pinto (2000) give a general review of the problem in the J. *Barry Cooke Volume* CFRD – *Recent Experience on Design, Construction and Performance of CFRD* – containing eight references.

Cruz published two papers – *Leakage on Concrete Face Rockfill Dams* (2005, Proceedings of the International Conference on Hydropower, Yichang, China) and *Stability and Instability of Rockfills During Throughflow* (2005, Dam Engineering magazine) – including 20 and 10 references respectively.

The interest in analyzing the flow in CFRDs can be summarized in the words of Cooke and Sherard (1987): "*One further advantage of compacted rockfill over dumped rockfill dams is their ability to withstand passage of flood water through and over the uncompleted dam*", but they also mention that "*for possible flow over the rockfill, reinforced rockfill is necessary*".

The recent failure of part of the downstream rockfill at the Arneiroz Dam II in Ceará (Brazil) in 2003 – repeating the disaster that occurred at Oroz Dam in 1961, downstream from Arneiroz II on the same Jaguaribe River – confirms the fact that dumped and even compacted rockfills can fail when overtopped. On the other hand, rockfills have resisted throughflow and over flow as discussed in sections 6.3 and 6.4 below.

In 1967 Olivier proposed a formula to compute the maximum flow admissible through rockfills based on his laboratory experiments. But so far we still do not have a practical way to solve the problem.

In the next sections, the basic questions related to throughflow in rockfills are discussed both in respect to theory and practice.

6.2 THEORIES ON FLOW THROUGH ROCKFILLS

Even though most authors state that the flow in rockfills is turbulent, Penman (1971) says that if the permeability of the rockfill is 10^{-3} cm/s or less, it is to be analyzed according to the theories of soil mechanics. Thus, the flow is laminar and controlled by Darcy's law, expressed by $v = ki$.

To ensure a permeability of 10^{-3} cm/s, the fine fraction of rockfill comprised of rock powder of sand size, rock fragments, and even soil would have to fill the voids between the rock blocks and therefore control the flow.

At Itauba Dam (Brazil – rockfill with clay core), it was common to see tiny water pools on top of the rockfill. A pratical test to decide whether to approve or reject the rockfill consisted of digging a shallow trench in the rockfill with the blade of a scraper, filling as it with water and taking a walk. If soon after returning from that walk the water was still there, the rockfill would have to be removed; but if the water was not there, the rockfill could stay.

A simple calculation can be made to estimate the rockfill permeability. If $v = ki$ and the flow is vertical, $i = 1$. In one hour the travelled distance would be $d = 3600\ v$ ($= 3600\ k$), and for a water level of 50 cm, the required permeability would be $k = 0.0138$ cm/s.

If k is 10^{-3} cm/s, after one hour water would still remain inside the trench. For a heavy rain of, let's say, 200 mm in one hour, the velocity of infiltration (or permeability) would have to be of 5.55×10^{-3} cm/s in order to drain the water.

Water is usually added during or before compaction of rockfill to accelerate the settlement. On such occasions it is possible to estimate the rockfill permeability. If a volume of 300 liters per cubic meter of rockfill is added, then the infiltration velocity should be of 8.33×10^{-3} cm/s to allow the water flow freely away in within one hour.

This analysis suggests that rockfills must have a minimum k in the order of 10^{-2} cm/s to avoid the occurrence of water impounded on the surface.

Whenever rockfills with fines are used in CFRDs, it is recommended they are placed in the central area, because if the amount of fines is excessive the shear strength will drop and the stability of the slopes could be jeopardized.

Marulanda and Pinto (2000) also claim that adequate transitions are necessary in order to avoid migration of fines to the upstream and downstream shells of the dam. Considering the grain size distributions of Figure 6.1, we see that the two curves on the left, which are the upstream transitions, are permeable materials in the order of about 10^{-1} cm/s to 10^{-2} cm/s calculated by an expression like:

$$k = cd_{10}^2 \tag{6.1}$$

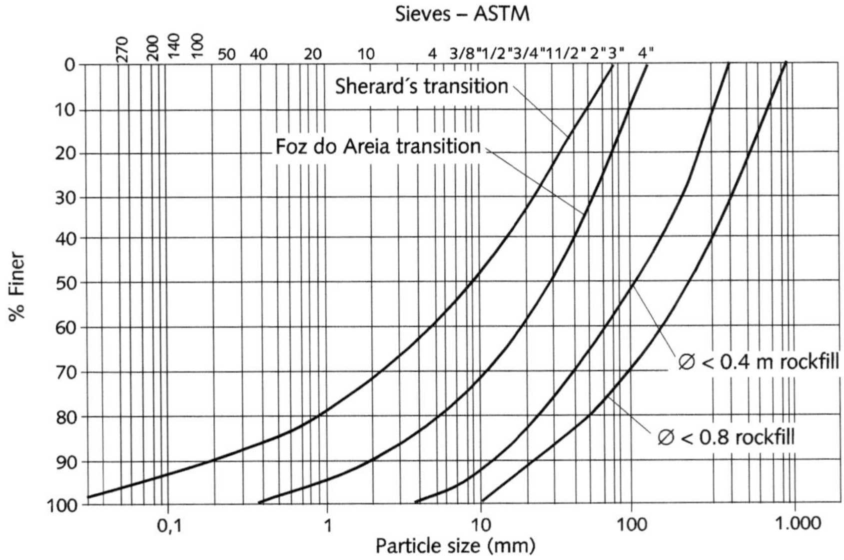

Figure 6.1 Rockfill and transitions typical curves (Marulanda & Pinto, 2000).

For the sand,

$$d_{10} = 0.02 \text{ cm} \quad \text{and} \quad k \cong 100 \times 0.02^2 = 4 \times 10^{-2} \text{ cm/s} \tag{6.2}$$

and for the transition,

$$d_{10} = 0.2 \text{ cm and } k \approx 40 \times 0.2^2 = 1.6 \text{ cm/s} \tag{6.3}$$

For the curves on the right, the permeability level has to be computed considering turbulent flow because d_{10} equals 1 cm and 2 cm.

The research developed on rockfills was aimed at two issues.

- An attempt to establish an equation for the water flow velocity, or even to define an equivalent to "permeability coefficient" or discharge velocity for rockfills.
- An attempt to define the maximum permissible flow through a rockfill that would not endanger the downstream slope stability. This question is discussed in section 6.3.

Back in 1933, Lindquist (cited in Cruz, 1979) noted that in seepage tests on glass beads of 2.7 cm diameter placed in a transparent tube flow was laminar only up to a gradient of 0.7 (see Figure 6.2).

A second test performed by Fancher et al. (1933, cited in Cruz, 1979) with sands relates the coefficient of resistance to flow (Darcy) with Reynolds number.

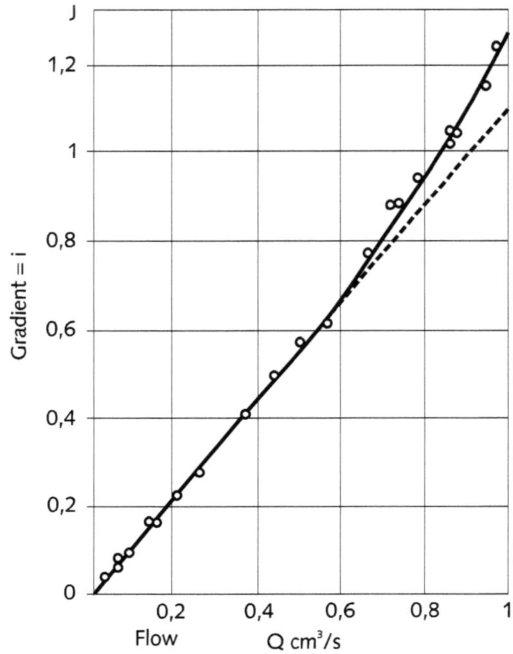

Figure 6.2 **Relationship of flow versus gradient.**

Fancher defined the λ coefficient as:

$$\lambda = \frac{d\,\frac{\Delta P}{\Delta L}}{2pv^2} \tag{6.4}$$

and the Reynolds number by:

$$R_e = \frac{pvd}{n} \tag{6.5}$$

The parameters are defined in Figure 6.3. To estimate *d*, Fancher used the expression:

$$d = \sqrt{\frac{\sum n_s d_s^3}{\sum n_s}} \tag{6.6}$$

where n_s is the number of particles with d_s (arithmetic average of the openings of two successive sieves).

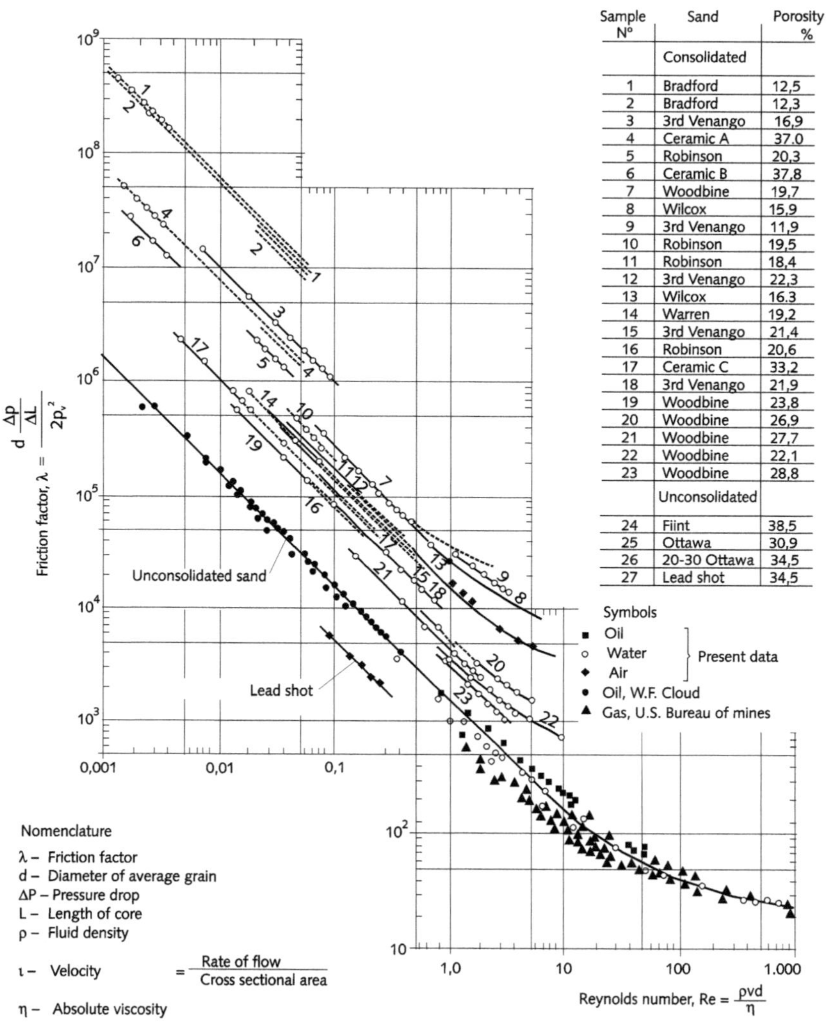

Figure 6.3 **Relationship between λ and R_e (Fancher et al., 1933).**

Usually λ is computed by:

$$\lambda = \frac{\frac{JD_h}{V^2}}{2g} \tag{6.7}$$

and R_e is computed by:

$$R_e = \frac{VD_h}{\mu} \tag{6.8}$$

where J – hydraulic gradient; D_h – hydraulic diameter; V – velocity; μ – kinematic viscosity.

What is interesting in Fancher's work is that the linearity between log λ and log R_e changes when R_e roughly approaches 1.0. However if the d value is taken as the void size instead of the particle size, the value number of R_e at the change in flow regime would be as low as 0.1.

Silveira (1964) demonstrated that the voids size distribution in sand is similar to the grain size curve, but d_v void size values were 1/8 to 1/10 of the grain size. It is well known that flow is laminar as long as the linearity between log λ and log R_e persists. After that, the flow regime becomes transient and thereafter it turns to a turbulent regime.

Twenty years after the Fancher and Lindquist contributions, Escande (1953) ran laboratory tests on rockfill and concluded that the flow was turbulent. Wilkins (1956) proposed the first formula to estimate the effective rate of velocity V_v, that is, the velocity in the voids of the rockfill:

$$V_v = C\eta^a m^b i^n \tag{6.9}$$

where: η – viscosity; m – mean hydraulic radius; i – gradient.

Leps (1973) revised these studies and shows that:

$$V_v = Wm^{0.50} i^{0.54} \tag{6.10}$$

where W is an empirical constant for each rockfill. It is primarily a function of the rock particle shape and the rugosity of water viscosity.

Wilkins, as well as others mentioned by Leps (1973), obtained a value of $W = 33$ for a granite gravel and $W = 46$ for polished spheres (glass beads) measured in inches and seconds. Wilkins, tests (1956, 1963) were carried out on samples of 20 cm to 62 cm in diameter, with particle size of 2.0 cm to 7.5 cm tested in permeameters with upward water flow.

New tests with rock block samples of 18 cm (7″) and 20 cm (8″) in diameter were run on cylinders of 90 cm (36″) and 213 cm (84″) long and gradients varying from 0.1 to 0.9. The effective velocity was confirmed as:

$$V_v = 52.5\, m^{0.5} i^{0.54} \text{(cm/s)} \tag{6.11}$$

The mean hydraulic radius according to Taylor (1948) is either the ratio of the void volume and the surface area of the particles, or the void ratio (e) divided by the surface area of the unit volume:

$$m = [e/(d/c)] \text{ for spheres.} \tag{6.12}$$

A new version of the Wilkins formula is proposed by Marulanda and Pinto (2000):

$$V = C\, i^{0.54} \tag{6.13}$$

where:

$$C = 5.24\, n\, R_h 0.5\ (m/s) \tag{6.14}$$

n – porosity (V_{voids}/total volume)

$$R_h = CV_s/A \tag{6.15}$$

where V_s – volume of solids;
A – surface area.
C is called the discharge coefficient, equivalent to "permeability" in a turbulent regime.

The ratio $\frac{V_s}{A}$ is a function of the particle diameter and its shape:

$$\frac{V_s}{A} = \frac{d}{6}\phi \tag{6.16}$$

where ϕ is a shape factor and d the equivalent sphere diameter for the same particle volume.

Leps (1973) computed the values of m (R_h) for rock particles from ¾″ up to 48″ and found the ratio:

$$m(R_h) = \frac{d}{8} \tag{6.17}$$

Comparing this value to that which Marulanda and Pinto proposed, one could take ϕ as 0.75. In fact the authors mention that ϕ must vary from 0.60 to 0.80.

For $\phi = 0.70$ they got the expression:

$$C = 1.79\, n\, d^{0.5}\, e^{0.5}$$

or

$$C = 1.79 d^{0.5}[n/(1-n)]^{0.50} m/s$$
$$a = 1/c^{1.85} \tag{6.18}$$

Marulanda and Pinto (2000) considered the grain size distribution of sands, transitions and rockfills (see Fig. 6.1) and, assuming for d the value d_{50} proposed by Leps as a representative diameter, presented values of C for different grain sizes (see Table 6.1).

Considering the values of C from Table 6.1 it is possible to estimate the discharge coefficient (a kind of turbulent permeability) of rockfills.

For the first two materials (zone 2) of Table 6.1 (sand and transition) the permeability in the laminar regime was estimated as 4×10^{-4} m/s and 1.6×10^{-2} cm/s. The first value is very far from the one in the table, but the second value is close to it. It is necessary, however, to correct the value for laminar flow k dividing it by n

Table 6.1 Rockfills "permeability" (Marulanda & Pinto, 2000).

Zona	d_{50} *(m)*	e*	*n*	*C (m/s)*	*a*
2	0.01	0.2	0.17	0.0138	2800
	0.02	0.2	0.17	0.0195	1500
3A	0.05	0.23	0.20	0.040	390
	0.10	0.23	0.20	0.056	200
3B	0.20	0.23	0.19	0.074	120
		0.28	0.22	0.094	80
3C	0.30	0.25	0.20	0.098	70
		0.30	0.23	0.123	50

*e – void ratios; $a = 1/C^{1.85}$

(porosity) for comparison reasons, because in the computation of *C*, it is considered of *K* porosity.

The new values would be 2.35×10^{-3} m/s and 9.41×10^{-2} m/s. In terms of velocity ($V = ki$), $V_v = ki/n$ in the laminar regime, and $V_v = Ci^{0.54}$ in the turbulent regime.

It is a fact that the water flow regime is a function of the gradient and at the change of regimes the velocities should be close. The gradient can be found by making the velocities equal.

For sand:

$$V_v = 2.35 \times 10^{-3}\, i = 1.38 \times 10^{-2}\, i^{0.54} \tag{6.19}$$

and $i = 46.9$. Thus the flow in the sand will be dominantly laminar.

For the transition:

$$V_v = 9.41 \times 10^{-2}\, i = 1.95 \times 10^{-2}\, i^{0.54} \tag{6.20}$$

and

$$i = 0.03 \tag{6.21}$$

So for the transitions the regime is basically turbulent.

Other authors, as Yang and Løvoll (2006), have adapted another expression for the velocity:

$$V = \sqrt{k_t i} \tag{6.22}$$

where k_t is the turbulent permeability in cm²/s² and the exponent adopted for *i* is 0.50, instead of 0.54.

As a field test a rockfill 6 m high was built with a material with $d_{10} = 3.0$ cm, $d_{50} = 12$ cm and $d_{max.} = 30$ cm, and the upstream water level was raised gradually. The flow was measured for each water level. Flow varied from 91 ℓ/s · m to 190 ℓ/s · m for water heights of of 4.07 m to 6.11 m, respectively.

The k_t permeability values estimated as a function of the data measured were 50 to 53 cm²/s². These values were compared against Bear's formula (1972, cited in Cruz, 1979):

$$k_t = \frac{1.7 d_{10} g n^3}{\beta_0 (1-n)} \tag{6.23}$$

For $d_{10} = 3.0$ cm, $\beta_0 = 3.6$ (shape factor) and $n = 0.30$ (assumed), $k_t = 53.6$ cm²/s², similar to the value measured.

The C value computed by the expression of Marulanda and Pinto (2000) is:

$$C = 1.79 d_{50}^{0.5} n \left(\frac{n}{1-n}\right)^{0.5} \text{ m/s} \tag{6.24}$$

$$C = 0.12 \text{ m/s} \tag{6.25}$$

To compare values of k_t and C it is necessary to get the square root:

$$k_t \sqrt{\frac{51.5}{0.30}} = 7.17 \text{ cm/s} \quad \text{or} \quad 0.0717 \text{ m/s} \tag{6.26}$$

Correcting the exponent of i and assuming $i = 0.25$, the value of C is 0.075 m/s, similar to the value measured at the field test.

Leps (1973), considering the failure in the Hell Hole Dam, has estimated the effective velocity that could have occurred when it failed. For a d_{50} of 20 cm and a gradient of 0.15, the velocity would be 29.2 cm/s = 0.292 m/s.

Computing the value of C for $e = 1.0$, $n = 0.50$ and $d_{50} = 0.20$ we get $C = 0.400$ m/s.

The effective velocity is very close to the one obtained by Leps:

$$V_V = \frac{C i^{0.54}}{n} = 0.286 \text{ m/s} \tag{6.27}$$

In summary, Leps (1973) reviewing Wilkins' work proposed for the velocity in the voids, the expression:

$$V_V = W R_h^{0.5} i^{0.54} = 53\ R_h^{0.5} i^{0.54} \text{ cm/s} \tag{6.28}$$

Marulanda and Pinto (2000) proposed the expression:

$$V = C\ i^{0.54} \tag{6.29}$$

where $C = 5.24$
$nR_h^{0.5} = 1.79\ d^{0.5}\ n(\frac{n}{1-n})^{0.50}$ m/s for the average velocity.

As the void velocity is equal to V divided by n:

$$V_V = \frac{C}{n} i^{0.54} = 5.24 R_h^{0.5} i^{0.54} = 1.79\ d^{0.5}\ e_{0.50}\ i^{0.54} \text{ m/s.} \tag{6.30}$$

6.3 CRITICAL ASPECTS FOR STABILITY

6.3.1 Flows

Rockfill structures might fail if submitted to a throughflow above a critical value, called the critical flow. According to Marulanda and Pinto (2000):

> "The instability of the rockfill slope initiates by the rattling of the rocks producing shallow slides at the emerging zone of the seeping water. The phenomenon tends to intensify with time as the flow concentrates in the initially scoured area. Steeper slopes are formed and thus deeper slides occur. The instability progresses upstream, reaching the crest of the fill and eventually causing the breaching of the dam."

A pioneering work by Olivier (1967) established the inclination of stable slopes for rock blocks dumped in running water. Figure 6.4 (Thomas, 1976) summarizes Olivier's work from 1967.

Leps (1973), reviewing the work of Olivier, points out the characteristics and conditions that govern the stability of the water discharging slopes:

- Rock properties: specific weight, dominant diameter, gradation and shape of the blocks;
- Relative density of the rockfill;
- Maximum hydraulic gradient;
- Inclination of downstream slope.

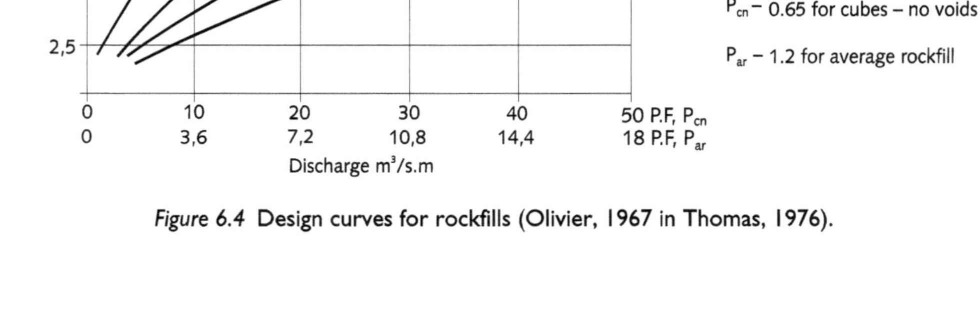

Figure 6.4 Design curves for rockfills (Olivier, 1967 in Thomas, 1976).

Olivier also proposed a packing factor to account for blocks interlocking that varies from 0.65 for a 100% relative density, up to 1.60 for relative densities of 20% to 30%.

As can be seen in Figure 6.4, the slope tested comprises a very large variation that does not concern CFRDs. In Table 6.2 the test results for slopes of 1.5(H):1.0(V), the steeper slopes tested, are summarized.

A second work on the same subject is mentioned by Thomas (1976). Tests were run by Hartung and Scheuerlein (1970) at the University of Munich. They proposed curves to estimate the permissible flow through rockfill considering the variables gradient, water height, particle size and an aeration factor.

Thomas (1976) computed the maximum permissible flows through a compacted rockfill considering three equivalent (or dominant) diameters based on the work of Hartung and obtained the values shown in Table 6.3.

We can find some relationship to the flow values obtained by Olivier for dense rockfill.

A first formula to compute the critical flow was proposed by Olivier. The metric version was developed by Collet (cited in Thomas, 1976) and is:

$$q = 0.2335(d)^{1.5}\left(\frac{\delta-\gamma_0}{\delta}\right)^{1.667} \cot g\beta - 1.167 \qquad (6.31)$$

where: δ – specific weight of the rock; γ_0 – specific weight of the water; β – slope angle.

The constant 0.2335 applies for crushed granite.

A second formula, also by Olivier (1967) and reproduced by Stephenson (1979, cited in Cruz, 1979), is:

$$q = \frac{Cg^{0.5}d^{1.5}}{n^{1/6}}\left(\frac{\delta-1}{1+c}\right)^{5/3}\frac{[\cos\beta(\text{tg}\varphi-\text{tg}\beta)]^{5/3}}{(\text{tg}\beta)^{7/6}} \qquad (6.32)$$

Table 6.2 Permissible flow.

		Permissible flow $m^3/s \cdot m$	
Downstream slope	*Dominant block size (m)*	*Loose*	*Dense*
1.5(H):1.0(V)	0.60 m	0.4	1.0
	1.20 m	1.4	3.7
	1.50 m	1.9	5.1

Table 6.3 Maximum permissible flow.

Equivalent diameter (m)	*Maximum flow $m^3/s \cdot m$*
0.30	0.40
0.60	1.10
0.90	1.80

where: n – porosity; β – slope angle; φ – friction angle of the rockfill; c – 0.25 for dumped rockfill and medium capacity.

6.3.2 Downstream slope stability

Another approach to analyse the critical throughflow and the stability of the downstream slope is proposed by Cruz (2005b) and described as follows.

CFRDs can be subjected to throughflow during a flood whenever the concrete face is not completed and the water level exceeds the cofferdam height. Both successful and unsuccessful cases have been reported in Leps, 1973; Cooke, 1984; Pinto, 1999; Marulanda and Pinto, 2000; among others.

In the present analysis only unreinforced rockfills are being considered.

Factors that seem to influence either the stability or the instability of rockfills are:

- Placement conditions, e.g. whether the rockfill was dumped or compacted;
- The average gradient in the rockfill;
- The exit height of phreatic line, and the gradient exit;
- The discharge rate.

Leps (1973) also includes:

- Rock block specific gravity;
- Rockfill dominant size;
- Rockfill block gradation and shape;
- Downstream rockfill slope inclination.

An approach to the problem is to consider the rockfill shear strength – that in itself accounts for many of the above factors – and the critical exit gradient of the flow, which if exceeded will lead to failure.

In fact any description of the failures observed in rockfills during throughflows always mentions that dams failed by progressive sliding and removal of rock blocks from the toe up to the crest.

The problem, therefore, is clearly the seepage forces that develop during throughflow and that are strong enough to uplifit and remove rock blocks from the slope, which leads to progressive failure.

A simplified analysis considering different positions of the upstream water level and a constant friction angle of the rockfill is shown in Table 6.4 for a dam 140 m high.

The safety factors were computed using Hoek and Bray (1974) stability charts, for a slope of 1.3(H):1.0(V). Data from Table 6.4 show that for a loose rockfill ($\varphi = 50°$), $\Delta H/H$ that leads to instability is in the order of 0.10. A compacted rockfill ($\varphi = 60°$) would support $\Delta H/H$ up to 0.25 or more. The analysis, however, is for illustration purposes only because the shear strength envelopes of rockfills are clearly curved, as is shown in Figure 6.5, and the assumption of a constant φ is too simplified.

Marulanda and Pinto (2000) suggest that compacted rockfills could tolerate average flow gradients up to 0.30 "*if precautions are taken such as providing large rocks at the emergent zone on downstream slope of the fill*". Furthermore, they say that:

Table 6.4 Safety factors for a rockfill slope of 1.3H:1V with throughflow.

$\Delta H/H$	H (m)	γ (t/m³)	FS $\varphi = 50°$	FS $\varphi = 60°$
0	140	2.10	1.58	2.31
0.125	140	2.10	0.92	1.23
0.250	140	2.10	0.79	1.08
0.500	140	2.10	0.74	1.01
1.00	140	2.10	0.66	0.91

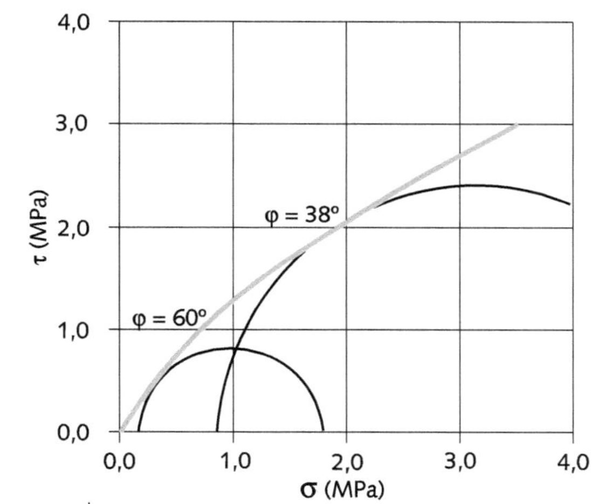

$\tau = A\sigma^b$

$\tau = 1{,}30\,\sigma^{0{,}9}$ MPa

$\varphi_{sec} = 1{,}3\sigma^{-0{,}10}$

σ MPa	φ_{sec}
0,08	60,3°
0,10	58,5°
0,25	56,0°
0,50	54,0°
1,00	52,0°
1,50	51,0°

Figure 6.5 Shear strength envelope of a basaltic rockfill (Maranha das Neves, 2002).

"the zoning of CFRDs is clearly favorable to stability in case of throughflow. The more impervious layers of the rockfill help to control the average gradient through zone 3C below the critical limit of 0.25–0.30."

Exit gradients in downstream zone

Figure 6.6 shows four flow nets drawn for isotropic permeability and laminar flow. Average gradients ($\Delta H/L$) varied from 0.16 up to 0.61. Exit heights taken from the flow nets ranged from 8 m to 58 m. The dam is 140 m high, and the external slopes

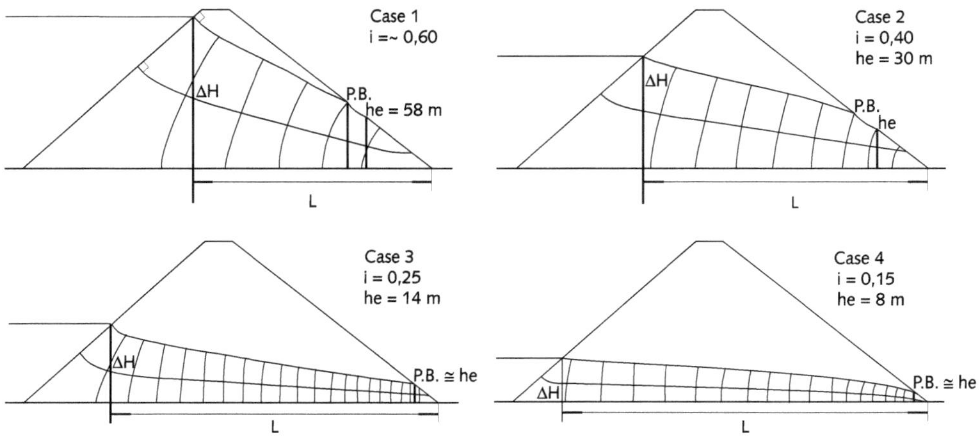

Figure 6.6 Flow nets for different water entrance levels (Cruz, 2005b).

are 1.3(H):1.0(V). Crest width length is 10 m and maximum water level is 130 m. The average exit gradient is approximately the same in the four cases because it is a function of the downstream slope inclination and it is equal to:

$$i = \sin\psi/\cos(\psi/2) = 0.643 \qquad (6.33)$$

where ψ is the downstream slope angle = 37.5°.

Another way of computing h_e, avoiding the trouble of drawing flow nets, is to consider Darcy's formula for flow:

$$Q = k \times i \times A = k\Delta H/L \times \Delta H/2 = k\ \Delta H^2/2\ L \qquad (6.34)$$

The flow at the exit height h_e is the same:

$$k\Delta H^2/2\ L = k\ \sin\alpha/\cos(\alpha/2) \cdot h_e \qquad (6.35)$$

and

$$h_e = \Delta H^2/2\ L \times [\cos(\alpha/2)/\sin\alpha] \qquad (6.36)$$

or

$$1.55\ h_e = \Delta H^2/2\ L, \quad \text{for } \gamma = 37.5°$$

Table 6.5 summarizes these data.

Results show that the computed values for exit heights and flows are similar to those obtained by flow nets. For practical purposes the values estimated by Darcy's law can be assumed.

Table 6.5 Exit heights h_e.

Case	ΔH (m)	L (m)	ΔH/L	Ave. exit i	h_e (m) Flow net	h_e (m) computed	Q – Flow net flow	Computed flow
1	130	212	0.61	0,64	58	62	40k	39k
2	100	254	0.39	0,64	30	30	22k	20k
3	70	290	0.24	0,64	14	13	6.7k	8k
4	50	320	0.6	0,64	8	6	5.2k	4k

Turbulent flows, however, seem to prevail over laminar flows in the case of rockfills. To compute flows, other considerations are necessary (see for example Leps, 1973; Marulanda & Pinto, 2000).

Figure 6.7 shows two flow nets drawn for laminar and turbulent flows. As one can see, the phreatic lines and the exit heights h_e are not so different. As a conclusion we may say that the previous analysis leads to a quite reasonable value of h_e, even when turbulent flow is considered.

6.3.3 Critical gradient

Figure 6.8 shows the downstream area of the flow nets, beginning at height h_e. This last "triangle" is always saturated and the exit gradients are $\sin\psi$ at the slope, $\mathrm{tg}\,\psi$ at the bottom and $\sin\psi/\cos(\psi/2)$ in an average direction of $\psi/2$. The seepage force F_p is $i\,\gamma_0\,V$; where $i = i_{ave}$, V = volume of the "triangle" and γ_0 the unit weight of water. The resistance force is a function of the submerged unit weight and the shear strength of the rockfill.

Silveira (1983) discusses the problem of critical exit gradients for granular materials and demonstrates that if there is vectorial equilibrium at the triangle of Figure 6.9 the angles are:

$$\delta = (\varphi - \psi),\ \alpha = (90 + \psi) \text{ and the third angle} = (90 - \varphi) \tag{6.37}$$

The submerged weight W_{sub} is equal to $\gamma_{sub}\,V$, being γ_{sub} the submerged unit weight of the rockfill. If W_p and F_p are divided by V, W_p will be replaced by γ_{sub} and F_p by $\gamma_0\,i_c$.

The critical gradient i_c can be computed by the expression:

$$i_c = \frac{\gamma_{sub}}{\gamma_0} \times (\cos\alpha - \mathrm{sen}\,\alpha\,\mathrm{tg}\beta) \tag{6.38}$$

where:

$$\beta = 90 - \alpha - \varphi + \psi \tag{6.39}$$

σ is the downstream slope angle.

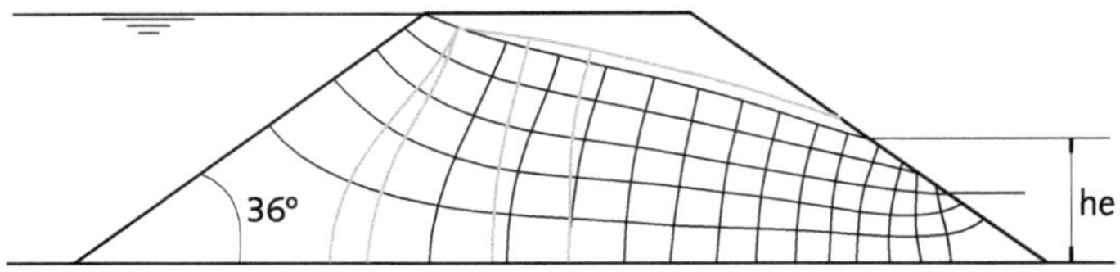

Figure 6.7 Laminar flow net with turbulent equipotentials superimposed (Thomas, 1976).

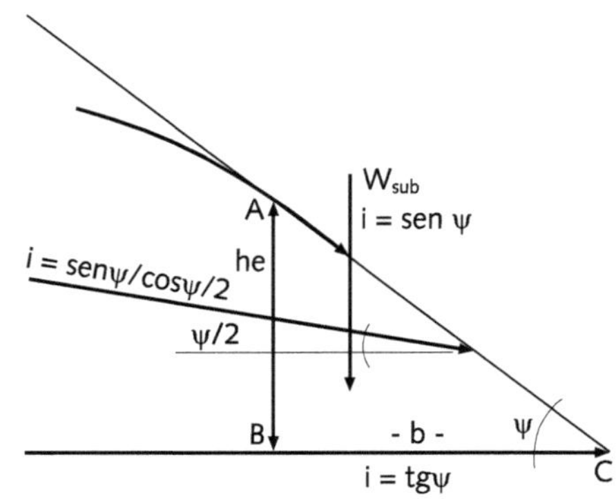

Figure 6.8 Flow gradients at the emerging zone downstream.

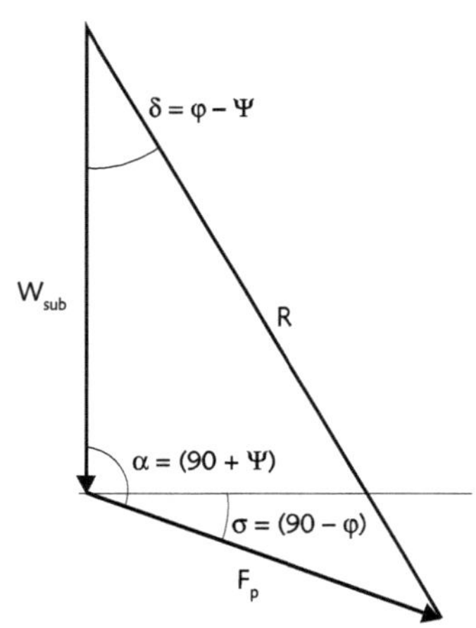

Figure 6.9 Submerged weight and seepage force at equilibrium.

Once the triangle of vectors is drawn, the seepage force and the critical gradient in any other direction can be obtained either graphically or by the same expression above, replacing α by the flow directions α_1:

$$i_c = \frac{\gamma_{sub}}{\gamma_0} \times (\cos\alpha_1 - \text{sen}\ \alpha_1\ \text{tg}\beta) \tag{6.40}$$

where:

$$\beta = 90 - \alpha_1 - \varphi + \psi \tag{6.41}$$

Stability and instability

The critical gradient i_c represents the limiting condition of stability. Whenever exceeded by the acting gradient of the flow, the rock blocks will start to be removed. From Figure 6.8, the acting average gradient in the "triangle" ABC is $i = \sin\psi/\cos(\psi/2)$. The critical gradient can be computed if φ is known. φ is a function of the average normal effective pressure σ' acting at the base BC of the triangle:

$$\sigma' = \gamma_{sub}\, h_e/2 \tag{6.42}$$

The shear strength envelope of the rockfill is:

$$\tau = A\sigma'^b \tag{6.43}$$

and

$$\varphi = \text{arctg}(\tau/\sigma'), \quad \text{variable with } \sigma' \tag{6.44}$$

If the triangle ABC is relatively small, and φ is computed for σ'_{ave}, one can calculate i_c and compare it with i. A kind of safety factor for flowing can be expressed by

$$\text{SF} = \text{i}_\text{c}/\text{i} \tag{6.45}$$

Practical applications

Referring back to Figure 6.5, the shear strength envelope has the following equation:

$$\tau = A\sigma'^b \tag{6.46}$$
$$\tau \cong 1.30\ \sigma'^{0.80}\ \text{MPa} \tag{6.47}$$

It is necessary to emphasize that the values of A are dependent on the units used. If the units are in kg/cm^2 or t/m^2, the values of A will be different; b is not affected by the units. From the flow nets of Figure 6.6, the average σ' values on the last flow "triangle" are equal to:

$$\sigma'_{méd} = (h_e\, \gamma_{sub})/2 \tag{6.48}$$

With the σ'_{ave}, it is possible to calculate:

$$\tau = \text{A}\sigma'^{\text{b}}_{\text{ave}} \tag{6.49}$$
$$\tau/\sigma = \text{arctg}\varphi_{\text{ave}} = \text{A}\ \sigma_{\text{ave}}^{(\text{b-1})} \tag{6.50}$$

Table 6.6 shows the values of σ'_{ave}, φ_{ave} and i_c.

The critical gradients i_c can be computed for the average direction of the flow, taken as:

$$\psi/2 = 18.7° \tag{6.51}$$
$$\alpha_1 = 90 + 18.7 = 108.7° \tag{6.52}$$

and

$$i_c = \frac{\gamma_{sub}}{\gamma_0} \times (\cos\alpha - \text{sen}\ \alpha\ \text{tg}\beta) \tag{6.53}$$
$$\beta = 90 - \alpha - \varphi + \psi \tag{6.54}$$
$$i_c = 1.15\ [-0.322 - 0.946\ \text{tg}(18.7 - \varphi)] \tag{6.55}$$

The acting gradient is 0.64 as shown before. So, the "safety factor" for the gradients is:

$$SF = i_c/i. \tag{6.56}$$

From Table 6.6, case 1 represents unstable conditions, case 2 is at equilibrium and cases 3 and 4 are of stable rockfill. The average gradients are 0.61, 0.39, 0.24 and 0.16.

Another example

A dam is 180 m high and the downstream slope is 1.40(H):1.0(V) (≅35.5°). The shear strength envelope for the downstream rockfill shell is:

$$\tau = 1.20\ \sigma^{0.90}\ \text{MPa} \tag{6.57}$$

The average emerging gradient on the slope will be:

$$\text{i} = \text{sen}\,\psi/\cos(\psi/2) = 0.61 \tag{6.58}$$

To reach stability the critical gradient should be equal or higher than the exit gradient:

$$i_c \le 0.61 \le \frac{\gamma_{sub}}{\gamma_0}(\cos\alpha - \text{sen}\ \alpha\ \text{tg}\beta) \tag{6.59}$$

Table 6.6 Critical gradient and SF values.

Case	h_e	σ_{ave} *MPa*	φ_{ave}	i_c	$SF = i_c/i$	$\Delta H/L$
1	60	0.35	58°	0.52	0.81	0.81
2	30	0.17	62°	0.65	1.01	0.39
3	13.5	0.08	65°	0.76	1.18	0.24
4	7.0	0.04	68°	0.89	1.39	0.16

$$\alpha = 90 + (35.5/2) = 107.8° \quad (6.60)$$
$$\beta = (\psi/2) - \varphi = 17.8° - \varphi° \quad (6.61)$$

For:

$$\gamma_{sub} = 12.0 \text{ kN/m}^3 \; \varphi_{nec} = 58.30° \quad (6.62)$$
$$\varphi_{nec} = \text{arctg}\, \tau/\sigma_{ave} = \text{arctg}1.20\sigma_{ave}^{-0.10} \quad (6.63)$$
$$\sigma_{ave} = 0.055 \text{ MPa} \quad (6.64)$$

and

$$h_e = (2\, \sigma_{ave})/\, \sigma_{sub} = 9.16 \text{ m} \quad (6.65)$$

Calculating ΔH, the maximum upstream water level, is a geometric problem. From Darcy's formula the flow is:

$$q = \Delta H^2/2L\; k = h_e\, i\, k \quad (6.66)$$

So:

$$\Delta H^2/2L = 9.16 \times 0.61 = 5.58 \text{ m} \quad (6.67)$$

But:

$$L = 2 \times 1.4 \times H_B + 10 - \Delta H \times 1.40 \quad (6.68)$$
$$HB = 180 \text{ m} \quad (6.69)$$
$$L = 264.8 - 1.40\, \Delta H \quad (6.70)$$

Solving the equation:

$$\Delta H = 47.0 \text{ m} \quad (6.71)$$
$$L = 198.8 \text{ m} \quad (6.72)$$

So, if the upstream water level reaches a maximum of 47 m high, the dam will still withstand the throughflow. The average gradient would be $\Delta H/L = 0.231$.

More examples

Pinto (1999) presents data from a physical hydraulic model made of successive layers of granular material between 0.80 mm and 8 mm in diameter. The downstream slope was 1.40(H):1.0(V).

When the water reached El. 23 (0.55H) signs of instability were evident. The average gradient was 0.25 $\Delta H = 23$ and $L = 92$. The exit gradient was $\sin\psi/\cos(\psi/2) = 0.61$.

For a γ_{sub} of 1.10 t/m³ the φ_{nec} is 59°, a value too high for compacted gravel. The exit height was roughly 12 cm, so the σ'_{ave} would be only 0.00054 MPa. For such a low value the shear strength of the gravel, considering the interlocking of the particles and the expansion during failure, could lead to a φ of 59°.

A field test was presented by Yang and Løvoll (2006). A rockfill dam 6.0 m high and 36.0 m in length was subjected to throughflow at progressive higher levels of flood until failure – see Figures 6.10 and 6.11.

The upstream and downstream slopes were 1.45(H):1.0(V) and the particules size distribution was d_{10} = 3 cm d_{50} = 12 cm and $d_{max.}$ = 30 cm. At failure point the flow was 30 m³/s or 0.83 m³/s · m. In Figure 6.11 we can see that part of the flow overtopped the dam. Under stable conditions the water level was close to the crest, and the flow through the rockfill was 0.19 m³/s · m.

The paper presents a seepage analysis considering a phreatic line similar to the basic parabola as presented by Leo Casagrande for laminar flow (see Figure 6.12).

Field measurements during the tests (Yang & Løvoll, 2007) are shown in Table 6.7.

Figure 6.10 Field test.

Figure 6.11 Field test during overtopping.

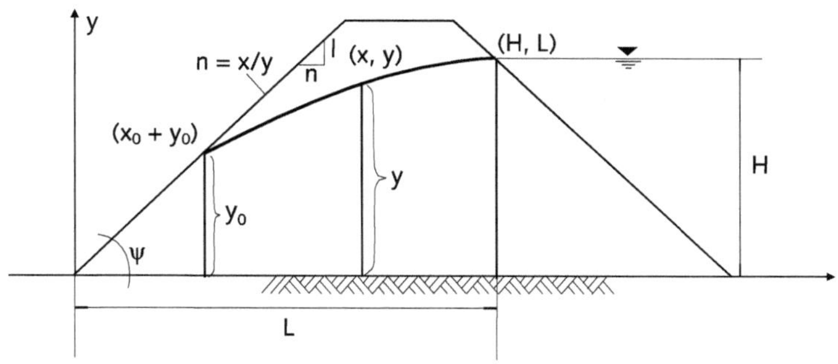

Figure 6.12 Calculation model.

Table 6.7 Field test data.

Test	H (m)	L (m)	i_{ave} (H/L)	i_{exit} (m)	$i_{critical}$	h_e (m) exit	Flow $m^3/s \cdot m$
T 1	4.07	15.5	0.26	0.60	0.69	1.18	0.091
T 2	5.35	14.3	0.37	0.60	0.69	1.89	0.115
T 3	6.11	13.2	0.46	0.60	0.69	2.42	0.190

The i_{crit} was calculated for a $\gamma_{sub} = 1.20$ t/m³, $\varphi = 60°$ and slopes of 1.45(H):1.0(V). It is surprising that the dike did not fail for an average i_{ave} approximately of 0.37 to 0.46. As the water exit height downstream is high, it is correct to compute i_{ave} as $\Delta H/\Delta L$. Its values would become ~0.21, 0.30, and 0.38. On the other hand, the exit i_{ave} is ~0.60, lower than 0.69 the value of $i_{critical}$.

A well-known case is the Hell Hole Dam, which was made of dumped rockfill (Leps, 1973; Cooke, 1984; Pinto, 1999) and failed in 1964 due to a flood during its construction. The maximum gradient at the start of the failure was 0.28 ($\Delta H/L$). The exit height h_e was estimated in 18 m. The downstream slope was 1.40(H):1.0(V) (35.5°). The average vertical pressure at the end "triangle" was $(18 \times 1.1)/2 = 9.9$ t/m² $\cong 0.10$ MPa.

For a dumped rockfill, the mobilized φ, even at low normal pressures should be of the order of 55° to a maximum of 60°. The critical gradient would be:

$$i_c = \frac{\gamma_{sub}}{\gamma_0}(\cos\alpha - \text{sen}\,\alpha\,\text{tg}\beta) \tag{6.73}$$

For $\alpha = 108°$, $\beta = 37.2°$ or 42.2°, $\gamma_{sub} = 11$ kN/m³, i_c would vary from 0.44 to 0.60, and once the exit gradient was $i = \sin\psi/\cos\psi/2 = 0.60$, the conditions of instability were imminent (Fig. 6.13).

6.3.4 The effects of anisotropy

Compacted rockfills are quite anisotropic in relation to permeability, and this condition affects the flow pattern within the rockfill dam. Let's assume a downstream shell

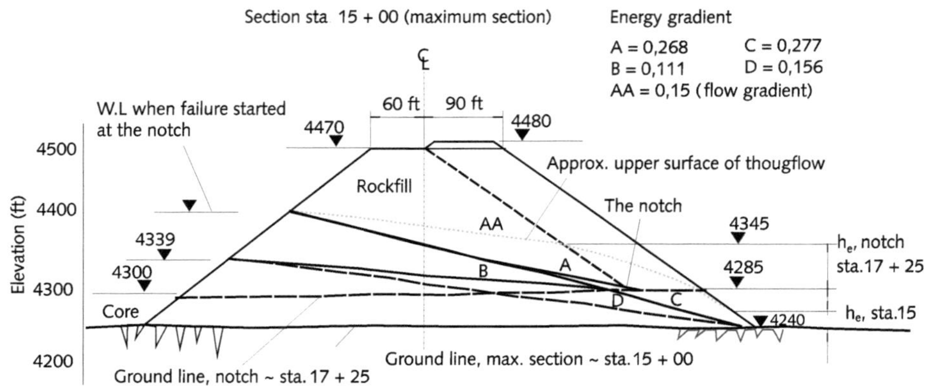

Figure 6.13 Hell Hole Dam, 12/23/1964 (Pinto, 1999).

compacted in 1.80 m thick layers and that the upper 0.45 m is 50 times less permeable than the lower 1.35 m layer. The equivalent horizontal permeability will be:

$$k_{eh} = (0.02k_h \times 0.45 + k_h \times 1.35)/1.80 = 0.755 \text{ k}_h \tag{6.74}$$

and the equivalent vertical permeability will be:

$$k_{ev} = 1.80/[(0.45/0.02k_h) + (1.35/k_h)] = 0.075\ k_h \tag{6.75}$$

So the ratio:

$$k_{eh}/k_{ev} = 10 \tag{6.76}$$

and the average permeability will be:

$$k = (0.755 \times 0.075)^{1/2} = k = 0 \times 24k_{\text{h}} \tag{6.77}$$

Figure 6.14 shows a flow net drawn for a k_h/k_v of 22, a more drastic ratio than that in above data. The exit height increased from 40 m to 60 m.

If the shear strength of the rockfill were $\tau = 1.30\sigma^{0.80}$ MPa, the critical gradient would drop from 0.59 to 0.52 due to increasing exit height. This analysis seems to indicate that an anisotropic permeability may lead to a more unstable situation for the rockfill under throughflow. A similar conclusion is presented by Pinto (1999):

> "The compacted rockfills, due to segregation of the material as a consequence of the construction procedure, have a larger horizontal permeability than vertical. This characteristic does not increase the stability of the rockfill under seepage. Its influence is negative."

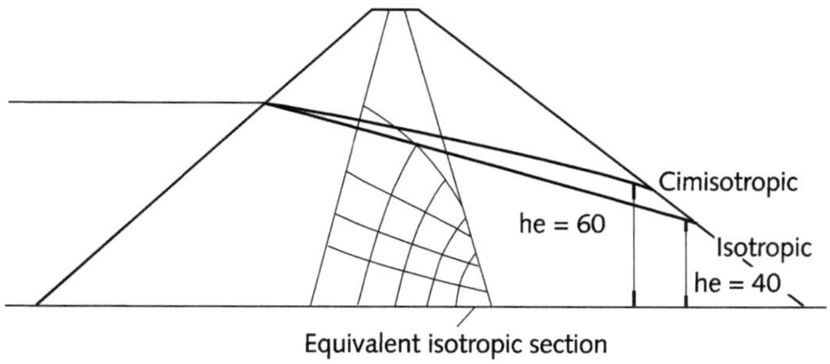

Figure 6.14 Anisotropic flow net for kh/kv = 22.

6.3.5 Discharge

For laminar flow the discharge can be computed either by a flow net or by Dupuit's approximate formula, as discussed above. When the flow is turbulent, the velocity can be estimated by Wilkin's formula:

$$V = C_{\mathrm{i}}^{0.54} \tag{6.78}$$

where C is the discharge coefficient, or the permeability or conductivity of the rockfill.

For average conditions of rockfill C is dependent on d_{50} and porosity.

$$C \cong 1.79\ d^{0.5}\, n\ [n/(1-n)]^{0.5} \tag{6.79}$$

If d values vary from 0.30 m to 1.00 m and n from 0.20 to 0.23, C values will be in the order of 0.10 m/s to 0.20 m/s. For exit gradients from 0.60 to 0.64 the velocities will range from 0.07 m/s to 0.16 m/s.

The discharge or flow will be:

$$q \cong \mathrm{V}\, h_{\mathrm{e}} \tag{6.80}$$

For the four cases in Figure 6.6 q values are shown in Table 6.8.

Olivier (1967) mentions permissible values of rockfill throughflow in the range of 0.40 to 1.50 $m^3/s \cdot m$ for loose rockfills and up to 1.00 to 4.00 $m^3/s \cdot m$ for compacted rockfills with d_{50} of 0.60 and 1.50 m, respectively.

The tests performed by Young have measured discharges bellow 0.20 $m^3/s \cdot m$. However, when it reached overtopping, with a discharge of 0.80 $m^3/s \cdot m$, failure has developed.

Flows measured in CFRDs (Cruz, 2005a) that are caused by cracks or flaws on the concrete face, in the plinth and in the foundation will be in the order of 2.0 ℓ/s · m to 5.0 ℓ/s · m, maximum values well below those estimates computed for unstable conditions.

Thus, we conclude that the stability of unreinforced rockfill to face throughflow depends on the exit critical hydraulic gradient – i_c – which can be computed if the shear strength of the fill is known or, at least, properly estimated.

Table 6.8 Discharges.

		q – Discharge m³/s · m		
Case	h_e *(m)*	*Min.*	*Max.*	*Obs.*
1	60	4.2	9.6	Unstable
2	30	2.1	4.8	Equilibrium
3	13	0.9	2.1	Stable if compacted
4	6	0.4	1.0	Stable

Whenever the exit seepage gradient exceeds the critical gradient, the external rock blocks will start to be removed from the downstream slope leading to the instability of the dam. Maximum average gradients within the rockfill should not exceed 0.25, but for design purposes gradients in the order of 0.10 to 0.15 are desirable.

6.4 SOME HISTORICAL PRECEDENTS

Successful cases are reported by Leps (1973) in his very comprehensive paper on flow through and over CFRDs and CCRDs.

Reinforcement of the downstream rockfill zone is recommended if a flood is predicted during the construction phase and the dam will be subjected to throughflow or overflow.

Some cases of regular rockfill without any reinforcement are worth reviewing, even though they have been published and republished by several authors in many different papers:

> "Hell Hole Dam (California, 1964) is an inclined core dumped rockfill dam, 123 m high, with a 471 m long crest and with a volume of 6,500,000 m³. During a flood, the throughflow in the rockfill was 540 m³/s, and the average gradient was 0.28. The dam failed by progressive sliding and by the removal of rock from the toe all the way up to the crest. The initial breach was in the order of 30 m, which progressively widened with a total loss of 5,100,000 m³ of material.
>
> The specific flow per meter of crest length was 1.14 m³/s · m. Considering, however, that the flow was closed in a narrow zone, the specific flow in this zone was much higher."

Cooke (1984), reviewing the Hell Hole failure, states that:

> "A similar event for compacted rockfill instead of dumped rockfill would have had less leakage and probably would have not failed."

Dix River Dam (Kentucky, 1924) is a dumped CFRD, 82.5 m high. During construction (when the dam was 30 m high and 225 m long), a flow of 86 m³/s passed through the dam. The specific flow was 0.38 $m^3/s \cdot m$ and the average gradient was 0.057. No sign of instability was reported.

Brownlee Dam (Idaho, 1956) is an inclined core, basaltic rockfill dam. A flood of 400 m^3/s passed through a 75 m wide channel, overtopping the dam. The downstream area was flooded, resulting in an average gradient through the rockfill of only 0.023. Damage was essentially restricted to the scour at the remotely located downstream cofferdam.

Another case is reported by Pinto (1999). The rockfill cofferdam (50 m high) of Itá Dam had been half built (the upstream soil protection was only at half its planned height) when 17,000 m^3/s flooded the area. A substantial through flow passed through the rockfill, with a maximum gradient of 0.45 (see Figure 6.15).

Attempts to reduce the flow were made by dumping saprolite and saprolitic soil on the upstream rockfill slope, and plastic sheets were placed on the upper part of the slope. Rockfill was dumped on downstream zones where the erosion process started. The cofferdam resisted the flood well and was completed successfully when the upstream water level lowered.

The throughflow can be roughly estimated by the formula:

$$q = k \cdot i \cdot As \tag{6.81}$$

where As is the area of the dam below the phreatic line where it reaches the downstream slope.

$$q = k \times 0.45 \times 16 = 0.72 \text{ m}^3/\text{s} \cdot \text{m up to } k = 10 \text{ cm/s} \tag{6.82}$$

A more refined calculation regarding turbulent flow would be:

Figure 6.15 Itá cofferdam during flood, water flow piping from downstream slope.

$$q = C \times i^{0.85} \times As \quad (6.83)$$
$$q = 0.098 \times 0.45^{0.85} \times 16 = 0.79 \text{ m}^3/\text{s/m'} \quad (6.84)$$

(Marulanda and Pinto, 2000).
C computed for $d_{50} = 0.30$ m and $e = 0.25$ (void ratio).

Cooke in 1984 referring to the modern period (1965–1982) of CFRD construction confirms that: "*One further advantage of compacted rockfill over dumped rockfill dams is their ability to withstand passage of flood water through and over the uncompleted dam.*" But Cooke also remind that "*for possible flow over the rockfill reinforced rockfill is necessary*".

6.5 LEAKAGE MEASURED IN CFRDs

Leakage weir gauges installed downstream of CFRDs may not register the total leakage that flows through the dam and its foundation. Water can partially escape through rock fractures, and avoid the measuring instrumentation. Despite the losses, instruments give a good indication of how much water is flowing through the concrete face slab, the perimeter joint, and eventual cracks or flaws on the slabs. See Figures 6.16 and 6.17.

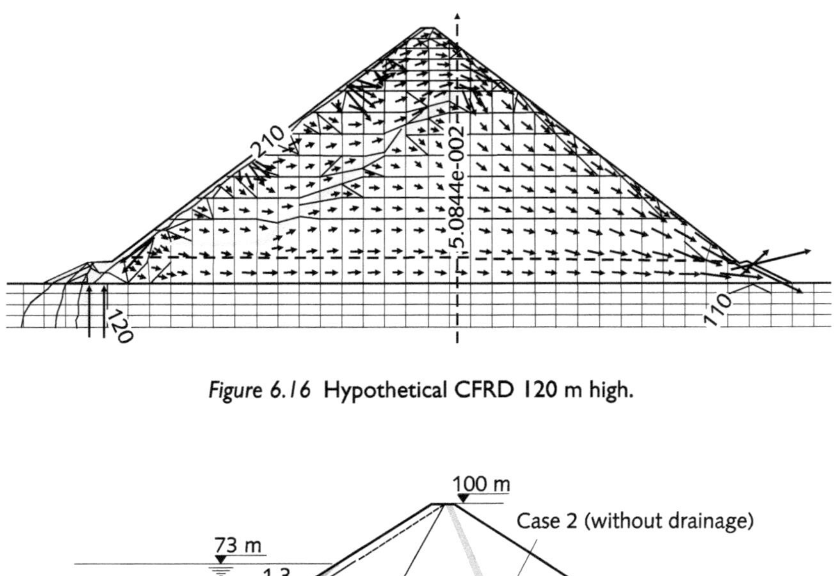

Figure 6.16 Hypothetical CFRD 120 m high.

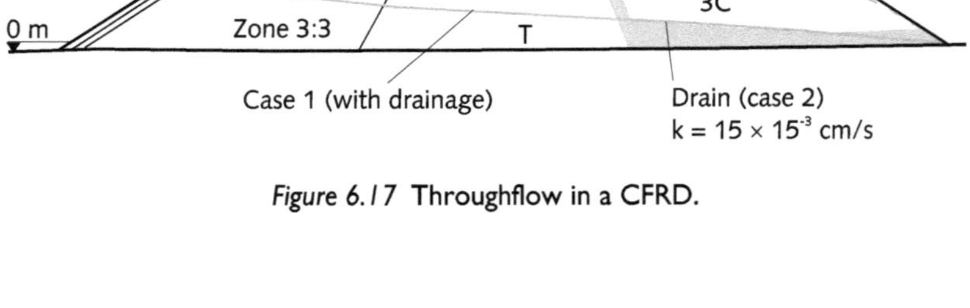

Figure 6.17 Throughflow in a CFRD.

Leakage data is usually expressed in liters per second and is given as a total value for the whole dam including the foundation.

Table 6.9 summarizes the measured flow in 26 CFRDs. Whenever the initial leakage exceeded 900 ℓ/s, investigations were made along the slab surface and repairs were done when cracks or flaws were found. In some cases, additional foundation treatments were also required.

It is worth mentioning that cracks sometimes appeared immediately after the first filling, but others showed up much later in response to progressive compression and accomodation of the rockfill. Seasonal fluctuations of leakage flow are reported in some cases. Some repairs were implemented after many years of active operation because no instabilities were reported, in spite of relatively high flows that percolated through the rockfill.

One exceptional case is the New Exchequer Dam built over the old Exchequer Dam in California, which we have already talked about in chapter 2. After the first partial filling (~110 m), the measured flow was 400 ℓ/s. In 1967, when the reservoir was almost full, the flow reached a record of 14,000 ℓ/s. Sub-aquatic investigations lead to the conclusion that the maximum flow occurred at the slab joints, and 15,000 m^3 of a soil with 25% gravel and 1.5% bentonite were dumped. After this treatment, the flow dropped to 230 ℓ/s.

Continuous treatments had been applied since 1967, and between 1976–1977 more effective treatments were done making use of the low water – as result of a dry year. In 1973 the flow increased to 2200 ℓ/s and even reached ~ 4000 ℓ/s (~560 ℓ/min · m considering the length of the crest of the dam).

Even with such a flow level, New Exchequer Dam did not become unstable. The loss was in generation. In 1985 a decision was made to conduct an effective treatment in order to reduce the leakage flow (Lepsc, Cashatt & Janopaul, 1985).

6.5.1 Foundation flows

An attempt to separate the flow through the rock foundation from the total flow can be made if it is assumed that the flow under the plinth is similar to the one under a gravity concrete dam because gradients, rock requirements in terms of permeability and treatments are similar, despite drainage design different features.

Table 6.10 summarizes measured flows in gravity concrete dams.

In these cases, foundations were always in basalt with average permeability of 0.5 to 5×10^{-4} cm/s. Gradients from the upstream toe to the draining gallery were in the order of 8 to 10 and the average flow was 0.03 to 0.20 ℓ/s · m.

If Porto Primavera spillway is taken into consideration, the total flow for the 320 m long struture will be 32 ℓ/s. For other foundation rock, as long as it is properly treated, the flows should not be much different.

Gradients under the plinth can be higher (15–20) whenever the rocks are sound, but they will drop to 7–10 if the rocks are less sound and an internal extension of the plinth is provided.

Thus, by taking measured flows through the foundation of concrete dams, and assuming a plinth length of 1.5 times the crest width of the dams it was possible to figure out the flow through the foundation of the CFRDs listed in Table 6.9. Total flow was usually under 100 ℓ/s. An exception is the Golillas Dam in which high foundation leakages are recorded.

By deducting the foundation flow from the total flow one can find the flow level through the slabs, perimeter joint, and cracks or flaws. Flow level values vary from

Table 6.9 Measured flow in 26 CFRDs.

Dam	Country	Year	Rockfil material	Foundation	Height (m)	Crest length (m)	Face area ($m^2 \times 100$)	L/H	A/H^2	Ev MPa	setl. (m)	Flow (l/s) Initial	After treatment	Today
Paradela[6]	Portugal	1956	Granite	Granite	112	540	55	4.82	4.38	72		1760	50	720
New Exchequer[8]	USA	1966	Meta andesite	Andesite	148	427	21	2.88	0.96	–	4.90	14000	230	–
Cethana1	Austrália	1971	Quartz	Riolite	110	213	24	1.93	1.96	–	–	7	7	–
Alto Anchicayá[5]	Colômbia	1974	Diorite	Schist	140	240	22.3	1.71	1.13	145	0.77	1800	>400	180
Bailey[1]	USA	1979	Sandstone	Sandstone	95	550	65	5.78	7.20	–	–	300	60	–
Foz do Areia[1]	Brasil	1980	Basalt	Basalt	160	828	138	5.17	5.39	32	3.78	260	60	60
Salvajina[1]	Colômbia	1984	Gravel	Sandstone	148	350	57.3	2.36	2.61	–	–	60	60	–
Golillas[3]	Colômbia	1984	Gravel	Sandstone	130	120	15.2	1.04	0.90	200	0.39	1080	650	200
Guanmenshan	China	1988	Andesite	Andesite	59	184	822	3.12	2.36	–	–	16	–	–
Chengping[1]	China	1989	–	Volcanic tuff	75	232	29.5	3.09	5.24	–	–	70	10	–
Zhushoqiao	China	1990	Slate-limestone	Slate-limestone	78	245	23	3.14	3.78	–	–	2500	–	–
Xibeikou[7]	China	1990	Limestone	Karst-limestone	95	22.2	29.5	2+33	3.26	–	–	1700	–	–
Longxi	China	1991	–	Lava tuff	59	141	7	2.38	2.01	–	–	2.6	–	–
Aguamilpa[4]	México	1993	Gravel	Rhyodacite	187	660	137	3.52	3.91	250	1.70	260	150	100
										50				
Segredo	Brasil	1993	Basalt	Basalt	145	720	92	4.96	4.37	45	2.20	390	45	50
Xingó[1]	Brasil	1994	Gneiss	Gneiss	140	850	112	6.07	6.22	37	2.30	200	135	–
Shirodo[2]	Nigéria	1994	Granite	Granite	130	1400	50	10.76	2.96	76	0.94	1700	100	–
Itá2	Brasil	1999	Basalt	Basalt	125	881	110	7.05	7.04	60	2.10	1700	380	200
Tianshengqiao	China	2000	Limestone	Limestone	178	1137	156	6.38	4.92	45	3.39	180	–	–
Machadinho[1]	Brasil	2002	Basalt	Basalt	127	700	77	5.60	4.93	40	1.60	900	700	600
Itapebi2	Brasil	2003	Gneiss	Gneiss-granite	120	583	67	4.85	4.65	–	1.55	902	127	37
Xiliushui	China	2004	–	–	146.5	190.6	–	1.30	–	–	–	113	–	–
Hongjiadu	China	2004	Limestone	Limestone	179.5	465	76	2.59	2.35	–	–	28	–	–
Barra Grande	Brasil	2005	Basalt	Basalt	185	666	108	3.60	3.15	–	3.40	1300	760	–
Campos Novos[7]	Brasil	2005	Basalt	Basalt	202	592	106	2.93	2.59	–	2.60	1500	1000	1300

1 No repairs were done. The flow was reduced naturally, possibly by silting.
2 Investigations detected cracks on the slab; treatment consisted of dumping fine soil and sand.
3 The flow was seeping through the lab joints and the foundation. The reservoir was lowered, treatment done, but the causes of the leakage are still unknown.
4 Fissures and cracks were detected on the slab. Some were repaired. The flow reduced over time by silting.
5 Leakage occurred along the perimeter joint and reduced to acceptable values after sometime.
6 A continuous sealing membrane covering the whole upstream face provided a permanent sealing of possible future cracks on the concrete.
7 See chapters 3, 8, 9 and 10.
8 See more details in the text.

Table 6.10 Measured flows in concrete gravity dams (Cruz and Silva, 1978).

Dam	*Structure*	*Height*	*Foundation rock*	*Gradient*	*Flow l/s · m*	*Permeability × 10^{-4} cm/s*
Ilha Solteira	Intake-Spillway	76	Basalt	5	0.08	0.7–1.9
		80		10	0.08	
Jupiá	Intake-Spillway	33	Basalt	10	0.04	0.5–1.4
		33		10		
Capivara	Spillway	29	Basalt	10	0.03	0.6–1.6
Promissão	Wall Spillway	38	Basalt	10	0.12	2–5
		40		8	0.18	
Porto Primavera	Spillway	29	Basalt	8	0.10	2–5

800 to 1600 ℓ/s before treatment. Residual flows drop considerably, showing the effectiveness of the treatment.

It makes no sense to consider specific flows (ℓ/s · m^2 or ℓ/s · m) for the initial values because in the majority of the cases flows are concentrated in cracks or flaws of the slab. However, in cases of residual or stabilized flows after treatment some figures are of interest. If it is assumed that all residual flow passes through the perimeter joint, flow varies from 0.02 to 1.8 ℓ/s · m. If it is also assumed that all residual flow passes through the slab, flow per area varies from 0.20 to 2.80 ℓ/s · m^2.

A lot more data is necessary before more definite or reliable conclusions can be drawn.

6.5.2 Finite element analysis

In order to understand a bit further the flow through CFRDs, a bi-dimensional model was prepared and parametric values of permeability were assumed for the rockfills, transitions, and rock foundation.

Figure 6.16 shows the considered CFRD. The dam is 120 m high with slopes of 1.3(H):1.0(V). A first analysis considers the dam without the slab. The total flow depends on the permeability of the rockfill and transitions. Values vary from 0.3 m^3/s · m to 1.0 m^3/s · m, compatible with throughflows in compacted rockfills.

The upstream transition layer inclusion (zone 2B) causes a significant depression of the phreatic line, as well as an effective control of the rate of flow through the sea wall which may go down to a range from 0.01 to 0.10 m^3/s · m, about a third to a tenth of the normal flow rate. Even so, along the 100 meters length of dam, flow rate will vary from 1,000 to 10,000 ℓ/s.

The next step took a quite impermeable concrete slab with $k = 10^{-8}$ cm/s. Table 6.11 summarizes the results.

Flows through the foundation are similar to those estimated for CFRDs as long as rock permeability is in the order of 10^{-4} cm/s; if it is of the order of 10^{-3} cm/s the flow increases significantly.

Slab flows are very low, and even lower than those measured on CFRDs.

A further step in the analysis was to introduce flaws on the slab in order to increase the permeability to 10^{-2} cm/s in one to three areas of the slab. Figure 6.16 shows three flaws on the slab, where there is a concentrated leakage.

Table 6.11 Permeability and flows.

Case	Rock foundation cm/s	Grout curtain cm/s	Slab cm/s	Rockfill cm/s	Flow foundation ℓ/s · m	Flow slab ℓ/s · m	ℓ/s · m²
6c	10^{-4}	10^{-5}	10^{-8}	1	0.031	0.013	0.0012
6	10^{-4}	10^{-5}	10^{-8}	10	0.023	0.032	0.0028
6f	10^{-4}	10^{-4}	10^{-8}	1	0.083	0.016	0.0013
6g	10^{-3}	10^{-3}	10^{-8}	1	0.47	0.011	0.0010

Equivalent permeability of 10^{-2} cm/s represents either a generalized fissure pattern, or rough open cracks of 0.5 to 5 mm spaced 0.5 m to 1.0 m (see Table 6.12).

Table 6.13 sums up flow simulations for CFRDs with flaws or cracks on the slab. The flow through the foundation remains of the same order of magnitude as it did in previous cases. The flow through the slabs on the other hand increases dramatically from values of 70 ℓ/s · m to 400 ℓ/s · m. Of course, there are not cracks all along the concrete face, but if cracks did develope a few meters from each other along the face, the total measured flows in CFRDs would be in accordance with the present analysis.

6.5.3 Anisotropic effects on CFRDs

When a well-graded rockfill is placed in layers of 0.8 m, 1.0 m, 1.6 m, and even 2.0 m thick, grain segregation is hardly avoidable. The larger blocks tend to settle at the bottom of the layer, and the amount of fines is seldom enough to fill in the voids.

During compaction fragmentation of the rock block occurs, which is more intense at the top of the layer. The pre-wetting of the rockfill or even watering during compaction contributes to the breakage of rock blocks and to the moving of fines. However, it is not enough to bring homogeneity to the grain size. We must add in to this process the movement of heavy construction equipment on the surface, which helps to form a thin flat layer that is good for traffic and tire durability.

The result is the formation of horizontal layers both more and less permeable, alternately. The permeability contrast between the upper and lower portions of a layer will increase with the thickness of the layer and with the widening of grain size distribution as well as with the compaction gradient from the layer top to the bottom.

Compacted rockfill is anisotropic material with regard to its permeability. If we divide the thickness of the layer by half, and estabilish k_T the upper half permeability and k_B the lower half permeability, the average permeability equivalent in the vertical and horizontal directions are given by the expressions:

$$k_h = \frac{k_T \frac{d}{2} + k_B \frac{d}{2}}{d} = \frac{k_T + k_B}{2} \tag{6.85}$$

$$k_V = \frac{d}{\frac{d}{2k_T} + \frac{d}{2k_B}} = \frac{1}{\frac{1}{k_T} + \frac{1}{k_B}} = \frac{k_T k_B}{k_T + k_B} \tag{6.86}$$

Table 6.12 Equivalent permeability.

Opening e (mm)	*Spacing b (cm)*	*Laminar flow (cm/s)*	*Turbulent flow (cm/s)*
0.5	50	8×10^{-3}	
0.5	100	4×10^{-3}	
5.0	50		2×10^{-2}
5.0	100		4×10^{-2}

Table 6.13 Analytical flow model.

Case n°	*Rockfill cm/s*	*Transitions cm/s*	*Slab cm/s*	*Flaw area cm/s*	*N° of flaw areas*	*Rock foundations cm/s*	*Grouted area cm/s*	*Flow through foundation ℓ/s · m*	*Flow through slab ℓ/s · m*
10	1	10^{-8}	10^{-8}	10^{-2}	2	10^{-4}	10^{-5}	0.012	126
9b	1	10^{-1}	10^{-8}	10^{-2}	3	10^{-4}	10^{-5}	0.009	147
9c	10	10^{-1}	10^{-8}	10^{-2}	3	10^{-4}	10^{-5}	0.018	400
8d	1	10^{-1}	10^{-8}	10^{-2}	1	10^{-4}	10^{-4}	0.033	73
8c	1	10^{-1}	10^{-8}	10^{-2}	1	10^{-4}	10^{-5}	0.015	73
10a	10	10^{-1}	10^{-8}	10^{-2}	2	10^{-4}	10^{-5}	0.019	316
9e	10	10^{-1}	10^{-8}	10^{-2}	3	10^{-4}	10^{-5}	0.018	404

If $k_B = 10\ k_T$, which is a probable hypothesis:

$$k_h = \frac{k_T + 10k_T}{2} = 5.5\ k_T \ \text{ or } \ 0.55\ k_B \tag{6.87}$$

$$k_T \ \text{ or } \ 0.1\ k_B \tag{6.88}$$

For a contrast of 50 x to 100 x, we have respectively:

$$k_h = 25\ k_T = 0.50\ k_B \tag{6.89}$$

$$k_V \cong k_T = 0.02\ k_B \tag{6.90}$$

and

$$k_h \cong 50\ k_T = 0.50\ kB \tag{6.91}$$

$$k_V \cong k_T = 0.01\ kB \tag{6.92}$$

From these analyses we can conclude that the horizontal flow depends on the lower layer permeability divided by 2, and that the vertical flow strongly depends on the upper layer permeability, almost independently from the permeability contrast.

6.5.4 Flow-related conclusions

The amount of data on leakage from CFRDs is very scattered and limited to a few historical cases, and much more data is needed in order to allow for clear conclusions on permissible or maximum flows through CFRDs.

A few case histories have been summarized in this chapter. Some preliminary conclusions may be drawn.

- When not protected by a slab, compacted rockfill dams can support extreme flow conditions of 0.5 to 1.0 $m^3/s \cdot m$. Whenever flows through rockfill or over rockfill are predictable, reinforcement of the downstream slope is strongly recommended.
- Maximum tolerable gradients are in the order of 0.15 to 0.18.
- Total flow through CFRD rock foundations, if sound and properly grouted, can vary from 30 to 100 ℓ/s.
- Whenever total flows measured in CFRDs exceeded 800 to 2000 ℓ/s, investigation and treatment is required. Total residual or stabilized flows vary from 30 to 400 ℓ/s in the majority of CFRDs.
- The analytical model confirms the low flow through the foundations of CFRDs and explains the relatively high flow through cracks or flaws on the concrete face.

6.6 DESIGN OF CFRDs FOR THROUGHFLOW CONTROL

6.6.1 Zoning

The classical zoning of CFRDs, set out in chapter 2, as well as the compaction procedures generate an anisotropic dam regarding permeability, and, increasing permeability from the upstream to the downstream area as a result of the lower compaction of zones 3B and 3C, or at least of zone 3C. These aspects, however, are not enough to ensure flood control and not sufficient to avoid the risk of failure in cases of throughflow as noted in section 6.3.

Two zones are built upstream from zone 3, not only to promote a transition between the concrete face and the rockfill, but also for flood control: the transition zone 3A and the cushion zone 2B. The transition zone 3A, with a finer grain size, should have lower permeability than the adjacent rockfill of zone 3B, which should contribute to reducing the flow and lower the phreatic line.

The main protection against the flow is, however, the cushion zone (2B) compacted in thin layers and with a content up to 40% of sand size particles and fines, with permeability ranging from 10^{-3} to 10^{-2} cm/s (see chapter 2).

The pumped out concrete with its cast recently placed to support the slab in zone 2B has a low permeability level, unless some fissures appear caused by the accommodations that accompany the rockfill displacements. These rockfill accommodations are usual, due to either the construction elevation or the rockfill wetting before the slab placement, or caused by a fortuitous cofferdam overflow when the river flood and is blocked by the cofferdam itself.

6.6.2 The ideal rockfill

Another benefit from compaction is noted by Cooke and Sherard (1987):

> "In conventional zoning of a CFRD, there is an increasing permeability principally in the thicker bottom areas of each of the four zones, and for the full width of the dam, as well as a desirable perched water table effected by the fine and semi-pervious surface of each layer. The perched water tables, assuming

a source of water, avoid a high phreatic line, with its consequent high pore pressures."

In an ideal design, excluding the case of overtopping, the upstream zones plus the extruded concrete would control the throughflow.

Figure 6.17 shows a CFRD with the cushion zone and 4 m of transition zone followed by rockfill with increasing permeability from the upstream towards the downstream (case 1).

The 100 m dam high is flooded before the face slab is built. Throughflow occurs. In case 1, the permeability increases towards downstream, so the phreatic line is lowered because the cushion zone has low permeability. No problems of stability are forseen because the exit height is below 1 m high, but if the 3C zone rockfill contains a high percentage of fines, the permeability decreases (case 2). Throughflow is now controlled by zone 3C and a large part of zone 3C becomes saturated. Stability decreases. The solution is to include an internal inclined drain, followed by an horizontal drain (see Figure 6.17and Table 6.14).

The phreatic line lowering is the result of the cushion layer presence followed by the transition layer. The gradients in the rockfill zones are significantly reduced. Without the two upstream zones the average gradient in the rockfill would be 0.20.

6.6.3 Deviations from the "ideal rockfill"

The main purpose of the CFRD zoning, as noted in chapter 2, is to control the displacements of the dam as a whole and, in particular, in the zones of the dam more affected by the filling of the reservoir. The smaller the displacements of the dam, due primarily to hydraulic loads acting on the face, and displacements due to rockfill creep, the smaller the risks of face fissures and cracks.

As the permeability of the rockfill is a function of many factors, such as grain size, distributions, void ratio, and block shape, the ideal dam would only be possible if the material found in dam zones was the same and the permeability at the downstream zones increased by reducing the degree of compaction.

Table 6.14 Throughflow in rockfill dam.

Zone	*Width (m)*	*Permeability (m/s)*		*Case 1*			*Case 2*		
		Case	*Case*	*h (m)*	*Δh (m)*	i_{ave}	*h (m)*	*Δh (m)*	i_{avev}
2B (cushion)	2	5×10^{-5}	5×10^{-5}	73	4	1.33	73	4	1.33
3A	4	5×10^{-4}	5×10^{-4}	24.9	1.50	0.25	67.1	1.10	0.65
3B	Variável	5×10^{-3}	5×10^{-3}	23.4	7.80	0.13	66.0	3.5	0.04
T	Variável	10×10^{-3}	5×10^{-3}	15.6	4.70	0.06	62.5	3.4	0.04
3C	Variável	15×10^{-3}	5×10^{-4}	10.9	10.0	0.10	59.1	53.0	0.53

Flow – 7.78×10^{-3} m³/s · m
i_{ave} – average gradient
Case 1 – $h_e = 0.90$ m
Case 2 – $h_e \cong 6.0$ m

But even in dams where the rockfill is of the same rock type, the construction requirements are that in the upstream zones the rockfill must have a higher shear resistance than in other zones. See, for example, Machadinho, Segredo, Campos Novos and Barra Grande dams (chapter 3). The rock is always basalt, but there are different requirements when it comes to compressibility but not to the permeability. There are dams in which different materials such as gravel and rockfill were used like in Aguamilpa (chapter 3). Moreover, there are scenarios in which because of an excessive amount of fines, an internal drainage system to control internal seepage has been proposed. This is a common practice in earth and earth-rockfill dams.

A simple analysis of the dams cross sections presented in chapter 3 is enough to conclude that no generalizations can be made regarding the internal flow in CFRDs.

6.6.4 Practical recommendations

1. It has been demonstrated in section 6.3.2 that the most vulnerable zone in terms of stability is the external downstream slope, at a width of 2/3 H to 1/3 H, where *H* is the water level. Whenever possible, the larger blocks should be accommodated in this zone because they are, in theory, more stable.
2. It has also been demonstrated that the critical exit gradient is a function of the friction angle of the rockfill. Thus, the higher the exit of the water at the downstream slope, the lower the safety factor regarding the flow – considering that shear strength of a rockfill decreases with increasing effective stresses. For this reason, the lower the phreatic line, the safer is the dam for through flow.
3. And finally, whenever there is a possibility that a CFRD could be subjected to throughflow or overtopping during the construction, the only guaranteed protection is to reinforce the rockfill as we discuss in section 6.7.

6.7 REINFORCED ROCKFILL

As already discussed in previous topics, dumped or compacted rockfills are structures that have limits in terms of bearing internal flow and are bound to fail if overtopped. The experience in Australia reported by Thomas (1976) is rich in examples of overtopping on reinforced rockfills.

Historically, the case of Orós Dam, made of rockfill with a central core (see Fig. 6.18) is well known. It failed in 1961 due to a flood. Over 1 million m^3 of fill was washed away in a few hours leaving behind a 200 m wide gap. At its peak the flow through the breach reached 10,000 m^3/s. The same happened to Arneiros Dam II, also made of rockfill with a clay core, in 2003 as is shown in Figures 6.19A, B, C. The two dams are built on Jaguaribe River, 50 km apart.

It is interesting to see how the compacted soil was to a large degree preserved in spite of the overtopping.

As far as is known, in none of the nine CFRDs built in Brazil was the downstream slope reinforced to resist overtopping. One known case is the auxiliary upstream cofferdam of the third phase of Tucuruí Dam, reinforced with "Telcon" grid ϕ 3.75 mm

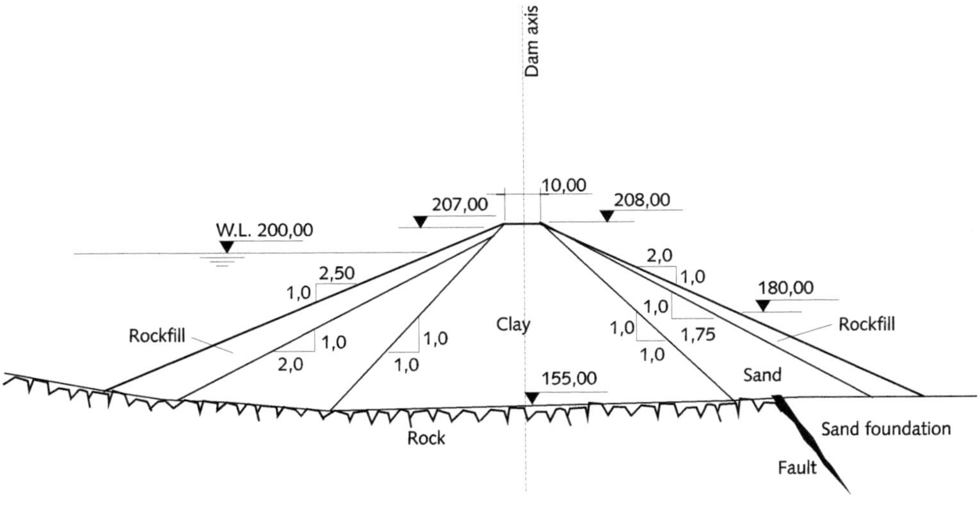

Figure 6.18 Orós Dam.

Figure 6.19 Arneiroz II Dam.

and 8 × 8 cm mesh, tied up by anchor bars of 20 mm every 2.0 × 2.0 m to concrete blocks 0.5 × 0.5 × 1.20 m inside the rockfill. The bars were 8 to 10 m long. See Figure 6.20.

Such reinforcement was placed to protect the cofferdam from high velocities close to the "sluice gates" as well as from vortices and turbulences observed in the hydraulic model (Eletronorte – UHE Tucuruí – Design Project of the Civil Structures – Consolidation of the Experience – Engevix – Themag, 1987).

More references to this subject can be found in Thomas (1976, Chapter 15, v. 2).

An example of a cell type reinforcement is shown in Figure 6.21. As can be seen, the protection is superficial and restricted to the crest of the dam.

> The cell is only 2.3 m long at the base and 0.81 m at the top. *"The vertical mesh at the back (upstream) side of the cell is a backstop and would prevent adjacent*

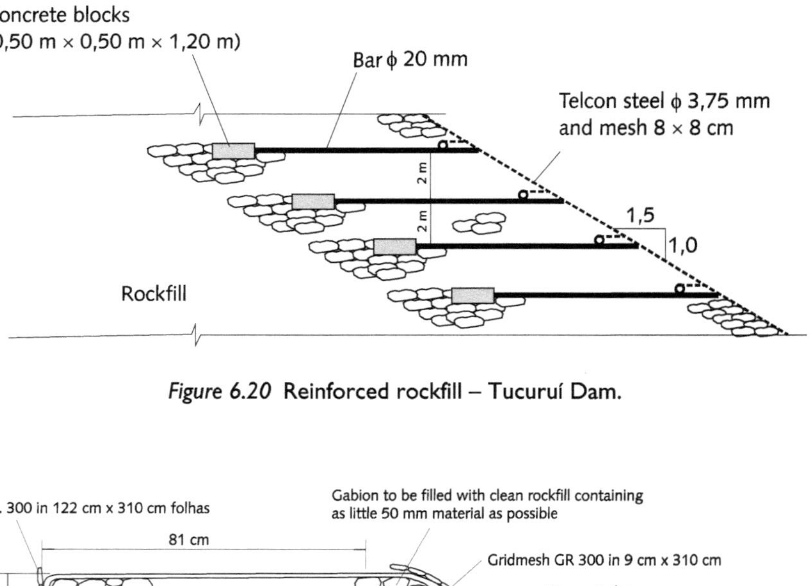

Figure 6.20 Reinforced rockfill – Tucuruí Dam.

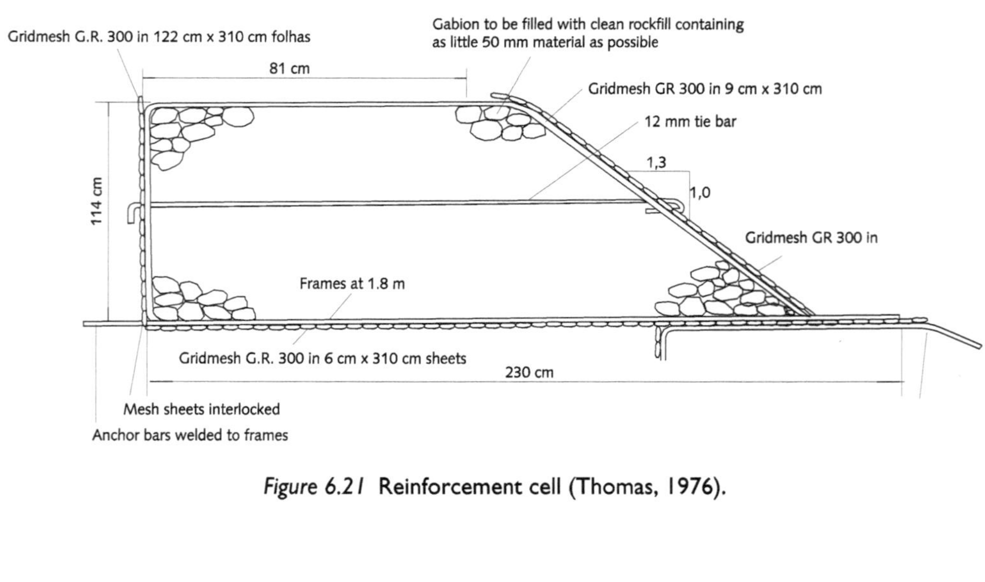

Figure 6.21 Reinforcement cell (Thomas, 1976).

rockfill from piping out should the rockfill within the cell wash out through a hole"

(Thomas, 1976).

An other historical case, already mentioned in chapter 2, is the canal left in the Tianshengqiao 1 rockfill to allow the passage of floods above 34,800 m^3/s (return period of 300 years) due to the limitations of the two tunnels (D = 13.5 m) to deliver such flows. The canal had slopes of 1.4(H):1.0(V), and it was 120 m wide at the base and 300 m long.

The entrance and the canal slopes were protected by gabions with rock stones of 20 cm – 30 cm, reinforced with anchor bars inside the rockfill (8 m long, ϕ 20 to 28 mm – see Figure 2.11). The maximum flood through the canal was 3,600 m^3/s, with a water level of 9 m and a 3,20 m/s velocity.

After the flood, a local inspection did not find any significant erosion problem either at the bottom or on the slopes. The rockfill protection was effective (Freitas, 2006).

Chapter 7

Foundation treatment

The design criteria for foundations in concrete face rockfill dams have been modified due to the progressive experience gained from building many dams during the last 35 years.

The following criteria, along with the discussion, represents the latest actual practice after analyses of the dams described in chapter 3 as well as other dams built in different continents.

7.1 PLINTH FOUNDATION

Traditionally the plinth is founded on hard, sound, non-erodible rock, which is consolidated by grouting. However, dam engineering experience has proven that it is possible to found the plinth in rock of lower quality, provided preventive measures are taken to avoid foundation erosion by reducing the hydraulic gradients and covering the potentially erodible zones with filters, gunite or pneumatic concrete.

During the construction of the 140 m high Alto Anchicayá Dam, in Colombia, the plinth sections submitted to higher hydrostatic pressures were founded on the more sound (competent) rock. The rock of the dam foundation was of schist with synclinal and antisynclinal folds as shown schematically in Figure 7.1.

The lower plinth was placed in the lidita (chert) and the higher portion, where the reservoir pressure was lower, over chloritic schists and limestones of lower quality, covered with filters downstream from the portion.

At Salvajina Dam, 148 m high, in Colombia, the plinth was founded on different rock formations (Sierra, Ramirez & Hacelas, 1985) designed with variable dimensions and gradients, as can be seen in Figure 7.2.

The concept of external and internal plinths was proposed by Barry Cooke for economic reasons in order to optimize excavations upstream from the plinth and still achieve the required gradients. The internal section of the plinth is built under the dam.

The choice of the foundation elevation used to be made by an experienced geologist who, based on the boreholes analysis, would indicate a tentative alignment. With the introduction of a geomechanic classification of rock foundations, new criteria to place the plinth have been established according to clear rules, and taking into consideration rock classification and reservoir hydraulic pressures.

An adequate correlation with the RMR (rock mass rating) classification (usually applied to tunnels) has been used in many modern dams by considering the

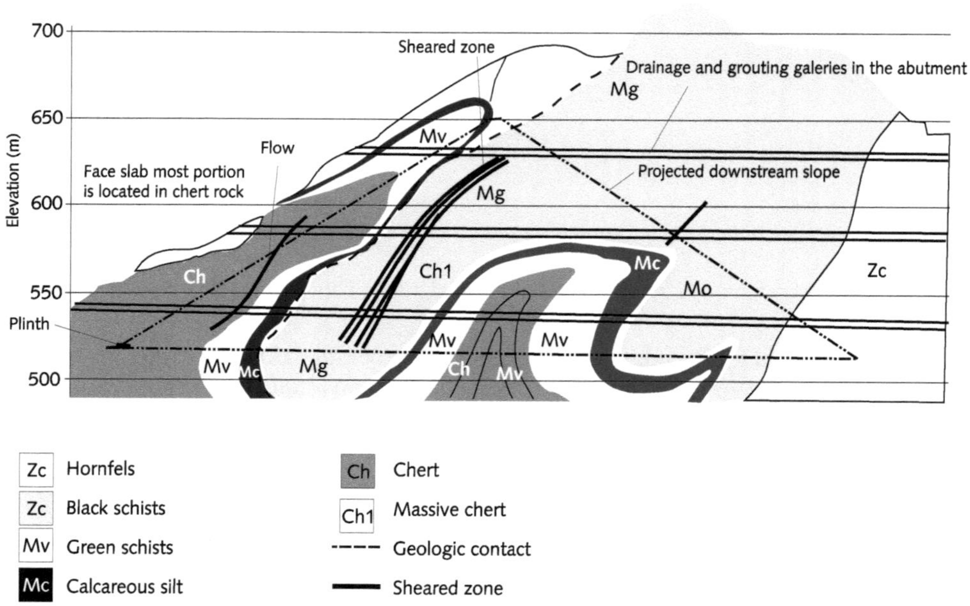

Figure 7.1 Alto Anchicayá Dam: left abutment geology (Materón, 1970).

type of rock and the appropriate gradient. At Berg River Dam, 60 m high, in South Africa, where the rock foundation is of poor quality, the Bieniawski RMR classification and the required gradients were considered in order to establish the length of the plinth within reasonable foundation excavations from the project point of view. Similar criteria were adopted at Bakun Dam, 205 m high, in Malaysia, at Merowe Dam, 53 m high, in Sudan, and at the Sia Bishe dams in Iran. These criteria were adopted:

- After the rock foundation level has been defined, a geological mapping of the clean surface on a stretch 25 m to 30 m long is performed.
- Litology, rock structure, water level, stratification system, shear zones, RQD, etc, are described at each stretch and photos of the area are taken.
- Bieniawski RMR is calculated based on the above data. Other rock classifications also can be used, such as those proposed by Hoek, Lagos, and others.
- A correlation between the RMR and the required gradient is used as indicated in Table 7.1.

The graph can be used for weathered rock foundations (Fig. 7.3).

Let's consider an example, where the rock classification leads to RMR = 55 and the reservoir pressure is 80 m.

For RMR = 55 from Figure 7.3 the required gradient is 11. The plinth length is 80/11 = 7.30 m. For an external plinth of 4 m, the internal plinth will be 7.30 – 4.0 = 3.30 m.

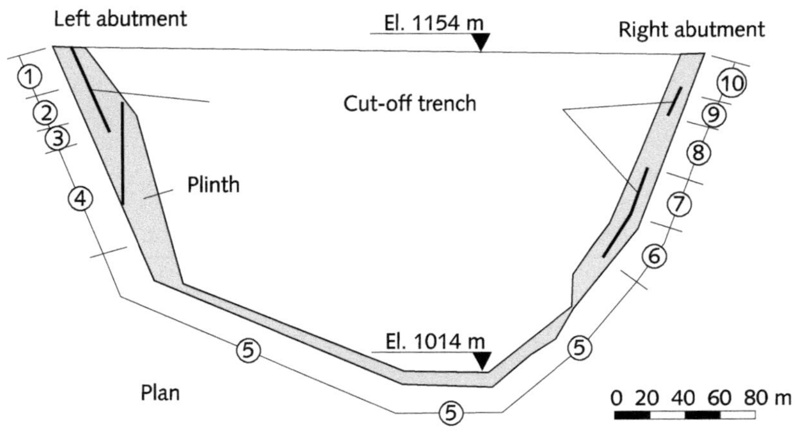

Sector	Elevation	Type of foundation	Description of rock
1	1154 - 1136	II - III	Friable or fractured sandstone
2	1136 - 1121	II	Fractured sandstone
3	1121 - 1113	II - III	Friable or fractured sandstone
4	1113 - 1067	II	Fractured sandstone with interbedded siltstones
5	1067 - 1014	I	Sound sandstone and siltstone slightly fractured
	1014 - 1062		
6	1062 - 1082	II - III	Fractured sandstone with interbedded wheathered siltstones
7	1082 - 1104	III	Friable sandstone
8	1104 -1126	II	Fractured siltstone
9	1126 -1138	III	Residual soil from weathering of diorite (ITO)
10	1138 -1154	II	Fractured sandstone

Figure 7.2 **Plinth sectors according to the rock type at Salvajina Dam (Sierra, Ramirez & Hacelas, 1985).**

At Salvajina Dam (148 m high), in Colombia, the plinth foundation was defined according to the river bed rock and the type and quality of the abutments.

At the river bed, the foundations were on sound sandstone and siltstone with few fractures. For that reason the plinth was fixed in conventional gradients of 18. On the abutments, the sandstone and siltstone were weathered. On the right margin besides the friable sandstone and fractured siltstone, a diorite dike completely decomposed by hydrothermal weathering was found.

The design criteria to define the plinth are presented in Table 7.2 (see Fig. 7.2 for a description of the geological stretches).

At the 50 m high Pichi Picún Leufú Dam, in Argentina, the plinth extension was established considering the degree of erodibility of the rock foundation. Table 7.3 summarizes the proposed criteria.

Table 7.1 Correlation between RMR and gradient.

RMR	*Gradient*
80–100	18–20
60–80	14–18
40–60	10–14
20–40	4–10
<20	Deeper foundation or use a cut-off wall

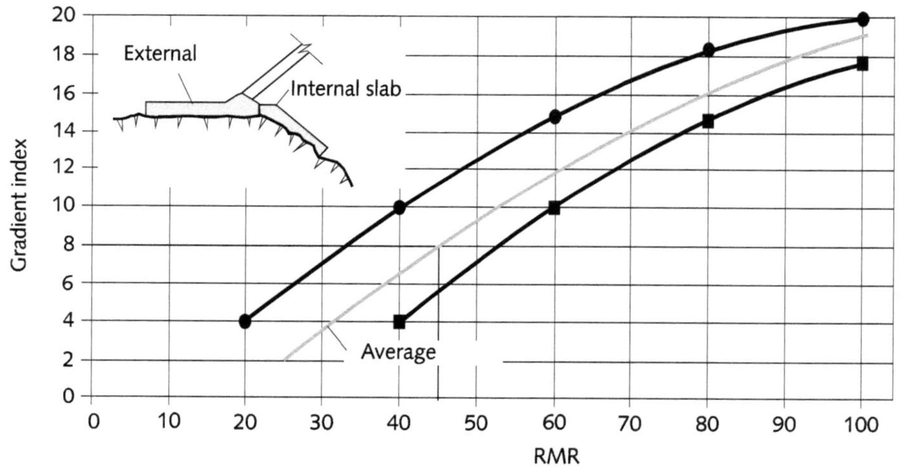

Figure 7.3 Plinth design with partial internal slab.

Table 7.2 Criteria of plinth design (Sierra, Ramirez & Hacelas, 1985).

		Maximum hydraulic gradient		
Type of foundation	*Description*	*Acceptable*	*Today*	*Width of plinth (m)*
Original design	Hard groutable rock	18	–	4–8
I	Competent rock	18	17.5	6–8
II	Intensely fractured rock	9	6.2	15–23
III	Intensely weathered sedimentary rock	6	3.1	15–18
IV	Residual soil from intensely weathered rock	6	1.3	13–14

Table 7.3 Gradient criteria according to foundation erodibility (Lagos).

A	B	C	D	E	F	G	H
I	Non Erodible	1/18	>70	I–II	1–2	<1	1
II	Slightly Erodible	1/12	50–70	II–III	2–3	1–2	2
III	Moderatedly Erodible	1/6	30–50	III–IV	3–5	2–4	3
IV	Highly erodible	1/3	0–30	IV–VI	5–6	>4	4

A – Foundation type; B – Foundation class; C – Gradient: plinth width/depth of water; D – RQD em %; E – Weathering degree: I – sound rock; VI – residual soil; F – Consistency degree: 2 = very hard rock; 6 – friable rock; G – Weathered macrodiscontinuities per 10 m; H – Excavation Classes: 1 – equires blasting, 2 – requires heavy rippers: some blasting, 3 – excavated with light rippers, 4 – excavated with dozer blade.

7.2 PLINTH STABILITY

The excavations for the plinth must be done with care in order to avoid over-excavation and damage to the rock. As usual, when the foundation is not adequate, it is more suitable to switch from deeper alignment in the ground to a better rock quality and longer alignments.

However, there are cases in which the plinth alignment requires the use of concrete walls. In such cases, it is necessary to analyze the forces acting in the rock wall and to check for sliding and tilting, taking the wall as a gravity structure.

In chapter 2, Figure 2.13 shows schematically the forces acting on the concrete wall block when the reservoir is full.

When the block is relatively low (less than 2.0 m), the stability can be improved by anchors installed directly over the plinth.

In the stability analysis one should disregard slab support, and the trust of the rockfill should be considered only partially because any movement of the block to mobilize the rockfill trust will breake the upstream slab as shown schematically in Figure 7.4. It is recommended to use a thrust for design purposes of 0.20 to 0.25 of the hydrostatic pressure.

The technical literature describes several examples of slab failure connected to plinths resting on high walls.

During the constructions of the 145 m high Mohale Dam, Lesoto, Africa, in many stretches of the plinth the stability analyses showed that tendons were required to guarantee stability against sliding. These were the criteria for these analyses.

- The friction angle between the concrete and the foundation φ was 45°.
- Cohesion at concrete-rock contact was estimated at 300 kN/m^2.
- No passive pressure of the rockfill was considered. Instead a trust coefficient of 0.25 H (of the hydrostatic pressure) was used, following an Australian criterion.
- The stability was computed using a reduction factor of 1.5 for φ and 3.0 for the cohesion:

$$FSD = \left(\frac{Ntg\varphi}{1.5} + \frac{cL}{3} \right) / T \geq 1 \qquad (7.1)$$

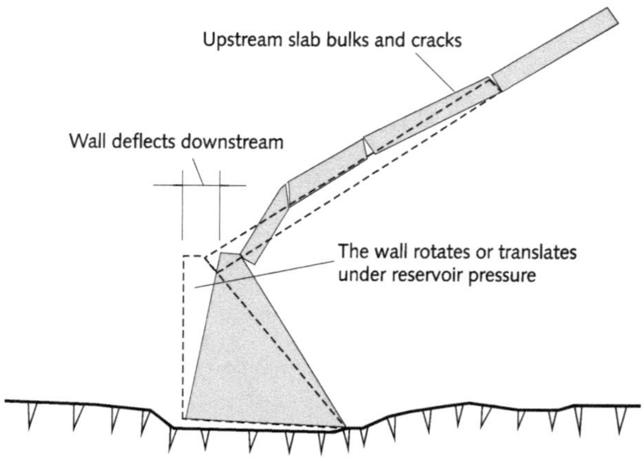

Figure 7.4 **Plinth on high walls (Cooke & Sherard, 1985).**

where:

N – net normal force on the failure plane.
φ – 45°
c – cohesion = 300 kN/m²
L – extension of the failure plane.
T – resultant shearing of forces on the failure plane.

Moreover, the uplift pressures over the failure plane after grouting were reduced by 20%. Tendons were installed to stabilize the wall.

A similar case occurred during the construction of Machadinho Dam, 125 m high, in Brazil. On the abutments where the plinth would be founded, an acid rhyodacite flow with frequent unstable fractures filled with soil was encountered making it unsuitable to place the plinth on it.

In some sectors, the soil was washed out leaving blocks with cavities. After the removal of these materials, and with the purpose of keeping the alignment of the plinth, it became necessary to build up concrete walls. The plinth was then founded on the top of these walls, whose gravity structures had upstream slopes of 1(H):5(V) and downstream slopes of 1.2(H):1.0(V). The walls were up to 17 m high.

Figure 7.5 shows a typical cross section of those walls.

Given the design premise that the wall should not move in order to avoid slab failure, it became necessary to: introduce an internal drainage system to reduce uplift pressures and, to intall three lines of tendons of 850 kN each anchored in sound rock.

Other adoped criteria and parameters were:

- Friction angle concrete/rock $\varphi = 40°$;
- Cohesion concrete/rock $c = 400$ kPa;

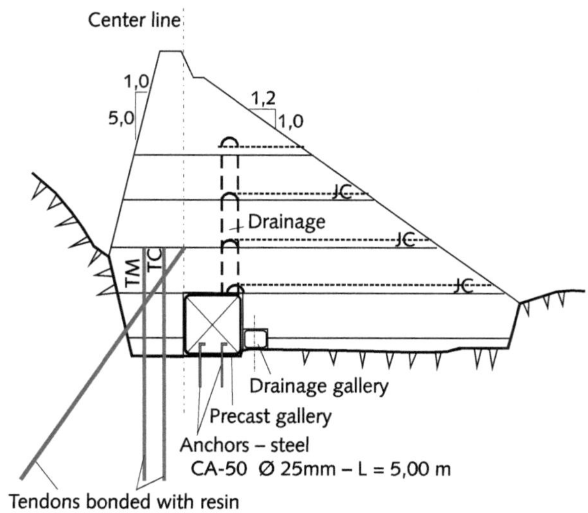

Figure 7.5 Gravity concrete structure at Machadinho Dam (Mauro et al., 2007).

External loads:

- Reservoir pressure
- Uplift pressures with a drainage system.

Stabilizing forces:

- Weight of the wall with $\gamma = 22$ kN/ m^3
- Rockfill with $\gamma_r = 19$ kN/m^3
- Water with $\gamma_0 = 10$ kN

The rockfill trust was taken as 0.15 of the hydrostatic pressure (similar to the Mohale case).

The performance of both Mohale and Machadinho dams has been satisfactory.

7.3 FOUNDATION TRANSITIONS

The 2B and 3A transitions, according to the internationally accepted zoning criteria described in chapter 2, are placed on a rock foundation after the removal of soft material, like colluvial or residual soils. If the foundation has cavities or negative slopes that obstruct an adequate compaction of the transition materials, the foundation should be excavated so to promote a satisfactory geometry that enables the proper placement of these materials.

The presence of banded rock with erodible material will require special treatment. At the foundation of the transition materials, erodible material is usually excavated to a depth twice its width and refilled with mortar or shot concrete. If the erodible band extends downstream, the treatment may follow with shot concrete up to 10 m downstream from the plinth surrounded by filters to avoid the migration of fines. These filters, if necessary, should reach up to 40% of the reservoir height.

7.4 ROCKFILL FOUNDATION

7.4.1 River bed

Whenever there are alluvial deposits on the river bed, they ought to be investigated in order to detect possible fine material pockets (silts, clays, very fine sands). These fine materials must be removed from the foundation, as they might liquefy during an earthquake. The alluvial deposits are usually excavated from the plinth and the foundations of zones 2B and 3A up to 30 m downstream from the plinth. Concavities in the bedrock filled with alluvial materials may be left in place as long as they remain confined by the rock itself.

When the alluvium deposits are very deep, or the valley configuration very large, the plinth can be founded directly over the river bed gravel, and be built with flexible elements, able to accomodate potential displacements. A diaphragm wall is also needed to cut down the alluvium strata permeability.

Articulated plinths coupled with diaphragm walls have been used in practice for the last 50 years. The technical bibliography mentions Campo Moro II Dam, in Italy, which had a similar plinth and was built in 1958. In China there are more than eight dams with articulated plinths over diaphragm walls on compressible materials. The Kekeya Dam, which has been in operation since 1982, exhibits an excellent performance. Hengshan Dam, 120 m high, also has a plinth over a dam built with gravels with impermeable core – it behaves excellently.

In Chile, Santa Juana (110 m high) and Puclaro dams (85 m high) were built over alluvium materials and also display an excellent performance. These project experiences open up the possibility of building new dams grounded on alluvium materials in the proximity of the Andes. Los Molles and Potrerillos dams were built in Argentina and the Caracoles Dam is nearing the end of construction. A new dam, Punta Negra, is starting to be built in San Juan. Also under construction is Limón Dam in Peru (Olmos project).

Figure 7.6 shows the articulated plinth of Santa Juana Dam in Chile.

7.4.2 On the abutments

The placement of the fills at the abutments follows these criteria.

a Upstream from the dam axis, the colluvial or residual soils are usually excavated to sound rock or a dense saprolite.
  At Salvajina Dam, Colombia, where there were thick deposits of colluvial material and residual soils, all soils were removed down to a depth in which the density of the foundation materials were similar to the fill to be built.
  At Itá Dam, in Brazil, the deep residual materials of basalt deposit were excavated until a saprolitic soil of SPT over 15 was exposed and left as the foundation surface.
b Soft materials on the foundation area upstream from the axis are excavated, in general until a material of density similar to the fill is displayed.

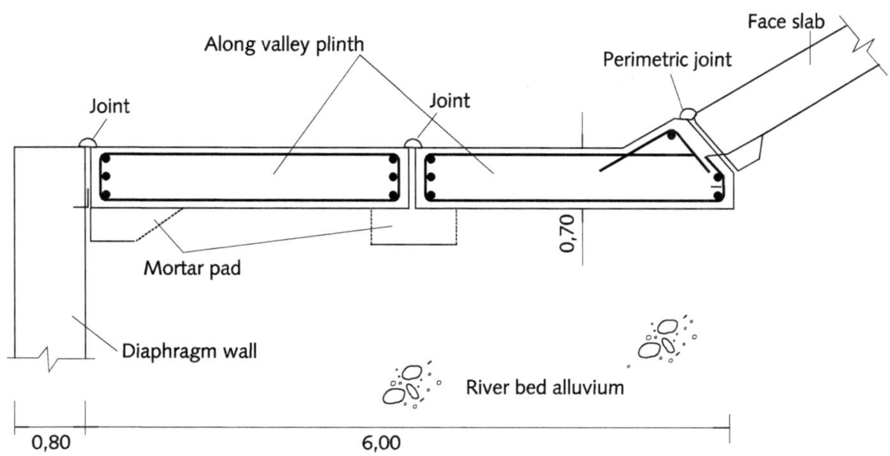

Figure 7.6 Articulated plinth at Santa Juana Dam, Chile (Noguera, Pinilla & San Martin, 2000).

c Downstream of the dam axis, the existing colluvial or residual dense straight materials can remain in straight contact with the fill to be built.

7.5 GROUTING

These criteria are usually applied to grouting operations.

a Two lateral consolidation lines with depths of 8 m to 15 m, depending on the rock type. Spacing between the grouting holes is usually 3 m.
b One central grouting line with depths between 1/3 H and 2/3 H, but with a minimum of 20 m to 30 m deep. *H* is the reservoir height.
c The primary grouting holes are separated by a 12 m spacing, the secondary at 6 m, and the tertiary at 3 m. In very fractured zones a fourth line drilled at every 1.5 m is necessary.
d In places in which erodible bands embedded in rock foundation exist, treatment should be extended down to 10 m downstream from the plinth, removing the weak material and replacing it with mortar or shot concrete before grouting.
e Grouting pressures vary from 1.0 kgf/cm^2 to 2.0 kgf/cm^2 at the top of the hole, increasing by 0.25 kgf/cm^2 for every meter of depth. In very fractured and permeable abutments, the number of grouting lines can be increased.

At Alto Anchicayá, Colombia, the plinth was founded on sedimentary rock and five grout lines were used, instead of the traditional three.

At Foz do Areia, during the river bed treatment, fractured horizons were found, down to 50 m deep and with artesian flows. It became necessary to make consolidation grout holes down to 20 m and the main grouting line went down to 60 m, with additional grout lines upstream and downstream from the plinth. The pressure at the top of the hole was increased to 5 kgf/m^2.

The GIN method has been used in dams like Aguamilpa, Mohale, Pichi Picún Leufú, and others still under construction. This method uses one single mixture after performing adjustments to ensure a minimum shrinkage, taking into account that penetration is limited by the grain size of the cement.

As a consequence, fine cements are more efficient.

The cement penetration is improved by adding super fluidificants, as they are able to reduce the viscosity of the mixture.

The W/C (water/cement) ratio is defined in the laboratory by considering:

- minimum sedimentation;
- high density;
- low viscosity;
- adequate cure time;
- high strength;
- wash resistance.

The grouting is controlled by three parameters:

a N° GIN = P.V, where *P* = pressure (in *bars*) and *V* = absorption of the mixture (in ℓ/m). This relation is a hyperbola; see Figure 7.7;
b Maximum pressure;
c Maximum flow.

The choice of GIN number depends on the geological conditions of the project. Grouting should begin with pressures that do not cause hydro-fracturations; although, in many projects the plinth has been raised because the suggested grout pressure was two to three times the reservoir pressure.

The choice of the grouting procedure depends on how experienced the designers are. The two methods described here give good results. However, the GIN method leads to a higher consumption of cement.

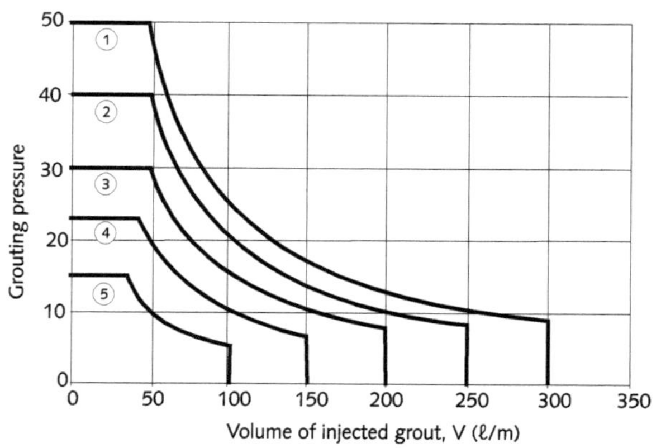

	Intensity	GIN = P.V. = bars x l	P_{max} (bar)	V_{max} (ℓ/m)
1	Very high	> 2500	50	300
2	High	2000	40	250
3	Moderate	1500	30	200
4	Low	1000	22.5	150
5	Very low	< 500	15	100

Figure 7.7 Grouting control with GIN method (Lombardi, 1996).

Chapter 8

Plinth, slab and joints

8.1 PLINTH

8.1.1 Design concept

The main function of the plinth is to control foundation seepage and hydraulic gradients. According to Cooke (2000b), *"the plinth, together with the perimeter joint waterstop, is the watertight connection between the foundation and the concrete face slab"*. The plinth is normally placed on sound, competent and groutable rock. The treatment is accomplished by a grout curtain. However, there are cases where plinths have been placed on weathered rock or saprolite – they are treated with gunite and an inverted filter on downstream stretches – and cases of alluvial foundation controlled by a diaphragm wall.

More recently, plinth structures have been built with an upstream slab that is used as a base for grouting. A built platform erected specially to construct the grout curtain has been an alternative at sites where the abutment is flatter. An internal slab may also be required in order to better control hydraulic gradients.

8.1.2 Width

The following factors have to be taken into account when designing the plinth:

- Hydraulic gradients;
- Geological characteristics of the foundation;
- Foundation geometry (topobatimetry).

From a construction point of view, a minimum width of 3.0 m (from the "x" point in Figure 8.1) must be specified in order to accomplish the grout curtain (in three lines). A concept that's been adopted in many dams is to establish an optimized minimum width on the upstream stretch continued by a downstream extension in order to reduce quite significantly upstream excavation and to improve gradient control.

Figure 8.1 shows the criteria used for the CFRDs of Barra Grande and Campos Novos. Table 8.1 shows the plinth slab dimensions at the Barra Grande Dam.

The concept of combining an upstream optimized slab with a downstream extension was applied initially at Itá and Itapebi dams (Brazil). The design specified widths of 6.5 m and 4.0 m respectively for a gradient of 20. The same concept has been used in Indonesia and Malaysia.

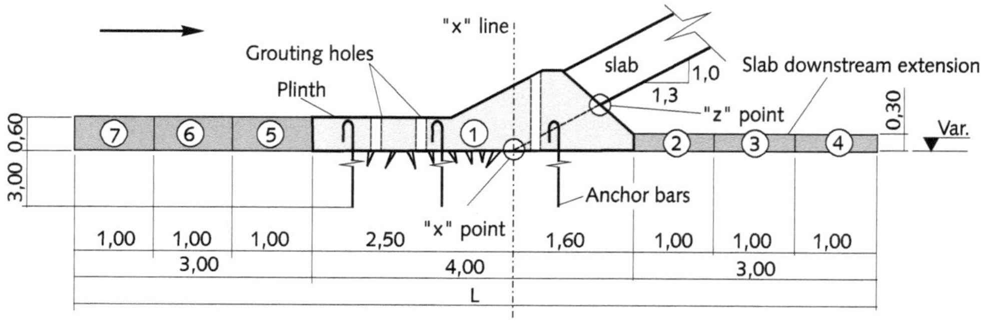

Figure 8.1 Dimensions of plinth slab of Barra Grande CFRD (185 m, 2005) and Campos Novos CFRD (205 m, 2005) (Engevix).

Table 8.1 Barra Grande: dimensions of the plinth slab (Engevix).

Situation	*Hydraulic head H (m)*	*Hydraulic gradient (H/L)* Min.	Max.	*Width – L (m)* Plinth	Internal slab	*Elevation EL W.L. max. ≅ 650*
1	H < 60	—	15	4.00	—	Crest at EL. 590
2	60 < H < 80	12	16	4.00	1.00	EL. 590 – EL. 570
3	80 < H < 100	13.3	16.7	4.00	2.00	EL. 570 – EL. 550
4	100 < H < 130	14.3	18.6	4.00	3.00	EL. 550 – EL. 520
5	130 < H < 150	16.3	18.7	5.00	3.00	EL. 520 – EL. 500
6	150 < H < 170	16.7	18.8	6.00	3.00	EL. 500 – EL. 470
7	170 < H < 185	17	18.6	7.00	3.00	EL. 470 – EL. 465

At Itá CFRD (125 m, 1999), as well as a significant reduction in excavation, there was a concrete reduction of up to 0.75 m^3/m due to a thinner internal stretch of plinth (Antunes Sobrinho et al., 2000).

Figure 8.2 shows a detail of the upstream excavation reduction that can be achieved when the concept of a downstream slab extension is adopted. Figure 8.3 shows a detail of the upstream optimized slab with a downstream extension for the plinth constructed on the steep abutment of Shuibuya Dam (China).

A modern method of plinth design has been achieved by taking into account the foundation rock quality (RMR Bieniawski classification) and the hydraulic gradients (Materón, 2007), as discussed in chapter 7, where the criteria and foundation treatments for the plinth are presented.

This concept has been applied in Itá (1999), Machadinho (2001), Monjolinho (2006), Barra Grande (2006), and Campos Novos (2006) in Brazil, as well as several other dams including Caracoles (Argentina), Bakún (Malaysia), Merowe (Sudan), Berg River (South Africa). It's allowed for optimization of excavation procedures and construction schedules.

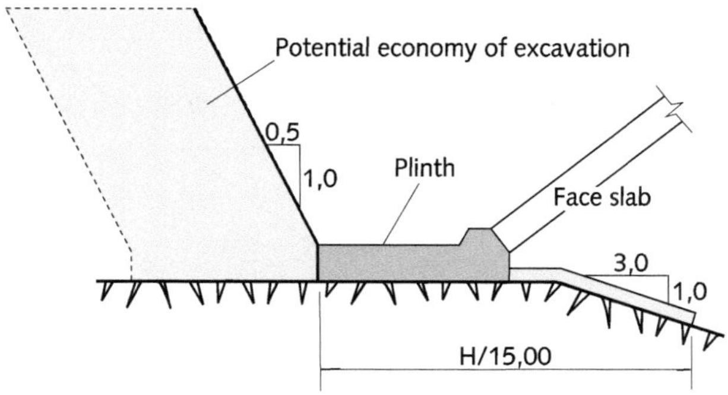

Figure 8.2 Plinth slab section: upstream and downstream segments (Marulanda and Pinto, 2000).

Figure 8.3 Shuibuya Dam: view of the plinth slab on the left steep abutment (upstream and downstream extension). (See colour plate section).

8.1.3 Thickness

The plinth slab thickness may vary for high dams (>120 m) from 0.90 m to 1.00 m at the the river bed section, reducing thickness to 0.60–0.40 m at the abutments. A downstream extension with a constant thickness of 30 cm has been commonly adopted. As a first attempt, the plinth thickness can be set as the same as the face slab thickness.

8.1.4 Plinth-slab connection

An empirical recommendation for a design to improve the sensitive connection between the plinth and the face slab consists in establishing the downstream face of the plinth so it makes an 90° angle with the face slab in order to attenuate the cracks and fissures on the slab contact area (Ramirez & Peña, 1999). A minimum height of 80 cm below the cooper joint seal is usually specified. This dimension may be brought down to 50 cm for dams under 40 m high (Figure 8.4).

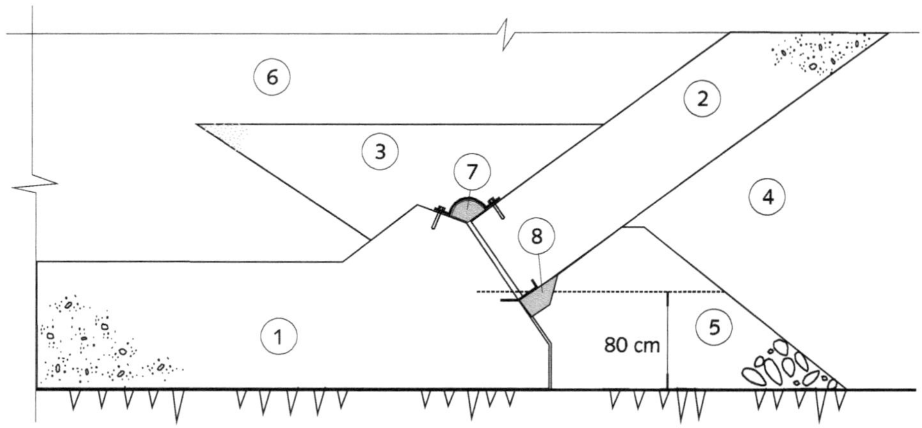

1 - Plinth
2 - Concrete face
3 - Silty sand
4 - 2A Zone
5 - 2AA Zone
6 - Random
7 - Mastic
8 - Asphalt cement

Figure 8.4 Plinth and perimetric joint details.

8.1.5 Features and practices

The construction sequence of the plinth and the main current practices are presented in chapter 12. Some features and practices drawn from the experience of Brazilian and foreign engineers are presented below.

At the river bed section and along the abutments, the specification should state that all alluvial, colluvial, and residual soil must be removed, as well as the weathered rock along the plinth, in an area 0.3H to 0.5H wide (H = hydraulic head of the reservoir).

As a basic design criterion, the structures of the plinth and zones 2A, 2B and 3A must be founded on sound, competent, and groutable rock. The foundation treatment for leakage control consists of building the grout curtain. Grouting should also reduce the permeability and the uplift pressures, and control the erosion on piping.

The design of the plinth geometry and its alignment along the abutments must aim towards reducting excavation and concrete treatment. However, in cases of extremely decomposed rock, additional excavations (over 10 m) and the construction of concrete blocks to support the plinth and contention walls might be necessary and it can significantly increase the amount of concrete used.

Under such conditions, the stability of the plinth wall must be carefully analyzed and assessed by the designers. In most cases where the plinth is founded on competent rock foundations, this criterion does not exclude the possibility of placing it on weathered rock not susceptible to piping or erosion, or on layers of gravel and sand in alluvial deposits.

8.1.6 Foundation on deformable structure – Hengshan case

The first stage in the Hengshan Dam (China) was to build a rockfill dam with a central impervious core. Afterwards, in a second stage built between 1987 and 1993, the dam needed to be raised up about 70 m, and a CFRD alternative was then chosen. The plinth structure was built on the crest of the old dam (first stage), which was treated by a 72.26 m long concrete curtain built in the dam's impervious core (Beijing, China, 1993). See also chapter 3 Figure 3.51.

The construction of the plinth at the river bed area may be accomplished by slipping metallic forms, hydraulically actived, or most commonly by using conventional wooden or metallic forms, as in Foz do Areia, Itá, Quebra-Queixo, Aguamilpa, Tianshengqiao 1 (TSQ1), Campos Novos, and Shuibuya. As a general practice, the use of slipping forms (metallic rails driven by a hydraulic system or even by a manual pulley system) is used on flat abutment areas. In wide valleys, for conventional plinth structures and concreting stretches over 100 m long, the slipping form production may reach values over 0.80 m/hour. Whenever it becomes a hustle for concrete mixer trucks to reach the concrete pouring fronts (in abutment sections), the use of pumped concrete is more feasible.

The plinth excavation and concreting works generally precede the rockfill placement and may begin either at the river bed section (after river diversion) or on the abutments. Plinth concreting may take place simultaneously at both abutments depending on progress in excavation, cleaning-up, foundation surface treatment operations (see chapter 7). The construction joints are implanted according, and in sequence, to each construction stage.

The critical path in a plinth construction schedule is building the grout curtain along the plinth's entire extension (river bed and abutment areas). The grout curtain is initially done at the river bed and at the low margin areas and it must be concluded before reservoir filling starts.

8.1.7 Transversal joints

For practical reasons, in modern CFRDs transversal contraction joints are no longer designed and built along the plinth. Construction joints are done only during different concrete work (see Figure 8.5).

Figure 8.5 Plinth: details of upstream and downstream and construction joint at the shoulder-deep part (Shuibuya CFRD, China, 2006). (See colour plate section).

8.1.8 Foundation treatment and regularization

After the removal of the superficial residual and decomposed soils, as well as the weathered and decomposed rock, the rock surface is cleaned by air blowers and water jets. Once the foundation is clean, the drilling and anchorage work installing the steel bars begins, followed by the placing of the reinforcement and copper water-stops according to the design. The concrete pouring activities start by spreading dental concrete (less than 1.0 m thickness) and back filling (over 1.0 m) over foundation depressions. The placement of the structural concrete should then continue.

In some stretches of the foundation, the regularization concrete may reach thicknesses over 2.0 m. Although good practice standards recommend a careful excavation in order to avoid over-excavation, in some places it is necessary to deepen the excavations to reach a competent rocky foundation. Construction of concrete walls may be necessary to support the plinth (chapter 7).

8.1.8.1 *Rock slopes excavations & regularization*

The conventional procedure consists of removing negative and abrupt slopes as well as the existing promontories in the downstream area (≤0.3 H away from the plinth). One of the purposes of this excavation and sharpness removal is to allow for a better compaction of the fine transition zone material (zone 2B) near the plinth structure. Another important purpose of the removal and rock surface regularization is to avoid the permanency of irregular and promontory rock areas, which may cause sharp variations in the rockfill thicknesses under the connection zone between slabs and the plinth structure. This effect may be significant in a zone of 0.3H or 0.5H from the plinth, particularly if the hydraulic head in this area is higher than 50 m. Sudden and abrupt changes in the rock surface topography may influence the deformations of rockfill in the plinth-slab junction. At Xingó, Brazil, slab cracks were recorded on the upper left shoulder after the filling of the reservoir. These cracks were soon associated essentially with existing protuberances on the rocky foundation at zones 2B and 3A next to the plinth (Souza et al., 1999).

8.2 SLAB

8.2.1 Slab design concept

These are the basic premises of a CFRD design.

- All the compacted rockfill lies downstream from the reservoir, protected by an impermeable surface (concrete face).
- The total hydrostatic load acts along the foundation of the plinth, upstream from the dam axis.
- The flow through the foundation, be it on rock or on alluvial deposits, is controlled from the upstream by either a grout curtain or by a diaphragm wall.
- The integrity and durability of the concrete slab, the plinth, the perimetric joint and the vertical joints (between slabs), are important factors for a good dam

performance. The basic concepts for the face slab design are listed as follows, not necessarily in order of importance:

- Sealing;
- Durability;
- Strength associated with deformability (elastic behavior).

Slab design criteria and construction specifications have been established over the last three decades based on CFRD performance and the experience presented in *Concrete Face Rockfill Dams – Design, Construction and Performance*, by J. Barry Cooke and James L. Sherard, published by the American Society of Civil Engineers (ASCE, 1985).

In chapter 2 of this book, "Design Criteria for CFRDs", Cooke and Sherard's main criteria are presented and discussed. However, the basic design premises of sealing the face slab as an "impervious barrier" has been challenged by the unexpected occurrence of cracks and ruptures during the reservoir impounding of Barra Grande, Campos Novos, and Mohale, resulting in an increase in leakage (over 1,000 ℓ/s) as a consequence of this unpredicted behavior.

Although in all of these cases safety has never been jeopardized, the designers, builders, owners, and advisors involved have reviewed the slab design criteria. The recorded fissures and cracks exceed in many cases the limits of opening allowance (<0.3 mm), a perfect condition for the corrosion reinforcement steel bars, mainly at the reservoir water level oscillation. Thus, durability must be factored in, because the lack of it leads to the need for repair work.

The design of the slab, and the construction techniques, must be such to absorb properly the deformations that will develop and to guarantee an efficient impervious barrier. A maximum leakage value can be set as a design criterion. If it is exceeded, the causes of the unpredicted leakage should be investigated and repairs should be undertaken in order to reduce leakage to within the specified values.

The slab sealing concept must be always linked to other properties of the concrete, such as strength and elasticity, given the differential deformation of the various zones of the rockfill during construction as well as throughout and after the reservoir impounding and dam operation.

8.2.2 New impermeability concepts

Although the use of concrete for the face slab is the most commonly applied solution, other available alternatives for water proofing may be applied.

a Geomembranes: the benefits of a geomembrane as a sealing system for dams are:
   - high tension strength;
   - high elongation capacity;
   - impermeability;
   - durability.

The synthetic membrane (geomembrane) is already being used in dam projects (concrete structures and CFRDs) as a sealing element. Initially it had been used for repairing and improving the sealing of concrete structures, like the Lost Creek Dam (USA) in 1997 (CARPI). In the Strawberry (USA, 2002) and in the Salt Springs

(USA, 2004/2005) CFRDs, a synthetic membrane was employed in order to improve the sealing condition of the slab after years of operation.

In Karáhnjúkar CFRD (Iceland, 2007), a geomembrane was installed in the construction of the lower levels of the slab (see Figures 8.6 and 8.7) (Scuero, Vaschetti & Wilkes, 2007).

In spite of its extreme efficiency in slab sealing performance and its easy installation, the use of geomembrane solution has been restricted by its high cost. There is currently an ICOLD sub-committee handling this matter specifically.

b Water proofing with asphalt concrete: the alternative of asphalt concrete (aggregates, sand and asphalt) has been used in medium and small-size dams in the European Union countries, but it has never been used in Brazil or in Latin America. The alternative of water proofing by asphalt has been used so far in channels, reservoirs (reversible mills – pump storage) and tailing fills only.

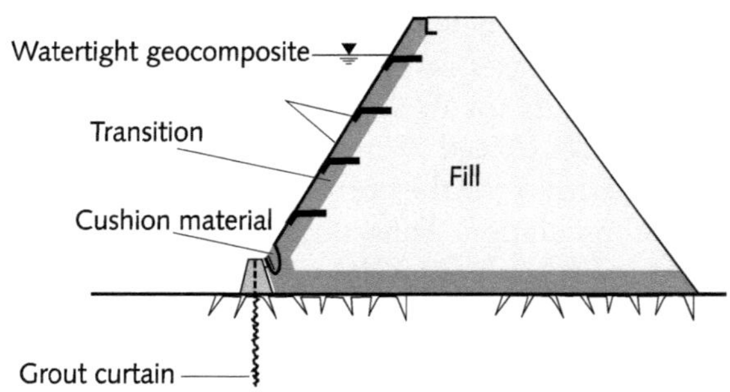

Figure 8.6 Geomembrane impermeability concept (CARPI).

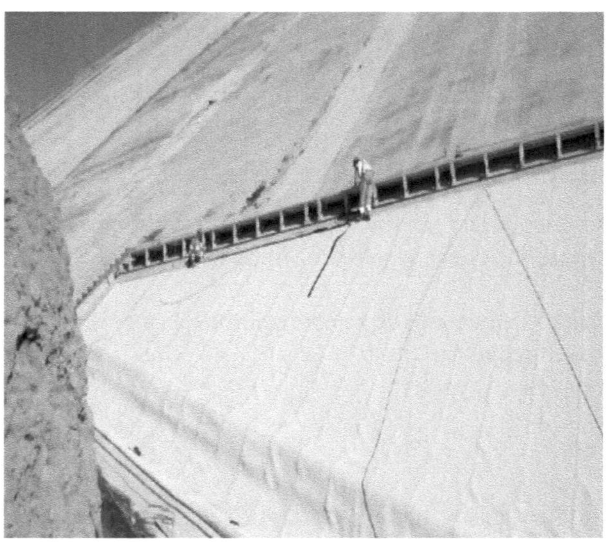

Figure 8.7 BEFC of Kárahnjúkar: geomembrane installation (CARPI) at slab 1st stage. (See colour plate section).

The asphalt concrete presents important features such as impermeability, flexibility regarding rockfill or foundation deformations, mechanic strength and longevity.

8.2.3 Slab thickness

The thickness of the concrete face of CFRDs over 80 m in height has been determined, empirically, by the following formula:

$$e = e_0 + kH \quad (8.1)$$

where:

e – thickness at a depth of H (m);
e_0 – minimum thickness of the slab at its top (m): it varies according to international literature from 0.30 to 0.35;
H – water head from reservoir level;
k – constant, varying from 0.0020 to 0.0065, according to each country's experience.

The value of $k = 0.0065$ has been applied to older dams (dumped and not compacted rockfill), and from 1999 to 2008, k values have ranged between 0.002 (Brazilian CFRDs) to 0.0035 (Chinese CFRDs).

Based on international experience (Cooke, 2000a), the recommended concrete face thickness is currently determined by the formula:

$$e = 0.30 + 0.002\ \mathrm{H} \quad (8.2)$$

In construction a minimum thickness of 300 mm has been adopted. Based on seepage and hydraulic gradient analysis (through the family of slab cracks), the ANCOLD (Australian Committee on Large Dams) has recommended that the hydraulic gradient through the face should be limited to 200 (Casinader & Rome, 1988). However, in some dams there have been estimated hydraulic gradients exceeding that value, e.g., Aguamilpa, where values around 215 were accepted.

A linear relation $e = 0.30 + kH$ has been applied for slab thickness for dams up to 100 m high and hydraulic gradients up to 220.

Figure 8.8 is a graph that shows the gradient values following a non-linear relationship. So, for gradients over 220 and CFRDs higher than 120 m, the function e = 0.0045 H must be used.

While developing the detailed design of Barra Grande and Campos Novos CFRDs, the criterion for depths over 100 m required a constant hydraulic gradient equal to 200, rather than a constant thickness, as usual practice dictates. The slab thickness has been determined by the relationships:

- e (m) = 0.30 + 0.0020 H(m) H < 100 m
- e (m) = 0.0050 H(m) H > 100 m (8.3)

8.2.3.1 *The Chinese experience*

At the moment, there are over 100 projects being planned or under construction in China. The CFRDs have been built with slab widths between 12 m and 18 m,

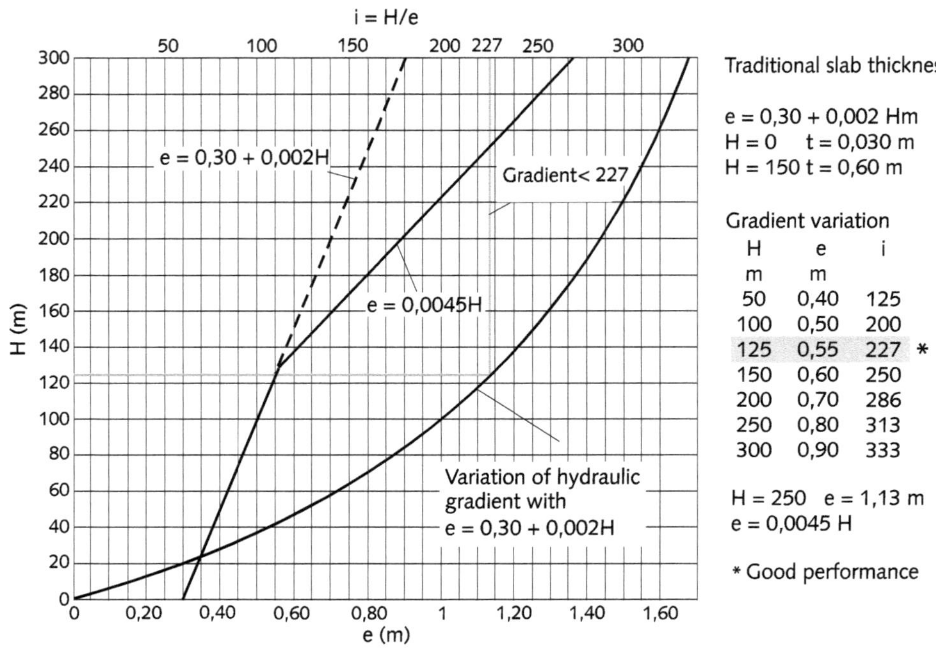

Figure 8.8 Gradients values × curve relation (Materón, 2002).

depending on the site characteristics (height, valley geometry). The slab thicknesses vary from 30 cm to 40 cm and are constant in CFRDs under 60 m high. For heights over 80 m, the equation applied has been:

$$e = 0.30 + (0.002 - 0.004)\ \mathrm{H} \qquad (8.4)$$

The specified concrete strength has been higher than 25 MPa. For regions with very low temperatures, the frost resistance grade required varies from F200 to F300. In a moderate climate this variation must not be below F50 to F100.

The concrete workability, as far as the slipping form works go, must stay in the range of 3~8 cm while being controlled in the feeding metallic chute. An air-entraining agent varying from 4% to 6% may improve considerably its impermeability, resistance to freezing, and durability characteristics.

According to Guocheng and Keming (2000):

"The main problem related to face concrete is fracturing as a result of temperature and dry contraction formed soon after concrete hardening. The measures taken to prevent these kind of cracks can be enhancing the crack preventing property of concrete itself and reducing the active force induced by external factors. Optimizing raw materials and mix proportion; using effective additives and admixtures, such as water reducing agents, air-entraining agent, fly ash, etc; selecting the most favorable season for pouring concrete and careful curing; minimizing the water-cement ratio and water content – all of these measures are effective in preventing the crack forming process. Some projects have used a slightly expanding agent to compensate for the contraction and to enhance crack prevention. But no common procedure has been achieved so far."

8.2.4 Joint sealing

In concrete dams, metallic joint seals (copper or another alloy) have not been used for over 50 years. They have been replaced by PVC joints with high resistance to tension (over 10 MPa), impermeability, high resistance to rupture, and durability.

The petrochemical industry has made available, in the last few decades, alternative compounds which are technically and economically more suited to civil construction. An alternative for replacing PVC and copper waterstops, resulting in technical, operational (installation), and cost effective advantages, has been discussed among professionals and experts, but it still needs to be proved in practice.

8.2.4.1 *Joint types*

Concepts and details of joints installed in several dams are described below.

1 **Perimeter joint** (external, between the plinth head and the slab)
The perimeter joint concept has been historically important and has played a key role in the design and safety criteria for CFRDs. The multiple joint sealing concept was initially developed for Australian dams and has been one of the most important features of the perimeter joint. The "multiple layers concept", using several combined materials such as mastic (IGAS), neoprene, PVC, copper waterstops and fine materials placed over the perimeter joint, has been used since the 1970s. Alto Anchicayá (1974) and Foz do Areia (1980) CFRDs are remarkable examples of the multiple layers concept.

A compressive material (wood or any other material), 12.5 to 20 mm thick, is normally placed in the interface between the plinth and the face slab. The main objective of this compressive material is to attenuate compression efforts along the joint due to face slab movements against the plinth structure as result of rockfill displacements during the construction and reservoir filling phases.

Here are some examples of CFRD joint design details.

- **Alto Anchicayá** (145 m, 1974) – concept of joint with multiple protection through the use of fine silty material over the joint and mastic (IGAS).
- **Foz do Areia** (160 m, 1980) – concept of joint with multiple protection (Fig. 8.9A).
  - Upstream fill of fine silty sand – zone 1A over perimeter joint;
  - Mastic application and either neoprene or PVC protection;
  - Installation of double joint sealing: central (copper or PVC) and bottom copper waterstop, over PVC pad and asphalt concrete pad;
  - Construction of a filter layer downstream (grain size, max. Ø 38 mm, zone 2A).
- **Segredo** (145 m, 1992) – concept of a triple protection (Fig. 8.9B).
  - Upstream fill over perimeter joint;
  - Mastic application and either neoprene or PVC protection;

- Placing of fine silty sand (max. Ø 0.5 mm) over the perimeter joint – zone 1A;
- Installation of joint sealing (simple) of copper background over PVC pad and concrete asphalt pad;
- Construction of a filter layer downstream (max. Ø 38 mm, zone 2A).

- **Aguamilpa** (187 m, 1994) – multiple protection concept (Fig. 8.10).
  - Upstream fill over the perimeter joint;
  - Placing of fly ash protected by a geotextile wrap and metallic plate (perforated);
  - Installation of double joint sealer: central (copper or PVC) and at the bottom (copper) over PVC band and concrete asphalt pad;
  - Placement of a filter layer downstream (max. Ø 40 mm, zone 2A).

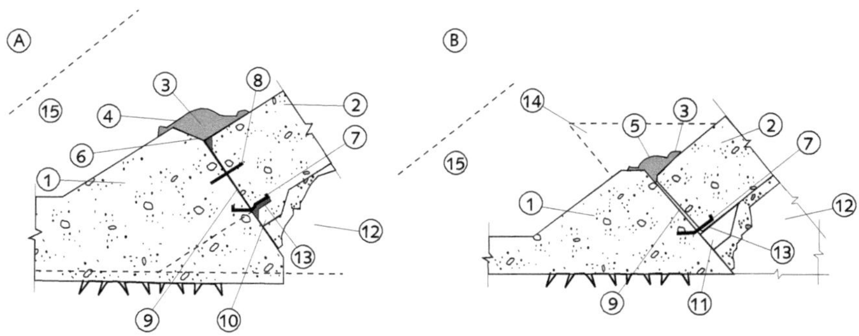

1 – Plinth
2 – Face slab
3 – Mastic
4 – Neoprene sheet
5 – Wide reinforced rubber band
6 – Neoprene cylinder
7 – Copper waterstop
8 – PVC waterstop
9 – Plywood
10 – Asphalt sand pad
11 – Cement mortar pad
12 – Special transition 2B-B zone
13 – Neoprene strip
14 – Silty sand strip
15 – Earthfill

Figure 8.9 A) Foz do Areia and B) Segredo perimeter joint (Pinto, Blinder & Toniatti, 1993).

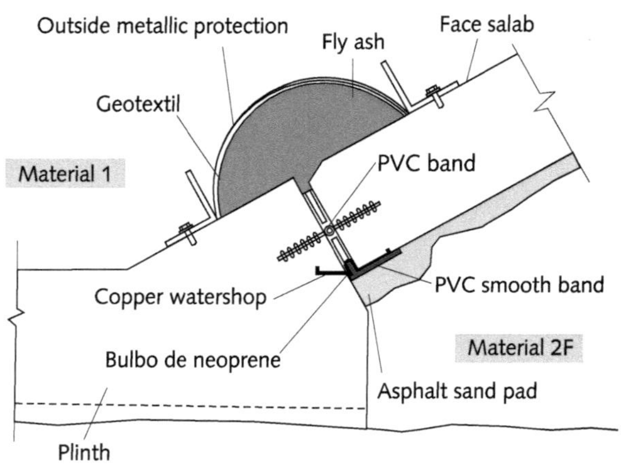

Figure 8.10 Perimeter joint: concept of multiple protection – Use of fly ash as an alternative to mastic (Aguamilpa, Mexico) (Gómez, 1999).

- **El Cajón** (190 m, 2006) – concept of multiple protection (see Fig. 2.15).
  - Upstream fill over perimeter joint;
  - Placing of fly ash protected by a geotextile and metallic plate (perforated);
  - Installation of double copper joint sealer: i) at the bottom on a PVC band and asphalt concrete pad; ii) copper waterstop at the upper part of the perimeter joint.
- **Xingó** (145 m, 1993/94) – concept of triple protection (Fig. 8.11).
  - Upstream fill over perimeter joint;
  - Mastic application and either neoprene or PVC protection;
  - Placing of fine silty sand (max. Ø of 1.0 mm) over the perimeter joint;
  - Installation of a copper joint sealing (simple) at the bottom over a PVC band and concrete asphalt pad;
  - Construction of a filter layer downstream (max. Ø 38 mm, zone 2A).
- **TSQ1** (178 m, 2000) – concept of multiple protection (Fig. 8.12).
  - Upstream fill over the perimeter joint;
  - Placing of fly ash protected by a geotextile wrap and metallic plate (perforated); solution applied on the vertical joints of the tension zones at the abutments;
  - Installation of copper double joint sealer: i) at the bottom over a PVC band and asphalt concrete pad; ii) a copper/PVC waterstop at the middle part of the perimeter joint.
- **Shuibuya Dam** – new concept of perimeter joint (Fig. 8.13).

In Shuibuya Dam (233 m, China, 2007) a new concept of multiple protection was adopted, including a corrugated internal joint protected by elastic material (GB), and a synthetic rubber cylinder – all protected externally by a PVC cover (Fig. 8.13 A). A similar concept has been recently applied at Bakún CFRD (205 m, Malaysia, under construction) and at Mazar Dam (185 m, Ecuador, under construction), on the right abutments (Fig. 8.13B). On the left abutment and river bed, the fly ash joint (El Cajón) concept was used.

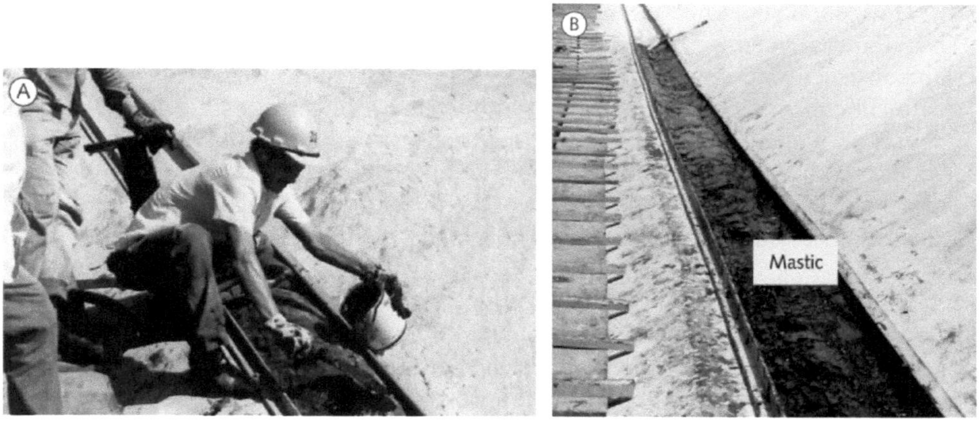

Figure 8.11 A) Perimeter joint and B) mastic placement in Xingó Dam. (See colour plate section).

Figure 8.12 TSQ1: A) perimeter joint with fly ash (covered by geotextile and metallic plate on the abutments); B) perimeter joint details. (See colour plate section).

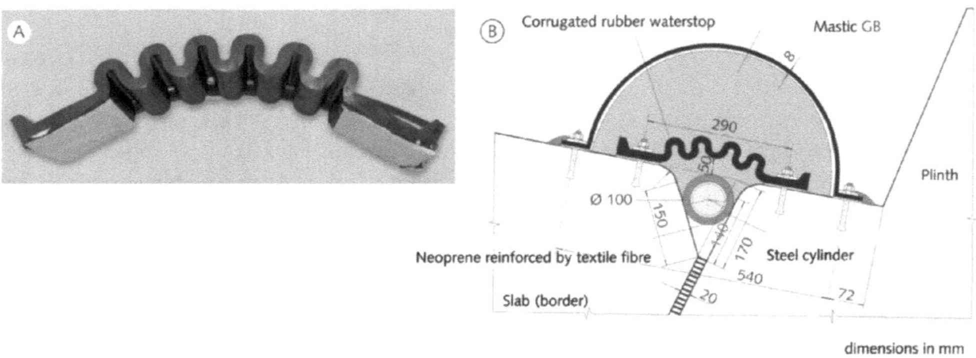

Figure 8.13 A) Corrugated joint adopted in Shuibuya (China) and B) Mazar (Ecuador). Courtesy of Consortium Mazar Management. (See colour plate section).

2 Horizontal construction joint

These joints are defined by the designer/constructor according to the various stages of slab construction (two or three stages for dams over 140 m). The construction joint treatment consists in removing some centimeters off the pre-existing concrete, cleaning up (air, water) and extending the reinforcement connections. The reinforcement steel bars continue overlapping the concrete of the next stage. This construction joint should not be considered as a contraction joint (with the installation of a joint sealer) – this is unnecessary and expensive, according to good engineering standards.

3 Slab-plinth connection joints

At a L distance (often ranging from 10 m to 20 m) from the plinth, an additional joint sealer (besides the copper joint sealer at the bottom of slab) has been installed in some projects to increase protection and to improve sealing (in case of cracks) throughout the plinth area. At TSQ1, an additional PVC joint sealer has been designed along slabs joints, at a distance of $L = 20$ m from the plinth (Fig. 8.14). In the central area

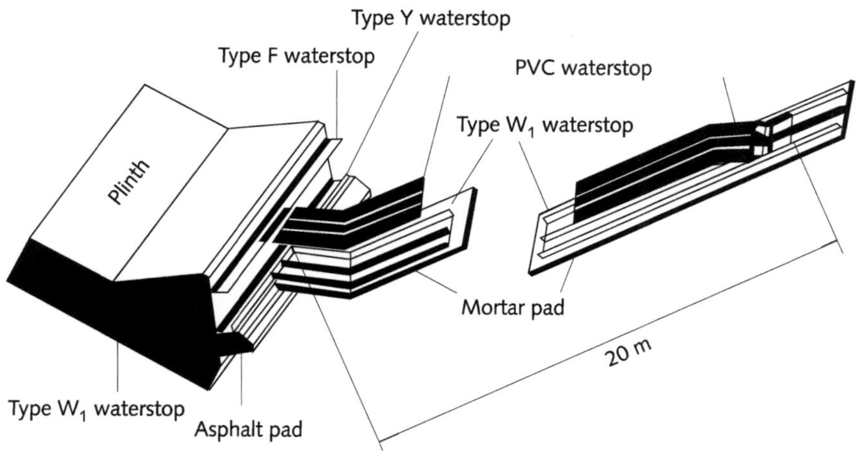

Figure 8.14 TSQ1: connection of vertical joints (bottom and center) with the plinth (band $L = 20$ m) (Wu et al., 2000a).

between joints the installation of this second joint sealer turns out to be a burden on construction sites. There are also the difficulties in maintaining the alignment of the PVC joint during the sliding forms displacement. Additional care has to be taken during concrete pouring to assure the homogeneity and quality between the two joint waterstops (copper at the bottom and PVC at the center).

In the case of Barra Grande and Campos Novos CFRDs, a joint sealer of mastic protected with a PVC band has been designed over the vertical joint top at a distance $L = 5$ m from the plinth (Fig. 8.15).

4 **Vertical contraction joints** between slabs

These are used where there is no reinforcement overlapping between two adjacent slabs. Areas where contraction joints are placed:

- Abutment areas, tension zones – As a common practice a copper joint sealer is installed at the bottom of the slab, and a second joint sealer (Jeene type, PVC) at the top, between the two slabs. This practice was adopted in the late 1990s and early 2000s in the CFRDs of Itá, Itapebi and Machadinho, in Brazil.
- Central area – At the central area where compression zones prevail, waterstops are not installed in the upper part, only the copper waterstop is installed at the bottom. Between the two slabs an asphalt paint is applied to one of the slab faces before concreting the adjacent slab. This is a practice adopted in Brazil since the 1980s.

After the cracks and ruptures recorded at Barra Grande and Campos Novos (2005) and at Mohale (2006) during reservoir impounding, and at TSQ1 (2003/2004) after three years of operation, the vertical joint concept in the central areas includes a compressive filler between slabs consisting of wood – or equivalent material – to absorb compression stresses (Materón, 2008). This "compressive filler concept" along the vertical joint was applied after repair works at Barra Grande, Campos Novos and

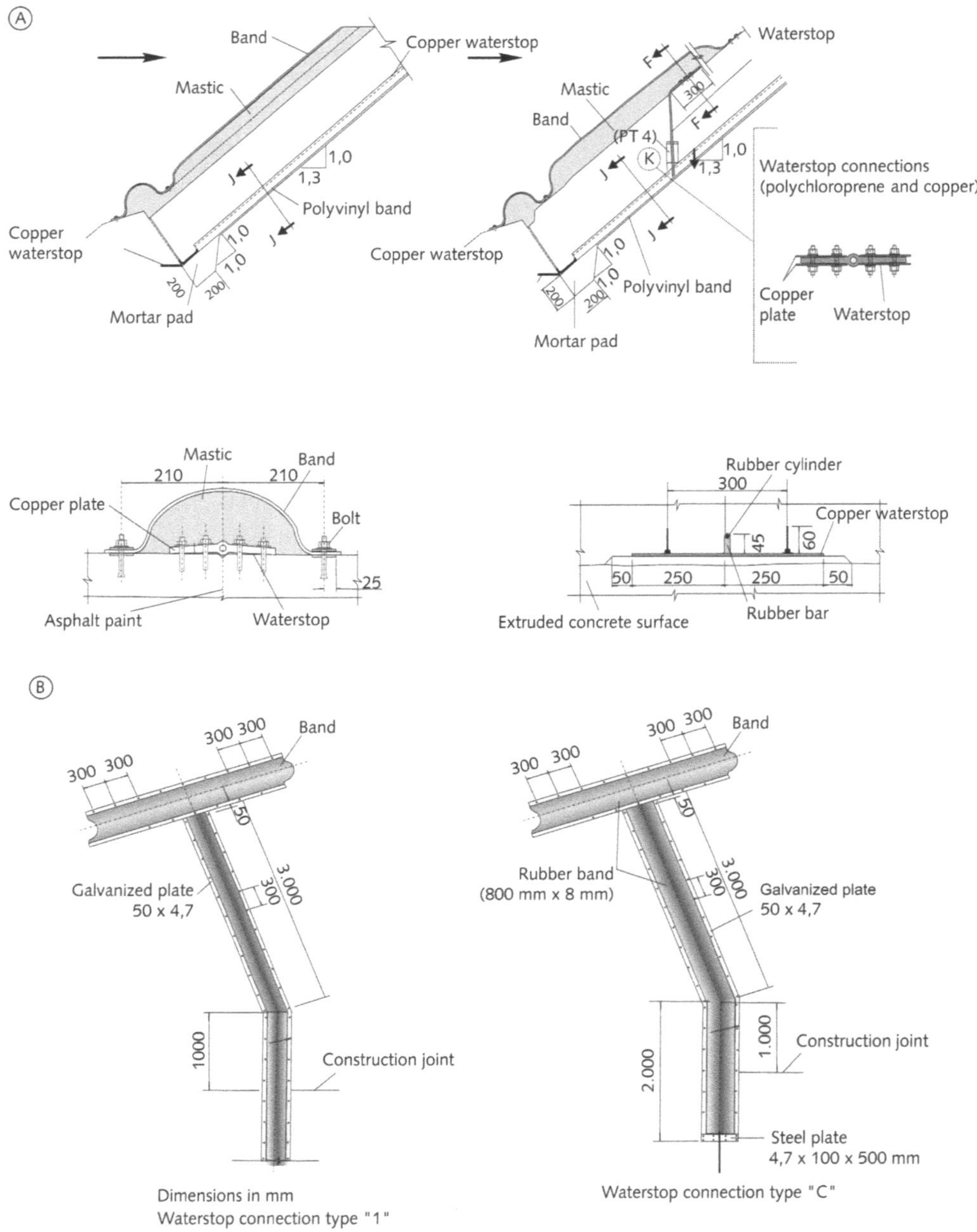

Figure 8.15 **Detail of the contraction joint waterstop of the vertical joint (A) and the plinth (B) adopted for Barra Grande and Campos Novos dams (Engevix).**

Mohale, and has been adopted in several current CFRDs, such as El Cajón, Kárahnjúkar, Bakún and Mazar.

Figure 8.16 presents examples of vertical joints used in CFRDs in Brazil before the inclusion of the compressive filler. The new joint used in Kárahnjúkar Dam is illustrated in Figure 2.18.

The following criteria must be met in the design and construction of vertical joints in areas of high compression:

- Maintain the design thickness of the slab above the surface of the extruded concrete;
- Imbed the mortar pad into the extruded concrete;
- Install a flexible filler between the two slabs (wood, neoprene or asphalt) to absorb the compression efforts;
- Reduce (maximum 2 mm) or eliminate the upper *V* notch;
- Apply anti-spalling reinforcement.

For narrow valleys, additional vertical joints (slabs 7.5 m wide) must be considered at the abutment areas to mitigate compression stresses (as, for example, at Shuibuya, El Cajón, Bakún and Mazar).

8.2.4.2 *Vertical joints: other concepts*

TSQ1 (China, 178 m, 2000): placement of fine non-cohesive silty sand (fly ash or coal ash) on the top and throughout the joint, enveloped by a geotextile and protected by a perforated metallic plate. A similar waterstop has been applied throughout the perimeter joint (Fig. 8.12).

El Cajón (Mexico, 190 m, 2006): fine non-cohesive silty sand (fly ash) on the top and throughout the vertical joint, enveloped by a geotextile and protected with a perforated metallic plate. Upper and lower copper joints, similar to the perimetral joint, complete the system of multiple protection (Fig. 8.17).

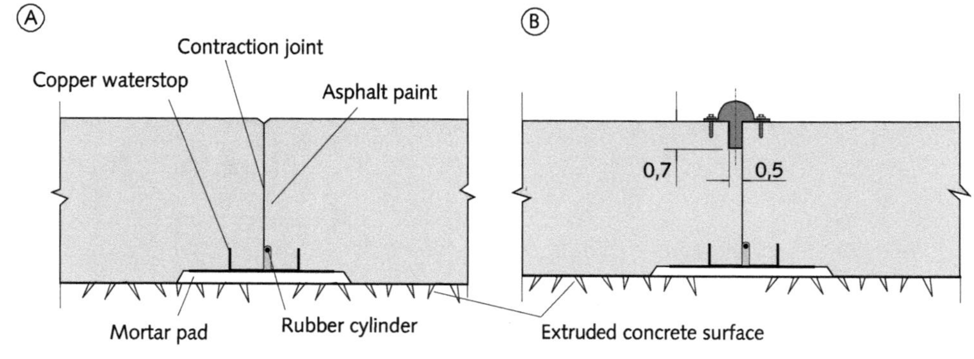

Figure 8.16 Vertical joints: A) Campos Novos, central area (compression zone); B) Barra Grande, abutments (tension zone) (Engevix).

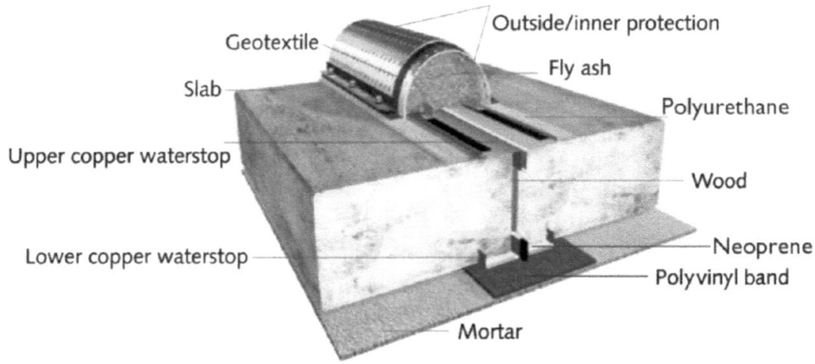

Figure 8.17 Detail of perimeter and vertical joints (zone of tension – abutments) (Mena Sandoval et al., 2007a). (See colour plate section).

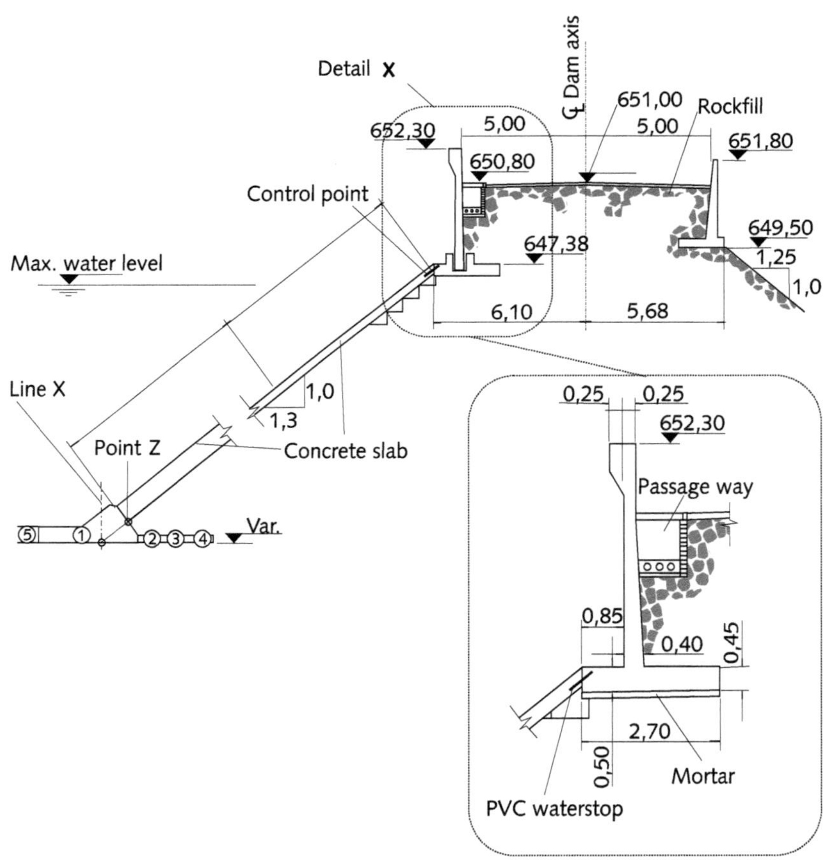

Figure 8.18 Barra Grande CFRD: parapet wall and expansion joint detail slab – wall foundation contact (Engevix).

Figure 8.19 Xingó CFRD: parapet wall under construction. (See colour plate section).

5 **Expansion joints** – connection between the upper slab end and the crest wall. Although placed some meters above the maximum normal water level of operation, an expansion joint is designed between the upper slab portion and the parapet wall base. A copper waterstop has been often used in these expansion joints. (Figs. 8.18 and 8.19).

8.3 REINFORCEMENT DESIGN

The percentage of steel bars in both directions has been fixed empirically in CFRD design. The main concern of designers is to guarantee sealing by minimizing cracks and maintaining the integrity of the slab when it is submitted to compression and tension as a consequence of rockfill deformations.

Some criteria that have been empirically adopted are:

- Application of 0.4% to 0.5% (vertical) and 0.30% to 0.35% (horizontal) of steel bars in each direction, in the form of mesh, with the exception of the region near to the plinth and abutments where 0.4% is generally specified;
- Elimination of overlapping steel bars over slabs along vertical joints;
- Installation of anti-spalling double (upper and lower) reinforcement;
- Installation of double reinforcement (0.4% in both directions) in a strip of 10 m to 15 m wide along the plinth (TSQ1, Barra Grande, Campos Novos).

Table 8.2 shows a chronological list of the design criteria for the face slab of important CFRDs (Materón, 2008).

8.4 CREST PARAPET WALL AND FREEBOARD

Improved design, as well as progress in construction techniques over the last decade with regard to the use of precast concrete, have contributed to time and cost reductions in the building of parapet walls.

Table 8.2 Face slab design of main CFRDs (Materón, 2008).

Name	Country	Year	Height (m)	Slope H to V: Upstream	Slope H to V: Downstream	Face slab thickness $e = e_0 + kH$	Reinforcement each way (%)
Cethana	Australia	1971	110	1.3	1.3	0.30 + 0.002H	0.6
Alto Anchicayá	Colombia	1974	140	1.4	1.4	0.30 + 0.003H	0.5
Golillas	Colombia	1978	130	1.6	1.6	0,30 + 0.0037H	0.4
Foz do Areia	Brazil	1980	160	1.4	1.25–1.4	0.30 + 0.0034H	0.4
Murchison	Australia	1982	89	1.3	1.3	0.30	0.65
Salvajina	Colombia	1983	148	1.5	1.4	0.30 + 0.0031H	0.4
Chenbing	China	1989	75	1.3	1.3	0.30 + 0.002.7H	0.3(H):0.5(V)
Segredo	Brazil	1992	145	1.3	1.2–1.4	0.30 + 0.0035H	0.3:0.4
Aguamilpa	Mexico	1993	187	1.5	1.4	0.30 + 0.003H	0.3 (H):0.35 (V)
Xingó	Brazil	1993	145	1.4	1.3	0.30 + 0.0034H	0.4
Itá	Brazil	1999	125	1.3	1.3	0.30 + 0.002H	0.3(H):0.4(V)
TSQ1	China	2000	178	1.4	1.4	0.30 + 0.035H	0.3(H):0.3(V) Double at 3rd stage slabs
Torata	Peru	2002	100	1.3	1.3	0.30 + 0.002H	0.4
Xekaman	Laos	–	187	1.3	1.3	0.30 + 0.002H	0.35(H):0.4(V)
Itapebi	Brazil	2002	110	1.25	1.3	0.30 + 0.0020H	0.35(H):0.4(V)
Machadinho	Brazil	2002	125	1.3	1.3	0.30 + 0.0033H	0.35(H):0.4(V)
Mohale	Lesotho	2002	145	1.4	1.4	0.30 + 0.0035H	0.4
Campos Novos	Brazil	2006	202	1.3	1.4	0.30 + 0.002H ($H \leq 100$ m) 0.005H ($H > 100$ m)	0.5 in both ways Double at 20 m from plinth
Barra Grande	Brazil	2006	185	1.3	1.4	0.30 + 0.002H ($H \leq 100$ m) 0.005H ($H > 100$ m)	0.3(H):0.4(V) central part
Messochora	Greece	2006	150	1.4	1.4	0.30 + 0.003H	0.5
Shuibuya	China	2007	233	1.4	1.4	0.30 + 0.003H	0.4
Kárahnjúkar	Iceland	2007	196	1.4	1.4	0.30 + 0.002H	0.3(H):0.4(V)
El Cajón	Mexico	2006	188	1.4	1.4	0.30 + 0.003H	0.4
Bakún	Malaysia	Under constr.	205	1.4	1.4	0.30 + 0.003H	0.3(H):0.4(V)

Water Power & Dam Construction – Yearbook 2008.

The parapet wall must meet the following criteria and guide-lines:

- Standing wave run up and dissipation;
- Freeboard should be calculated from parapet wall top level;
- Parapet wall must be extended along the crest all the way to reach both abutments;
- Estimating of nominal dam crest elevation should take into account dam settlements from post construction effects, past impounding, and long-term accomodation due to creep effect;
- Wall concreting works can be carried out at the site or by using precast elements.

After completing the wall concrete work, upstream and downstream fine rockfill is placed and compacted ($Ø_{max.}$ 30 cm), filling up to the ultimate dam crest elevation.

8.5 FISSURES, CRACKS, AND FAILURES – TREATMENTS

The current important cases of cracks and failures on slabs that have occurred in important CFRDs, such as TSQ1 (2003/2004), Barra Grande (2005), Campos Novos (2005), and Mohale (2006), have been discussed in international papers presented at Portugal (Lisbon, 2006/2007), China (Yichang, 2007) and Brazil (Florianópolis, 2007).

An overview of crack and fissure issues (Cruz & Freitas, 2007) elaborated by experts is being presented here.

Fissures

It is agreed upon that fissures are openings on the slab surface up to 0.3 mm wide that generally have been recorded before the reservoir impounding and are mainly caused by concrete retraction induced by thermal tensions and/or daily temperature variations. They often do not cause any significant damage to the CFRD slab and may be tolerated without any surface treatment. These fissures have been recorded on practically all CFRDs.

Cracks

Cracks can be defined as larger openings, over 0.3 mm wide, that normally occur during the construction or the reservoir impounding through compression or tension induced stress from the rockfill deformations and protuberances in the foundation. In some cases, these cracks are observed only years after the reservoir fills as at Xingó, Aguamilpa and TSQ1. At Xingó Dam (1994) cracks have been recorded on the left abutments slab after two years of operation (1996). Openings of approximately 10 mm have been caused by a non-uniform foundation geometry (deflection point) between the slabs L4/L5 and the perimeter joint. On slab #6, a horizontal opening has reached 15 mm on the surface. At Aguamilpa Dam, there were horizontal cracks on the central part of the upper slab, with a maximum opening of about ½ inch (~1.25 cm). At TSQ1, during the partial filling of the reservoir at the end of construction phase (2000), a total of 930 fissures and cracks were monitored at the second stage of the slab (between El. 680.0 and El. 746.0) and at the upper part built in a third stage (between El. 767.0 and El. 780.0). About 80% of them have presented openings smaller than 0.3 mm (fissures), 5.0 m in maximum length, and were predominantly horizontal. The remaining (18%) have presented openings that vary from 0.3 mm to 1.0 mm, and a small percentage (<2%) have showed openings of 1.0 mm to 1.5 mm. Unlike the fissuring process, the cracks (and failures) on the slab generally result in a significant increase of seepage flows that have not been foreseen during the design phase, both during reservoir filling and operation.

Failures

Besides cracks that often occur due to tension or compression stress, failures are currently being recorded that are associated with excessive compression on the central part of the face slab, with consequent concrete damage and spalling, as well as exposing the reinforcement. Slab failures were observed at TSQ1, Barra Grande, Campos Novos, and Mohale CFRDs. These failures are often accountable for an increase of seepage discharge, and have contributed to significant dam leakage (over 1000 ℓ/s).

1 Construction features

The construction of the rockfill dam requires detailed planning because after the river closing or diversion, the dam can face at least one or more flood periods before reaching the crest level.

In the case of cofferdam overtopping, and if the slab has not been concreted, water proofing works should be performed on the upstream face in order to avoid free seepage through the rockfill. Gunite concrete has been used as an alternative material to provide an impervious surface. In addition, it is required that the rockfill itself must be able to face a 500-year flood period.

The concrete curb, on its own, does not offer a satisfactory sealing condition for floods of 300 or 500 years period of recurrence. In face of these restrictions, the convenience of dividing the slab construction in two or even three stages, as at Barra Grande, Campos Novos and TSQ1, is attractive.

Thus, the question to be faced is how to protect the dam from free seepage during construction, in which it would be exposed to one or more flood periods. The issue is more severe in the case of organizing slab concreting in a single stage and, consequently, scheduling it at the end of rockfill construction. Putting it this way, organizing slab concreting in stages is the solution and not the problem. Naturally, there must be gains in productivity in concentrating the slab concreting in one stage. These gains, however, must be compared with those costs of supplying temporary protection.

Brazilian experience seems to indicate that in CFRDs up to 130 m high, slab concreting in a single stage has the advantage. It has been proven, in terms of concrete technology, that there are no restrictions for concreting a slab of these dimensions in a single stage. Machadinho CFRD is a successful Brazilian experience.

Regarding rockfill behavior (and settlement progress with time), delaying the slab construction would have the advantage of allowing a longer time for rockfill deformations.

Fill and slab construction aspects are considered further in sections 12.8 and 12.9 respectively, in chapter 12.

2 Slab slipping forms main preparation

The main preparation steps that precede slab concreting are:

- Preparing the mortar pad on the curb (extruded) concrete surface;
- Installing and aligning the copper waterstops;
- Installing lateral forms and vertical joints (compressive filler) between the slabs; the compressive filler must be placed over the copper joint (bottom) and reach the upper part of the lateral form;
- Levelling the top of the form throughout the slope;
- Placing of the reinforcement steel bars and instrumentation;
- Final check list and inspection of the joints between slabs and between the slab and plinth (perimeter joint);

- Positioning the slipping form framework on rails;
- Implementation of concrete curing system (aspersion or wetting);
- Positioning of metallic chute or tubes to convey the concrete;
- Concreting works start (Fig. 8.20).

3 **Slipping form facilities**

- Hydraulic rail slipping system: Foz do Areia, Segredo, Xingó and TSQ1 – (first stage of the slab).
- Steel cable traction (electric winches on the upper part of the dam): TSQ1 (second and third stages of the slab), Itá, Machadinho, Itapebi, Quebra-Queixo, Barra Grande and Campos Novos. These facilities are currently being used in several countries, such as in Ecuador on the Mazar CFRD, 166 m, under construction – Figure 8.21.

4 **Curing**

Curing is a procedure of extreme importance that ensures concrete quality by controlling the concrete slab temperature and the process of crack development. The curing process must start immediately after the slab slipping form works. In some cases, curing begins at the downstream part of the slab after the concrete surface finishing service.

Chemical cure has been used in some cases, in order to simplify the hoses and pumping facilities needed to assure water supply along the whole slab surface during concreting. However, according to experts, the efficiency of chemical cure should not be used as an argument to dispense with conventional curing through wetting.

5 **Extruded concrete**

The transition zone (zone 2A) slope protection against erosion during construction has progressed significantly since the construction of Itá CFRD in Brazil (1999).

The main slope protections that were in use before Itá:

Figure 8.20 Metallic chutes for concrete convey down to the slipping form. (See colour plate section).

Figure 8.21 Mazar CFRD: detail of electric winches and steel cable traction system. (See colour plate section).

- Shotcrete facilities, a solution seldom used in Brazil;
- Concrete mortar, applied manually (in restricted areas);
- Asphalt emulsion (Foz do Areia, Segredo, Xingó and TSQ1), a solution adopted until the late 1990s.

The extruded curb, also known as the "Itá Method", was initially used at Itá Dam (1999) and since 2000 it has been the method adopted in all CFRDs worldwide. See chapter 12.

8.6 DRAINAGE NEAR THE PLINTH

A drainage system must be designed and implemented in areas downstream from the plinth on the river bed before starting the fine transition layer placing and compaction. The water addition used during compaction of zone 3B, as well as rain, results in a water flow to the upstream lower areas close to the plinth that can cause a raised water level and hydrostatic pressures (see, for example, Mazar and El Cajón CFRDs). At Xingó CFRD this pressure caused the failure of the lower part of zone 2A (before the slab concreting) as a result of a drainage system malfunction.

The drainage design must include the construction of a well (concrete tubulation, 1.0 m to 1.20 m diameter) from the foundation level up to the slope surface elevation. This well must be connected to a system of metallic pipes (over 6 inches diameter) and must stay operational during the whole period of construction even after concreting of the first slab stage. Before the upstream fill construction starts, the drainage system (well and tubulation) must be grouted.

Practical procedures for implementing this drainage system are in chapter 12.

Chapter 9

Instrumentation

"Monitoring of every dam is mandatory, because dams change with age and they may develop defects. There is no substitute for systematic and intelligent surveillance."

(Peck, 2001)

9.1 INTRODUCTION

CFRD instrumentation must be oriented to either focus on specific questions or to attend the design criteria. According to Cooke (2000a), CFRDs contain inherent safety features such as: i) all the zoned rockfill embankment is located downstream from the reservoir water; ii) the water load on the concrete face enters the foundation upstream from the dam axis; iii) uplift and pore pressure do not intervene; iv) the high reliability and shear strength of rockfill; v) the high seismic resistance of rockfill; vi) the rockfill zoning is favorable to throughflow.

All these inherent safety features have already been stated by Cooke and Sherard (1985). An updated review, however, is introduced in chapter 2. In addition, the 1970s evolution of CFRDs using compacted rockfill and the construction of the Cethana Dam have played an important role in consolidating engineering standard practice. So far, only one CFRD (Gouhou, China, 71 m) has failed (August, 1993) after reservoir impounding (Yuan & Zhang, 2004). A dam failure in the United States was recently reported by Qian (2008) (see chapter 10).

However, cracks and failures on face slabs and major leakages have been recorded at Alto Anchicayá (1974), Shiroro (1984), Golillas (1984), Aguamilpa (1993), Xingó (1994), Itá (1999), Itapebi (2002), Barra Grande (2005), Campos Novos (2005), and Mohale (2006). Even when considering that dam safety has not been affected in any of these cases, dam engineers have put a lot of effort into solving these unpredicted incidents. The aim here is to protect the concrete slab from future mishaps and to reduce dam leakage and avoid excessive water loss.

The remarkable CFRD performance of Cethana (110 m), Alto Anchicayá (140 m), Foz do Areia (160 m), Aguamilpa (187 m), Tianshengqiao 1 (TSQ1, 178 m) and of Barra Grande (185 m), Campos Novos (202 m), Bakún (205 m), Shuibuya (233 m) and La Yesca (220 m), can be attributed in part to the important role played by monitoring dam behavior. This contributes to the progress of CFRD design, safety and construction technology progress and to the planning of the very high CFRDs (over 300 m high) of the future.

9.2 MONITORING PARAMETERS

CFRD design and construction has steadily progressed from dumped to compacted rockfill placed in controlled horizontal layers, and this has become the usual engineering practice. Instrumentation and monitoring during construction has become a mandatory procedure, mainly because of the increase in height of CFRDs since the 1980s as in the case of Alto Anchicayá and Foz do Areia.

Thus, in addition to laboratory tests on samples and building test fills done on site, instrumentation has been used to monitor CFRD performance and to obtain important rockfill stress and deformability parameters. Such parameters describe the relationships between rockfill construction and the filling phases, and are very helpful for finite element analysis, both for comparison with design assumptions and for new and future design and construction of larger CFRDs.

Here we consider the main parameters to be focused on by instrumentation.

9.2.1 Dam movements

Vertical and horizontal rockfill displacements during the construction phase and reservoir impounding are currently the most important variables to be monitored. In addition, their affect on face slab displacements has concerned engineers since cracks and failure issues were recorded in TSQ1 (2003, 2004), Barra Grande (2005), Campos Novos (2005), and Mohale (2006).

CFRD displacements during construction due to foundation and rockfill zone settlement occur quite frequently. When founded on rock, displacements are the result of rockfill zone settlement only. For CFRDs set on granular alluvium foundations, seepage control is a major concern as foundation settlement is not altogether significant. All silt clay or sandy clay alluvium layers must be previously removed. This has been proved to be a good engineering practice.

Rockfill settlement during the construction and reservoir impounding must be carefully monitored for it has a major influence on the face slab.

Instrumentation must monitor the following aspects (Freitas, 2004):

- Slab displacements and deflections induced by differential rockfill settlement caused by abrupt geometry of the plinth on abutments;
- Differential settlements between the upstream (3B), the central zone (T), and the downstream area (3C) induced by different construction specifications for each zone such as layer thicknesses, vibratory roller passes, wetting during compaction and rock lithology;
- Rockfill construction stages or different speeds at abutments and /or speeding up construction of downstream zone (e.g. TSQ1);
- Different moduli of deformability between upstream and downstream zone materials (e.g. Aguamilpa Dam);
- Rockfill grain size distribution placed along the zones;
- Shape of the valley.

However, the relevance of these factors on dam behavior is still controversial and there is no agreement between different authors.

According to Silveira (2006), in order to assure a good monitoring program, different types of instruments must be installed to measure the same determined variable (settlement, deflections, etc.) The instruments are described below.

9.2.2 Monitoring rockfill displacements

Internal rockfill displacements are monitored through settlements gauges installed during the construction. Steel plates or magnetic rings detect settlement which is measured at the surface by reading equipment. A steel tube is anchored in the rock foundation as a reference benchmark.

The main settlement gauges installed in CFRDs are hydraulic cells (Swedish-box type), extensometers, electric cells, KM type (very common in Brazilian dams), USBR torpedo model – developed by the Bureau of Reclamation, USA – and magnetic rings (Geokon type). Inclinometers protected inside the rockfill zones have also been installed in several cases.

1 **Hydraulic cells** (Swedish-box type) are installed horizontally along inner rockfill zones (Fig. 9.1). There are three pipes connecting each cell to the instrumentation house: one reading pipe connected to the reading panel (instrumentation cabin), one air pipe, and one drainage pipe. Hydraulic pipes crossing the rockfill must be placed taking into consideration that differential settlement along rockfil zones can damage pipes or even cause the tubulation system to collapse. The tubulation system is connected to the reading panel and to the control cabin located downstream. The water level difference between the cabin water level and each cell allows for the monitoring of the rockfill settlement. During the construction and installation of the hydraulic pipes, the system must be well protected from

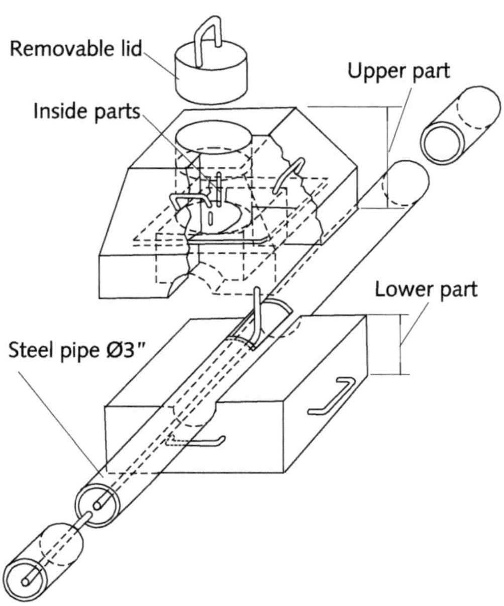

Figure 9.1 Hydraulic cells similar to the ones installed in Xingó, Itá and Itapebi CFRDs.

trucks and heavy construction equipment running over the site and it must be able to withstand high rockfill settlement without water leaking. Air bubbles in the hydraulic circuit in each cell must be avoided and carefully removed. An exhaustive program of tests must be implemented by circulating water throughout the pipe system, and by air injection and suction by vacuum. Blockage in the drainage pipe seems to be the main problem that can occur. Differential settlement during construction probably induces waving with deep concavities and vacuum in the pipes. Thus, water and air bubbles trapped in several points along the drainage pipe must be controlled by the instrumentation team. Hydraulic cells (Swedish-box type) have been used in nearly all CFRDs.

2 **Extensometers** are used to measure displacements along a vertical single axis. The magnetic extensometer consists of a series of magnets that are installed along an access tubing. The magnets are anchored at specified depths. Measurements are done by sliding a probe through the access tubing to detect the depth of the magnets (Durham Geo Slope Indicator Co.).
Extensometers and hydraulic cells located at upstream areas are temporarily installed with their Invar wires and pipes stretched to the temporary cabins during construction. In cases when the upstream and downstream rockfills are raised up almost simultaneously, the Invar wire and hydraulic pipes must be connected directly to the permanent instrumentation cabin.

3 **Electric settlement cell** consists of a reservoir, liquid-filled tubing, and the settlement cell, which contains a pressure transducer. One end of the tubing is connected to the settlement cell, which is embedded in the fill or installed in a borehole. The other end of the tubing is connected to the reservoir, which is located away from the construction area. The transducer measures the pressure created by the column of liquid in the tubing. As the transducer joins together with the surrounding ground, the height of the column increases and the settlement cell measures higher pressure. Settlement is calculated by converting the change in pressure to millimeters or inches of liquid head. In comparison to hydraulic cells, electric cell and tubing systems suffer less interference from rockfill construction activities (Durham Geo Slope Indicator Co.).

4 **KM type cell** is comprised of galvanized steel rods connected to steel plates and protected on the outside by a steel tube. It can be placed in both directions, vertically or horizontally along the rockfill mass (Fig. 9.2). When a KM type cell is installed for settlement recording, the steel rod connected to each plate (each at a different rockfill elevation) can freely move and is extended up to the rockfill surface inside the protection steel tube. The steel rod settlement is measured in relation to the reference steel tube anchored in the rock foundation. Brazilian field experience has proved that up to 12 plates can be installed in each rockfill dam position. For practical purposes each steel rod must be painted in a different color to make it easy for plate identification and monitoring. In Brazilian CFRDs, the plates were installed vertically in order to record horizontal rockfill movements (Fig. 9.3).

5 **Horizontal displacement meters:** Invar wire extensometers are installed to monitor horizontal displacements. All Invar wires are linked to the permanent instrumentation houses.

6 **Total pressure cells:** in general, pressure cells are used to verify assumptions made in the design by determining the distribution, magnitude, and directions of total

Figure 9.2 Vertical KM gauge (Silveira, 2006).

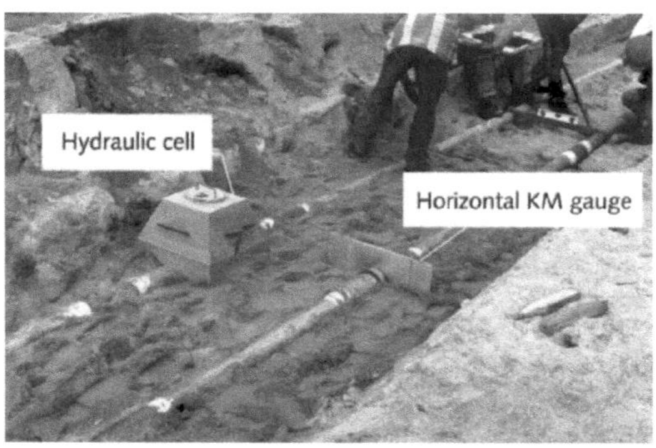

Figure 9.3 Xingó CFRD: settlement (hydraulic) cells and KM (horizontal) displacements gauge placed together (Silveira, 2006). (See colour plate section).

stresses and contact pressures between distinctive rockfill zones (Fig. 9.4). Cells can be installed and oriented to read pressure in several directions. It has been useful to install pressure cells in five or six directions in order to determine the principal stresses. Interaction between the curb concrete and the cushion zone can also be monitored by pressure cells, as can interaction between the extruded curb and the face slab.

7 **Inclinometers:** are used to monitor and detect displacements in any direction on slopes and in rockfill zones. Inclinometers are useful to establish whether a movement is constant, accelerating, or it is responding to the remedial measures taken.

Figure 9.4 El Cajón: total pressure cell (Sandoval et al., 2007b).

Inclinometer casing is permanently installed in an open borehole that crosses through zones movement is suspected. Inclinometer casing can also be embedded in rockfill, buried in a trench, or even cast into concrete (such as the face slab). Important features include the diameter of the casing, the coupling mechanism between successive tubing lengths, groove dimensions and straightness, and the casing strength (Fig. 9.5).

9.2.3 Surface movements

Levelling marks are distributed along the upstream parapet wall, along the crest, and on areas of the downstream slope. They are used to monitor the external rockfill surface displacements (Fig. 9.6).

Benchmarks founded on both abutments, bedrock are used as level reference points.

9.2.4 Pore pressure

Uplift and pore pressure on rockfill are not considered an issue either for CFRD design or for monitoring.

Electric or standpipe piezometers are being installed in upstream plinth areas to monitor the grout curtain efficiency during impounding. In addition, piezometers must be installed downstream from the plinth areas in order to control the face slab drainage system at the lower river bed areas.

The water added during rockfill compaction (about 200 ℓ/m^3) while elevating upstream zone 3B, plus occasional rainwater, may naturally flow down towards the

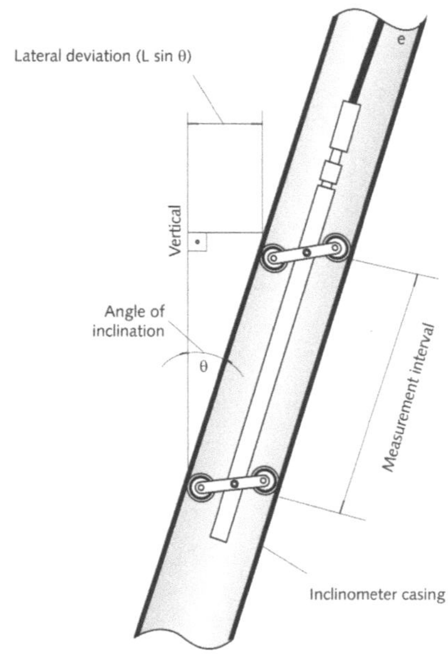

Figure 9.5 Inclinometer: reading detail (Durham Geo Slope Indicator Co.).

lower upstream areas close to the plinth and cause water level raise and hydrostatic pressures (e.g. Xingó, Mazar and El Cajón CFRDs).

9.2.5 Leakage control

Seepage measuring weirs are installed at the CFRD downstream slope toe to monitor seepage through foundation and rockfill. An impervious structure (compacted clay or concrete) must be built to ensure that all seepage flow throughout the rockfill or foundation is being collected and conducted to the measuring weir.

As for seepage through the foundation, the surfacing water should be easily collected and conveyed to the drainage system. In some cases, foundation treatment at downstream areas is carried out to avoid significant foundation flow escape from the monitoring system. Triangular and rectangular weirs are built and seepage measuring is recorded (Fig. 9.7).

Unpredictable high leakage (over 800 ℓ/s) from CFRDs has been recorded in several dams during impounding stages: Alto Anchicayá (1,800 ℓ/s), Shiroro (1,800 ℓ/s), Golillas (1,080 ℓ/s), Itá (1,730 ℓ/s), Itapebi (900 ℓ/s), Barra Grande (1,100 ℓ/s) and Campos Novos (1,300 ℓ/s). Remedial treatment was performed in these dams by dumping silty sandy material on concrete repair work. This managed to reduce the leakage substantially down to a range of 100 to 300 ℓ/s in almost all of the dams.

Therefore, leakage control is an important element to analyze face slab performance since it is sensitive to cracks or failure processes.

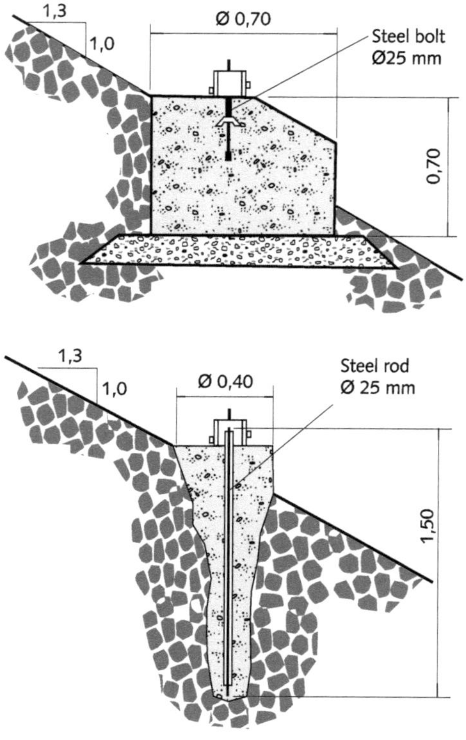

Figure 9.6 Rockfill surface marks on downstream slope (Silveira, 2006).

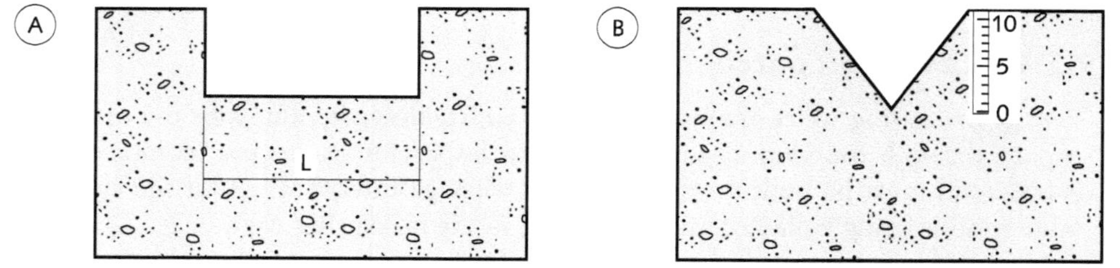

Figure 9.7 Seepage weirs: (A) rectangular; (B) triangular (Silveira, 2006).

Figures 9.8 and 9.9 show the TSQ1 (at the end of construction) and Campos Novos seepage weirs, respectively.

9.2.6 Slab deflections and strain X stress control

Monitoring the face slab displacements have deservedly got the attention of design and construction engineers after concrete slab cracking spalling and failures have been recorded in several CFRDs during impounding. Slab monitoring must be installed during the slab construction and not restricted to the impounding phase.

Figure 9.8 TSQ1: seepage weir before reservoir impounding (Oct., 2000). (See colour plate section).

Figure 9.9 Campos Novos: seepage weir in operation (Oct., 2007). (See colour plate section).

Joint meters to check the stress between slabs in the central slab area combined with electrolevels along the face slab can provide interesting information. In several CFRDs inclinometers embedded in the face slab have been used to monitor slab deflections.

1 **Electrolevel** – A cheap and simple gauge, electrolevels have been recently installed in current CFRD slabs to monitor their deflections. It consists of a glass capsule partially filled with an electrolytic fluid, and it is also commercially known as a *gravity sensing electrolytic potentiometer*. Electrodes penetrate the capsule and are used to measure electrical resistance through the fluid. Angular rotations of the electrolevels are monitored by the change in the electrical resistance between the electrodes that form a half Wheatstone bridge.

The individual readings for each electrolevel are plotted against distance along the sloping slab. This set of measuring values is used to establish a polygonal curve in a curve fitting procedure.

Electrolevels have been monitoring the face slab of dams such as Xingó, TSQ1, Hongjiadu, Barra Grande, El Cajón (Fig. 9.10) and Campos Novos. In El Cajón CFRD, besides electrolevels, three sets of inclinometers were embedded in the face slab.

2 **Inclinometers** monitoring deformation of the face slab of CFRDs have been successfuly installed. Inclinometer casings are embedded and cast into concrete, as shown in El Cajón (Figs. 9.11 and 9.12).

3 **Strain gauge and non-stress strain meters** are also recommended to be embedded in the face slab to measure concrete autogenous strains during rockfill construction and impounding.

Steel bar stress meters installed in the face reinforcing steel, and generally installed in pairs, are recommended to monitor longitudinal stress parallel to the slab axis. In addition to strain gauges and non-stress strain meters, thermometers embedded in the slab concrete are a common practice.

Figure 9.10 El Cajón: electrolevel installation on the face slab. (See colour plate section).

Figure 9.11 El Cajón: inclinometer casing to be cast into slab concrete. (See colour plate section).

4 **Three-orthogonal joint meters** – These instruments, arranged in groups of three, one meter in each direction, have been currently installed to monitor the perimeter joint movements. The movements monitored are: (1) perpendicular displacement of the perimeter joint, i.e. joint closing or opening between plinth and slab; (2) parallel displacement of the perimeter joint, i.e. settlement and bulging movements of the slab; and (3) tangential displacements in the plane of the perimeter joint, i.e. downward and upward movements of the slab along the perimeter joint.

Most of the 3-D joint meters are underlying the silty sand transition and upstream rockfill. Figures 9.13 and 9.14 show three-orthogonal joint meters at Xingó and El Cajón CFRDs, respectively.

Figure 9.12 El Cajón: inclinometer casing after installation at the dam crest. (See colour plate section).

Figure 9.13 Three-orthogonal joint meter – front view at Xingó Dam. (See colour plate section).

Figure 9.14 Three-orthogonal joint meter at El Cajón Dam. (See colour plate section).

5 **Single joint meters** are installed on the concrete face to monitor the opening of the vertical joints close to both abutments. Vertical joint opening and closing above the water level can be measured by pairs of stainless pins.

9.2.7 Permanent instrumentation houses

Inside the permanent instrumentation houses built on downstream slope berms are all the panels and reading equipment for the hydraulic cells (Swedish box), total pressure cells, electric piezometers, and extensometers that are installed. Safe access facilities such as stairs and catwalks must be in place for the instrumentation staff. Electrolevels and inclinometers readings can be recorded at the dam crest area on special panels or metallic boxes embedded in the wall of the parapet structure.

9.3 MONITORING AND MAINTENANCE CARE

CFRD instrumentation must perform rockfill and slab monitoring during construction, impounding and the operation phase. CFRD long-term monitoring and behavior analysis is mandatory because several unpredicted cracks and high amounts of leakage have been recorded almost one year after the first impounding (Xingó, Aguamilpa, Alto Anchicayá) as well as because of the failure of Gouhou CFRD. Therefore, it is paramount that during the operation phase a trained and skilled team must be in charge of dam monitoring. The monitoring measurements must be transmitted to the designer who is responsible for looking for any abnormal behavior.

Vertical displacements of the rockfill and foundation are highly concentrated during the construction stage, achieving 80% to 90% of the total settlement in the design. Thus, it is important that trained and skilled staff are available for instrument installation, calibration and maintenance, readings and data analysis as well as to seek corrective intervention whenever necessary from the beginning of the construction work.

These are the main parameters and features to be monitored for high CFRDs.

- Rockfill settlement should be recorded by hydraulic or electric cells, mainly in several elevations and different points, and particularly be monitored on upstream parts (close to cushion zone and zone 3B), central areas (T zone) and downstream zones (3C and 3D) where the largest settlements have been recorded, according to international records. It is utterly important that each settlement cell reading should be supplemented by some essential information: i) elevation and station of the cell; ii) rockfill elevation above the cell; iii) cell readings and related instrumentation cabin elevation surveying data (bench topographic marks on abutment rocky areas) on the same day, to obtain a more "accurate absolute settlement" of each cell and of the rockfill.

  During the impounding phase, in addition to settlement cell readings and instrumentation cabin surveying data, daily reservoir water level complemented by rain data from local meteorological stations must be recorded.

  It is mandatory that data from each cell should be fully recorded from the moment of the cell's installation, including during construction and reservoir impounding as well as through the dam's operation in order to have a historical series on each cell. The recorded reading on each cell are part of a "Instrumentation Data Book". This procedure must be kept for all and every dam instrument.
- Crest and downstream slope surface movements must be monitored and analyzed. Crest displacements downstream towards and from abutments to the central part of the dam during reservoir filling and in the operation period are key points for the dam performance analysis.
- Perimeter joint monitoring, mainly on the abutments, must deserve special attention by designers and instrumentation team members in the field.
- Face slab movements and stresses measured by joint meters and strain gauges, particularly at the abutments and central part of the slabs, play a key role for crack and concrete failure detection. These are accountable for high leakage levels. Electrolevels have been widely used in the high CFRDs in South America (Brazil) and China (TSQ1 and Hongjiadu). Today, electrolevels are a cheap, easy to install and handle, simple and reliable displacement monitoring system and their use should be encouraged for future high CFRDs.
- Seepage weir data from downstream areas must be monitored daily. Seepage readings up to 300 ℓ/s are acceptable after complete reservoir impounding. Dam leakages above this level must be reduced by slab remedial treatment, as a good engineering practice.
- Daily visual inspection of the dam crest, face slab above the reservoir water level, abutments and downstream low areas must be performed by a skilled team during and after reservoir filling. Unfortunately, this once usual and mandatory practice is often neglected.
- Seepage and natural ground water level on the abutments must be monitored and included within the dam monitoring instrumentation design.

9.4 FINAL CONSIDERATIONS

In instrumentation and dam monitoring activities only skilled staff must be engaged. Private companies, research institutes at universities, government laboratories and even instrumentation manufacturing houses have become training and expertise centers. These technological centers are an important aid to support instrument installation, monitoring and maintenance facilities during construction and in operational phases. These issues must be considered in order to ensure good monitoring and accountability for dam performance control. Instrumentation measures as well as all related data cannot be lost, as this would jeopardize behavior analysis and safety control.

By the same token, the readings data book and instrumentation analysis and the comparison with the dam design criteria are equally fundamental.

Installation of seismological stations at the reservoir area is a good engineering practice. The current experience (2008) recorded at Ziping CFRD (150 m high), located at Sichuan Province in China, showed that no significant damage of the CFRD structure and of the concrete slabs was recorded even after an earthquake of magnitude 8.0 on the Richter scale.

Table 9.1 presents the types and the quantities of instruments installed in some main CFRDs.

Table 9.1 List of instrumentation of main CFRDs.

Name/ Country/ Height/ Completed	*Rockfill* $m^3 \times 10^3$	*Slab area* m^2	*Rockfill zones 3B; 3C Slopes H:1V*	*Installed instruments*		*References*
				Slab	*Rockfill*	
Cethana Australia 110 m 1971	1,400	30,000	Quartzite US 1.3 DS 1.3	23 bench marks on face slab in 3 cross sections and abutments 3 inclinometers casings 8 joint meters (Carlson type) on perimetric joint 32 strain meters and temperature gauges (Carlson type)	4 hydraulic settlement cells 15 surface marks on dam crest 18 surface marks on downstream slope 1 leakage weir	Fitzpatrick et al., 1973
Alto Anchicayá Colombia 140 m 1974	2,400	22,000	US 1.4 DS 1.4	60 strain gauges 22 joint meters	(*) hydraulic settlements cells (*) surface marks 1 leakage weir	Regalado et al., 1982
Kotmale Sri Lanka 97 m (1st stage) 1984	—	60,000	Charnokite US 1.45 DS 1.4	5 joint meters at perimetric joint 3 inclinometers lines	16 pneumatic piezometers at foundation 38 hydraulic settlement cells 5 horizontal meters 33 surface marks on dam crest and downstream slope 1 leakage weir	Kulasingle & Tandon, 1993
Cirata Indonesia 125 m 1987	3,600	—	Andesite, breccia US 1.3 DS 1.3	3 inclinometer lines (*) joint meters on slab and perimetric joint (*) stress meters 2 inclinometers (one on each abutment)	(*) hydraulic settlement cells (*) bench marks at dam crest and downstream slope (*) piezometers at foundation (*) standpipe at abutments 1 leakage weir	Pinkerton, Siswowidjono, & Matsui, 1985
Golillas Colombia 125 m 1978	1,300	14,000	Gravel US 1.6 DS 1.6	110 marks at vertical joints and along parapet wall joint	39 hydraulic settlement cells 15 surface marks 8 standpipe piezometers at foundation 1 leakage weir	Amaya & Marulanda, 1985

(Continued.)

Table 9.1 (Continued.)

Name/ Country/ Height/ Completed	*Rockfill $m^3 \times 10^3$*	*Slab area m^2*	*Rockfill zones 3B; 3C Slopes H:1V*	*Installed instruments* Slab	*Installed instruments* Rockfill	*References*
Foz do Areia Brazil 160 m 1980	14,000	139,000	Basalt US 1.4 DS 1.25–1.4	30 strain meters (two directions) 12 reinforcement stress gauges 18 joint meters 14 electrical thermometers	40 hydraulic settlement cells 35 surface marks 1 leakage weir	Pinto, Materón & Marques Filho, 1982
Shiroro Nigeria 125 m 1983	3,900	50,000	Granite US 1.3 DS 1.3	89 joint meters 72 strain meters 22 thermoresistors 3 inclinometers (*) extensometers (*) piezometers	3 inclinometers 72 surface marks on dam crest and downstream slope 6 horizontal displacement gauges 2 sets of horizontal extensometers 2 leakage weirs	Bodtman & Wyatt, 1985
Salvajina Colombia 148 m 1983	3,395	57,500	Gravel US 1.5 DS 1.4	44 joint meters 49 strain meters	45 hydraulic settlement cells 7 sets of total pressure cells 4 pneumatic cells at foundation and rockfill contact areas 6 pneumatic cells (foundation) (*) bench marks	Sierra, Ramirez & Hacelas, 1985
Segredo Brazil 145 m 1992	7,200	87,000	Basalt US 1.3 DS 1.2–1.4	55 strain gauges 23 joint meters 3 thermometers	(*) hydraulic settlement cells (*) horizontal meters 1 leakage meter	Penman, Rocha Filho & Toniatti, 1995
Aguamilpa Mexico 187 m 1993	13,000	137,000	Gravel US 1.5 Rockfill DS 1.4	36 joint meters (electric extensiometers) 19 tri-orthogonal meters 45 joint meters in vertical joints and para-pet wall 4 inclinometers	3 inclinometers 40 hydraulic settlement cells 7 pneumatic piezometers 41 standpipe piezometers at abutments 3 sets of 6 total pressure cells plus 7 sets of 3 total pressure cells 6 extensometers 1 leakage meter	Macedo, Castro & Montañez, 2000

Xingó Brazil 150 m 1994	12,700	135,000	Granite gneiss US 1.4 DS 1.3	10 electrolevels on left abutment 10 electrical joint meters 5 tri-orthogonal meters 6 electric thermometers 7 strain meters	23 hydraulic settlement cells 23 horizontal extensometers (KM type) 24 surface marks 1 leakage weir 2 magnetic settlement cells	CHESF – Companhia Hidro Elétrica do São Francisco
Itá Brazil 125 m 1999	8,900	110,000	Basalt US 1.3 DS 1.3	25 electrolevels 6 electric thermometers 9 joint meters 3 tri-orthogonal meters on perimetric joint	26 hydraulic settlement cells 3 magnetic settlement gauge 7 horizontal extensometers (KM type) 19 surface marks 1 leakage weir	Engevix Engenharia
TSQ1 China 178 m 2000	17,700	173,000	Limestone US 1.4 Mudstone DS 1.4	64 electrolevels 26 joint meters 12 tri-orthogonal gauge (perimetric joint) 17 thermometers 20 tensiometers (reinforcement) 9 strain meters 11 non stress strain meters	50 hydraulic settlement cells 31 horizontal extensometers 21 total pressure cells 33 electric foundation piezometers 15 standpipe piezometers on abutments 1 leakage weir	Wu et al., 2000b
Itapedi Brazil 110 m 2002	3,900	67,000	Granite gneiss US 1.25 DS 1.3	13 electrolevels 9 electric join meters 12 electric meters along perimetric joint	21 hydraulic settlement cells 6 horizontal extensometers (KM type) 3 magnetic settlement meter (KM type) 17 surface marks 1 leakage weir	Engevix Engenharia
Machadinho Brazil 125 m 2002	6,200	77,000	Basalt US 1.3 DS 1.3	39 electrolevels (3 sections) 15 joint meters (tri-orthogonal) 12 electric joint meters (one direction)	29 hydraulic settlement cells (Swedish box) 4 settlements extensometers (35 plates) 3 horizontal extensometers (KM Type: 14 plates) 38 surface marks 1 seepage weir	CNEC Engenharia
Quebra-Queixo Brazil 75 m 2002	2,100	49,000	Basalt US 1.25 DS 1.20	10 electrolevels 7 electric joint meters 9 electric joint meters (perimetric joint)	25 hydraulic settlement cells 5 horizontal extensometers 3 magnetic settlements meters 15 surface marks 1 leakage weir	Xavier et al., 2003

(Continued.)

Table 9.1 (Continued.)

Name/ Country/ Height/ Completed	Rockfill $m^3 \times 10^3$	Slab area m^2	Rockfill zones 3B; 3C Slopes H:1V	Installed instruments		References
				Slab	Rockfill	
Campos Novos Brazil 200 m 2006	12,100	106,000	Basalt US 1.3 DS 1.4	17 electric joint meters 4 tri-orthogonal joint meters 25 electrolevels	28 hydraulic settlement cells 6 multiple horizontal extensometers 11 surface marks 4 magnetic settlement meters 1 leakage weir	Martins et al., 2006
Barra Grande Brazil 185 m 2006	12,000	108,000	Basalt US 1.3 DS 1.4	18 electric joint meters 5 tri-orthogonal joint meters 33 electrolevels	28 hydraulic settlement cells 6 multiple horizontal meters 18 surface marks 2 magnetic settlement meters 1 leakage weir	Engevix Engenharia
El cajón Mexico 190 m 2006	10,900	113,300	Ignimbrite US 1.4 DS 1.4	6 vertical inclinometers 96 hydraulic settlement cells 9 electric settlement cells 344 surface marks 12 total pressure cells 64 extensometers 3 inclinometers	6 inclinometers 40 surface 103 extensometers 90 standpipe piezometers 10 electric piezometers 12 total pressure cells 5 accelerographs (grout curtain) 6 seismological stations 1 leakage weir	Mena Sandoval et al., 2007a

Kárahnjúkar Iceland 196 m 2007	8,500	93,000	Sandy gravel (3B) and basalt (3C) US 1.3 DS 1.3	32 joint meters between slabs 14 perimetric joint meters 60 crack meters 29 strain meters	36 hydraulic settlement cells 20 piezometers 33 surface marks 6 seepage weir gauges	Johannesson, Perez & Stefansson, 2009
Shuibuya China 233 m 2007	15,640	103,000	Limestone US 1.4 DS 1.4	4 sets of 3D strain gauges 30 sets of 2D strain gauges 74 rebar strain meters (reinforcement) 15 thermometers 2 inclinometer sections 13 perimetric joint meters (11 at abutments and 2 river bed); 6 strain meters between face slab - parapet wall joint 46 strain meters between vertical joints 20 strain marks between face slab and rockfill	73 hydraulic settlement cells 73 horizontal extensometers (total 73 points) 28 piezometers in foundation 1 leakage weir 56 surface marks 63 standpipe piezometers at both abutments	IWHR – Institute of Water Resources and Hydropower Research
Mazar Equador 166 m under construction	5,000	45,000	Quartzite US 1.4 DS 1.4	29 (one direction) joint meters 2 (two directions) joint meters 8 (three directions) joint meters 6 non stress meters 12 topographic marks 80 electrolevels	16 piezometers 16 electric settlements cells 60 hydraulic settlement cells 1 leakage weir	Consorcio Gerencia Mazar

(*) Quantity not mentioned by the author
US – upstream
DS – downstream

Chapter 10

CFRD performance

10.1 INTRODUCTION

According to the *Water Power and Dam Construction Yearbook 2008*, today in the world there are 404 CFRDs (over ~30 m in height); 172 of those are in China and 11 in Brazil.

If the performance of CFRDs were reviewed only by focusing on the accidents that led to failure, according to a research conducted by Qian (2008) on dam failures of all types since 1860, only two failures on CFRDs would be registered out of a total of 48 dams. The first failure recorded occurred in 1993 at Gouhou Dam (71 m, China), built with sandy gravel. It was caused by internal flow six years after its completion. The second failure recorded occurred in 2005 at Taum Sauk Dam (29 m, USA) due to overtopping that happened 42 years after construction. This dam is not included in the *Water Power* list.

Failures apart, which are discussed in chapters 4, 5 and 6 of this book that deal with the mechanics of rockfill, static and dynamic stabilities and internal flow, another topic relating to CFRD performance is the eventual accidents, which are discussed in other chapters, at Mohale, Paradela and New Excheguer among others. Due to today's easy communication and the large number of congresses, symposia and workshops, on top of magazines and publications that circulate in the world of dams and in particular in that of CFRDs, virtually any new data, or any accident occurring in a dam, will quickly be in the e-mail inbox of specialists, consultants and designers. Thus, within a couple of days, reports on new accidents are available to national committees of dams everywhere, and in a few months, a group of papers on the problem will be published at national and/or international conferences. Explanations about what's happened start circulating, then repair details are published. One or two years after the accident, the owners, designers, and consultants are publishing on the measures that were taken and demonstrating that the behavior of the dam is in agreement with the design requirements for CFRD safety.

However, data on the behavior of these and other dams over time is quite rare, and such information is available only to a more restricted number of people who have access to the instrumentation and to the dam itself. This is a pity because CFRDs are designed to last, one reason why they deserve so much care.

Portugal is one of the keen countries in this respect. Periodic assesments are done on Portuguese dams by experts of the Civil Engineering National Laboratory (the LNEC).

Back to the subject of performance, it seems that once the problem of one dam is solved, and the dam is operating properly, the accident is relegated to CFRD history, even if the causes of the accident were not clearly explained, and/or emergency repairs have not been soundly sustained on a technical or scientific basis. On the other hand, on new projects or even on CFRDs under construction emergency solutions are often adopted. This is a result of both the empirical and practical approach that prevails in CFRD design and construction.

Accidents and incidents related to CFRD performance are discussed further in this book, and should be better understood for the benefit of forthcoming dams.

Once failures and the accidents mentioned earlier are discarded, the question of performance related to leakage and displacements is reported in chapters 3 (case histories) and 6 (throughflow). Chapter 6, more specifically, reports the throughflow measured in 29 CFRDs and the treatments adopted in order to control those flows (see Table 6.9). The instrumentation to register displacements, stresses, opening and closure of joints is described in chapter 9.

In this chapter, the performance of some well instrumented CFRDs is disclosed in order to evaluate displacements and settlement related to a dam's height as well as the deformability modulus related to the shape of the valley.

However, it is necessary to recognize that the CFRD performance commented on at the end of each dam description in chapter 3 is influenced primarily by throughflow (or leakage) and secondarily by the displacements that have been measured in the rockfill embankment and on the concrete face slab. The reasonable, good or excellent classification given in respect of the performance is directly associated with these two aspects that are indeed interrelated as cause and effect: "to larger displacements correspond larger flows".

On the other hand, it is necessary to acknowledge that dams are steadly becoming higher and higher. According to Qian (2008) there are six dams in China in feasibility studies that are likely to be over 300 m high (Maji, Lianghe Kou, Songta, Gushi, Shuangjiangkou and Rumei). These dams if designed as CFRD certainly will yield much larger displacements, and if the same design slopes are to be maintained, there will be lower safety factors.

Should we continue to assess the performance of such a CFRD in terms of throughflow and displacements or should we review and broaden this performance concept?

Data on CFRD performance, as well as proposed correlations, are reproduced in this chapter from bibliographical references. Some repetitions in data are to be expected and some old data has been updated.

Seven types of information are most discussed: (1) settlement (vertical movements); (2) horizontal displacements measured in the upstream and downstream directions. There are no data on displacements parallel to the axis, i.e., from the abutments toward the valley, which, as can be seen in chapters 3 and 11, is of major interest when analyzing the accidents that concrete slabs have suffered; (3) vertical deformability modulus computed using actual settlements; (4) slab deflections, measured in the normal direction to the slab; (5) deformability modulus related to slab deflections; usually called transversal deformabilly modulus; (6) long-term displacements or creep; and (7) leakage.

10.2 SETTLEMENT

Figures 10.1 to 10.4 show the evolution of the vertical displacements measured in Swedish boxes in Campos Novos Dam during construction and with water elevation during the reservoir impounding.

One can observe that the cells of the two lower elevations indicate settlement developing almost simultaneously with dam heightening. In the two upper elevations there is some time lag.

After the constructive settlement, creep starts at a slower deformation rate. As the water level rises, new settlement of a small magnitude occurs followed again by creep displacements at a rate quite similar to the previous one before the reservoir impounding, or with a small increase exhibited in the upper cells. It is a typical behavior observed in many CFRDs.

Another way to visualize the settlement is shown in Figures 10.5 and 10.6, which presents, respectively, the total settlement at the end of construction of Foz do Areia Dam, and the partial settlement that occurred along and due to reservoir filling.

At the central zone where the maximum settlement (358 cm) was measured, the additional settlement due to the water load on the slab was only about 10% (30 cm). Near the slab, the ratio of the new impounding settlement to the construction settlement was higher, as would be expected considering the water weight on the face slab. Figure 10.7 shows the curves of equal vertical movements after the filling of the reservoir.

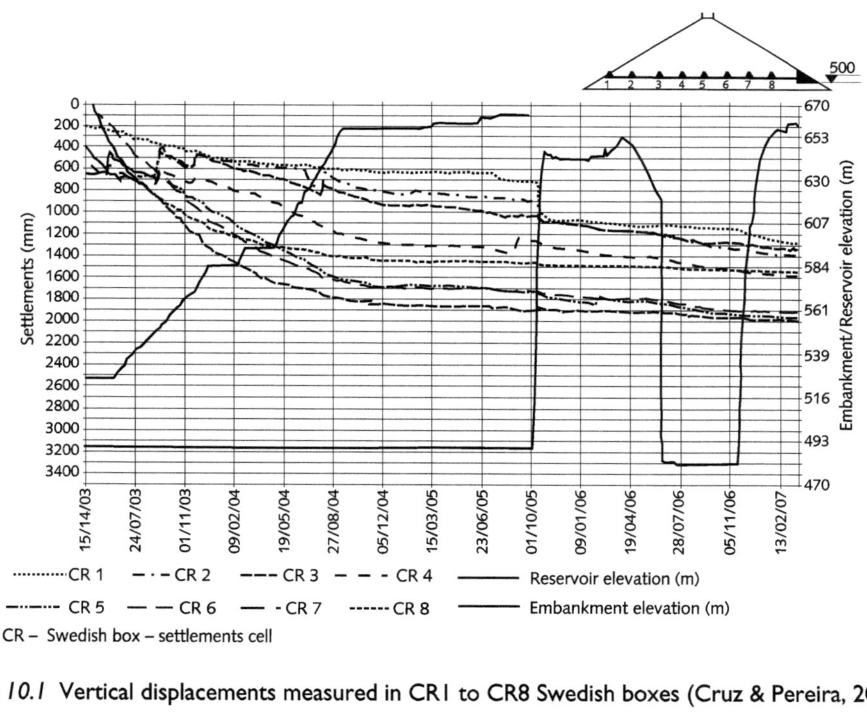

Figure 10.1 Vertical displacements measured in CR1 to CR8 Swedish boxes (Cruz & Pereira, 2007).

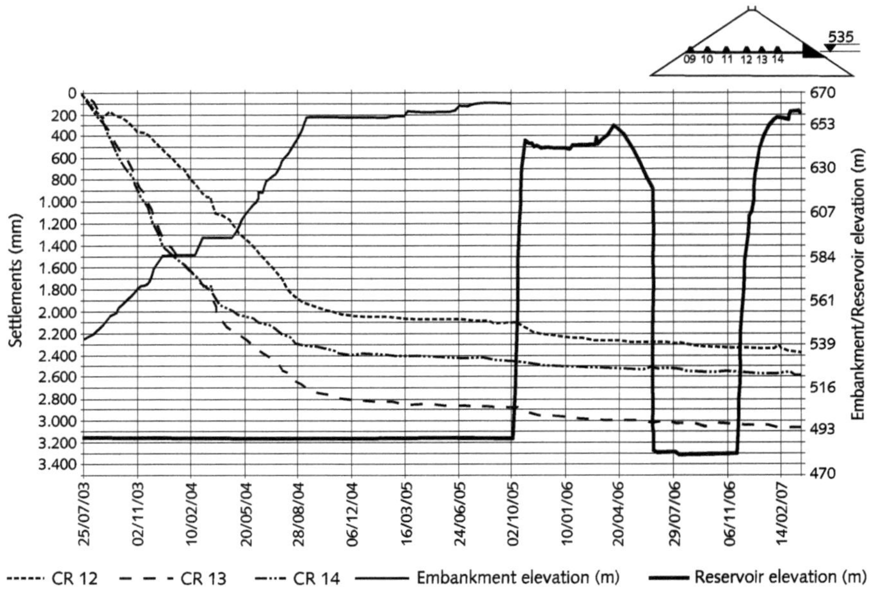

Figure 10.2 Vertical displacements measured in CR9 to CR14 Swedish boxes (Cruz & Pereira, 2007).

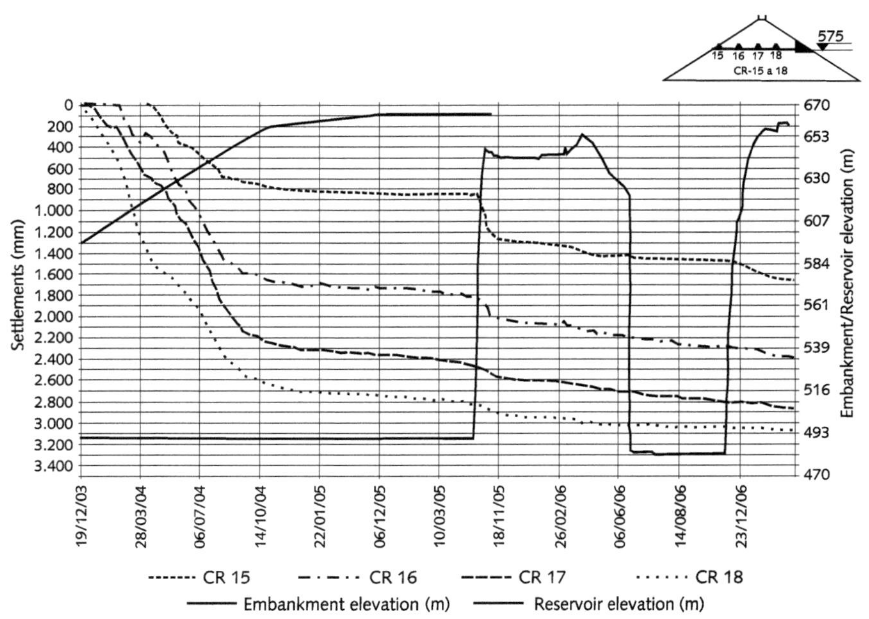

Figure 10.3 Vertical displacements measured in CR15 to CR18 Swedish boxes (Cruz & Pereira, 2007).

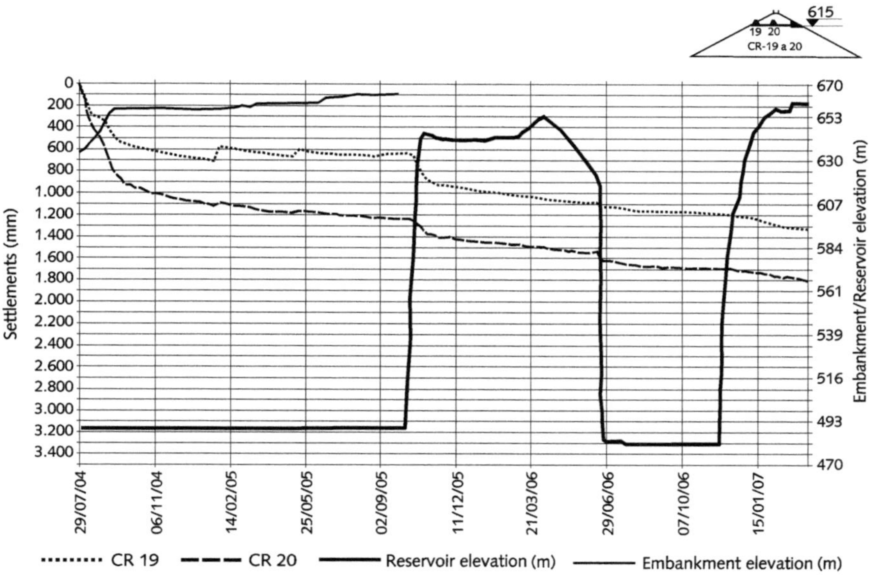

Figure 10.4 Vertical displacements measured in CR19 and CR20 Swedish boxes (Cruz & Pereira, 2007).

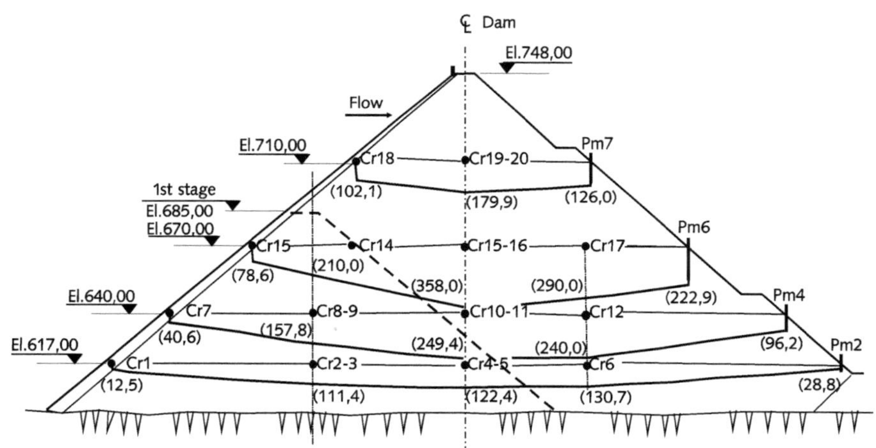

Figure 10.5 Foz do Areia Dam: settlement before reservoir filling (cm) (Pinto, Materón & Marques Filho, 1982).

A third graph presents the evolution of the settlement as the reservoir rises in Mohale Dam (Lesotho, Africa) Figure 10.8. The crest settlement increases with the reservoir level until the maximum water level is reached and continues to increase as the progressive accommodation of the rockfill and slab fissuring proceed.

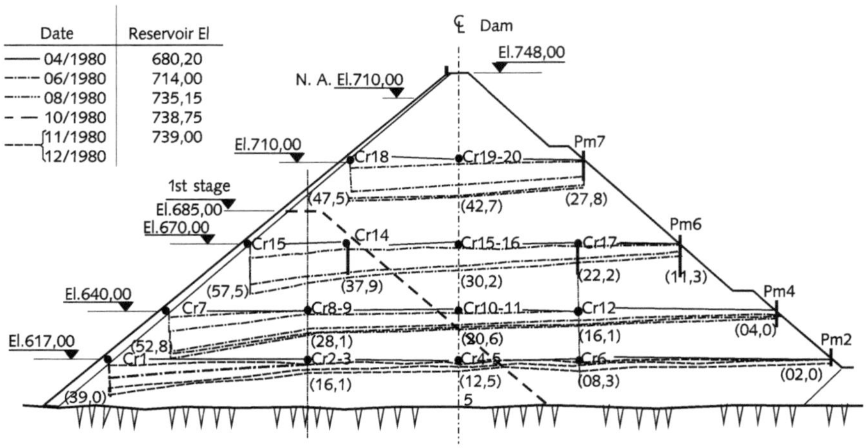

Figure 10.6 Foz do Areia Dam: vertical settlement with the reservoir filling (cm) (Pinto, Materón & Marques Filho, 1982).

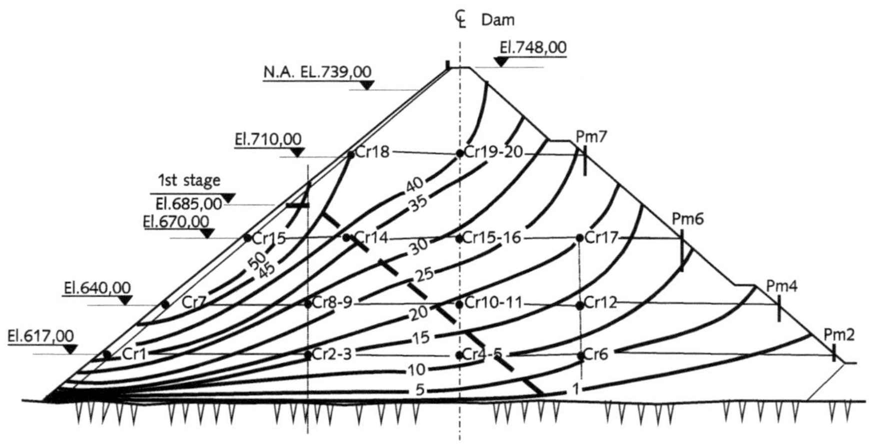

Figure 10.7 Foz do Areia CFRD: equal settlement (in cm) curves after reservoir filling – September, 1980 (Pinto, Materón & Marques Filho, 1982).

10.3 CORRELATIONS BETWEEN SETTLEMENT, DAM HEIGHT AND VALLEY SHAPE

It is usual to express the settlement at the end of construction as a percentage of the dam height.

Figure 10.9 shows these parameters for 21 dams built with rockfill from different rocks and gravels.

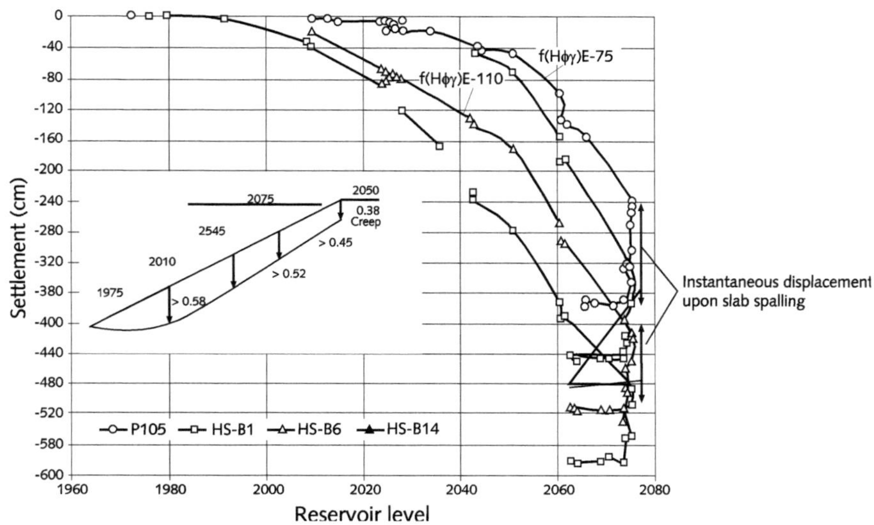

Figure 10.8 Settlement of Mohale Dam crest during reservoir filling and upon slab spalling (Johannesson & Tohlang, 2007b).

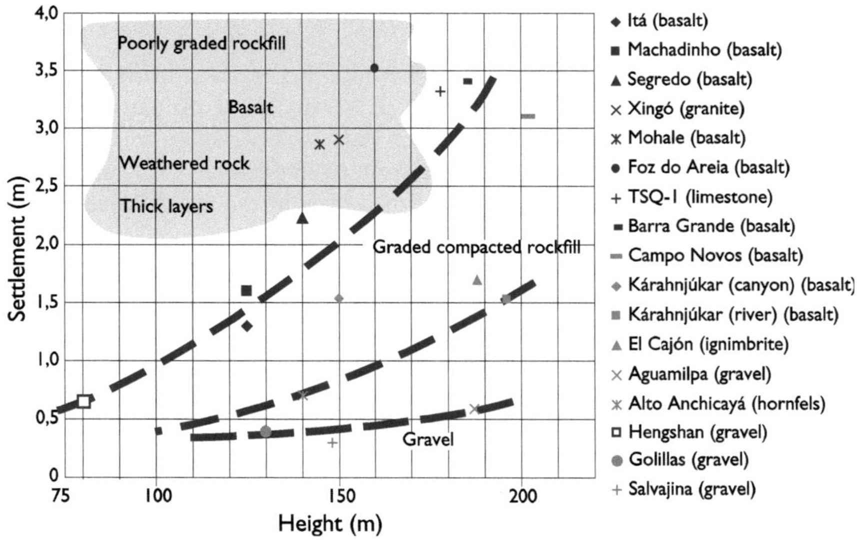

Figure 10.9 Evolution of crest settlement with height.

If the dam is homogeneous, i.e. build with the same type of rockfill, the larger settlement at the end of construction occurs approximately at the middle height and can be computed by the expression:

$$R = \frac{\gamma H^2}{4E} \tag{10.1}$$

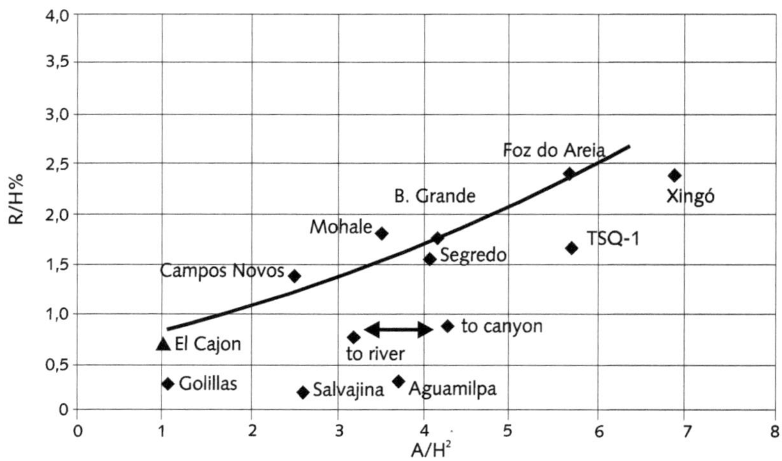

Figure 10.10 Percentage settlement with the valley shape (Johannesson, 2007).

where E is the vertical deformability modulus of the rockfill.

But as E varies with the type of rockfill, it is not surprising that two dams with the same height might show quite different percentage settlements. Gravel embankments are less deformable.

Chapter 4 shows that deformability modulus E decreases with increasing pressures and this is the reason why the ratio R/H versus H for the same rockfill type is not linear as shown approximately in Figure 10.9.

A second aspect to be considered in relation to settlement is the valley shape. Due to arching phenomena, the stress distribution acting in the rockfill mass is quite different in a narrow valley than it is in an open valley.

Pinto and Marques Filho (1998) proposed that the valley shape could be expressed through the ratio A/H^2 (A = face slab area and H = dam height). Johannesson (2007) shows in Figure 10.10 the correlation between R/H and valley shape A/H^2 for several dams. In an open valley, such as Foz do Areia and Xingó, the percentage settlement is higher than it is in narrow valleys, like those of Golillas, El Cajón, and Campos Novos.

According to Cooke (Cooke & Sherard, 1987), the effects of arching in narrow valleys can reduce with time and, as a consequence, the settlement developed by creep may be larger than it is in open valleys.

10.4 HORIZONTAL DISPLACEMENTS

Horizontal displacements are measured almost exclusively in the upstream-downstream direction, due to limitations of measuring instruments. The inclinometers used in earth dams show these displacements in two directions, but when installed in rockfills the readings lack precision because they have to be surrounded by relatively thick sand and transition layers to protect the casing.

In chapter 9 we call attention to the importance of measuring the actual displacements, which means in 3D especially in narrow valleys. In such valleys the displacements of the rockfill embankment are directed towards the valley, leading to high stresses in the slab. Pinto (2007) calls attention to this fact and shows in a graph of $E/\gamma H$ versus A/H^2 that those dams that had problems with the face slab (Campos Novos, Barra Grande and Mohale) are below a kind of "safe line" expressed by the equation $E/\gamma H = 120 - 20\ A/H^2$.

TSQ1 Dam also had problems with the slab. Figure 10.11 shows that the TSQ1 point on the graph is slightly below Foz do Areia Dam, and it is on the limit of the "safe line".

Figures 10.12 to 10.15 show horizontal displacements measured practically in the same places as the settlement readings of Figures 10.1 to 10.4 in Campos Novos Dam. In the lower elevation the displacements during construction were almost always towards upstream; nonetheless, the plates were also located upstream from the axis.

At the second level displacements were towards upstream and downstream. The same finding has been recorded at the third level. At the upper level both plates indicated displacements towards downstream.

This behavior can be predicted by mathematical modeling analysis as discussed in chapter 11.

During the reservoir filling there is a reversion of the displacements that now move towards the downstream direction. It is interesting to observe that in the upper elevations, again another reversion of the displacement was detected during the reservoir drawdown.

As was the case with settlement, the horizontal displacements develop almost at the same time as the water rises and then a process of creep takes place.

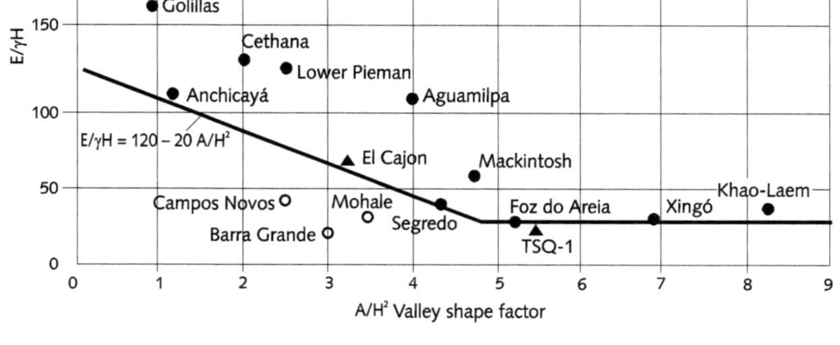

Figure 10.11 Factor A/H² relation (Pinto, 2007).

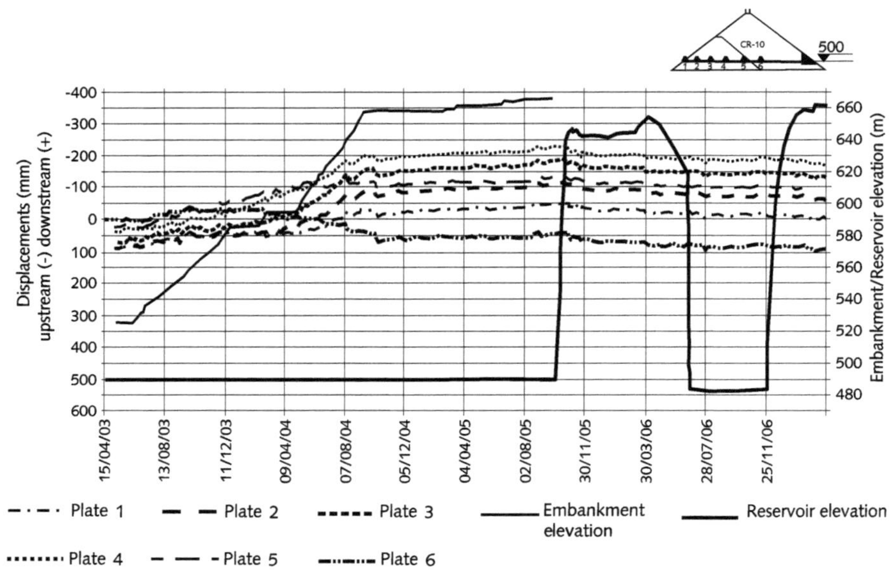

Figure 10.12 Horizontal displacements in Campos Novos CFRD (Cruz & Pereira, 2007), h = 500.

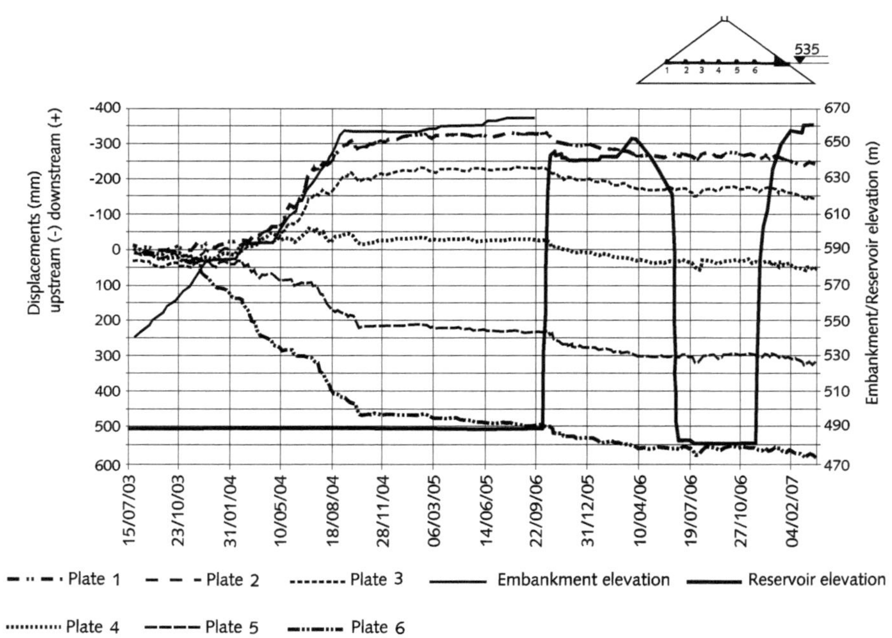

Figure 10.13 Horizontal displacements in Campos Novos CFRD (Cruz & Pereira, 2007), h = 535.

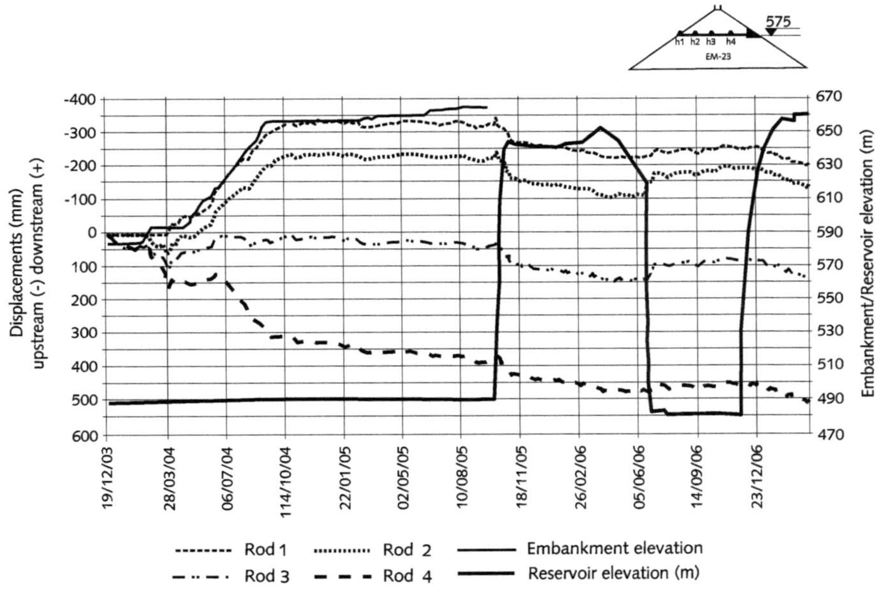

Figure 10.14 Horizontal displacements in Campos Novos CFRD (Cruz & Pereira, 2007), h = 575.

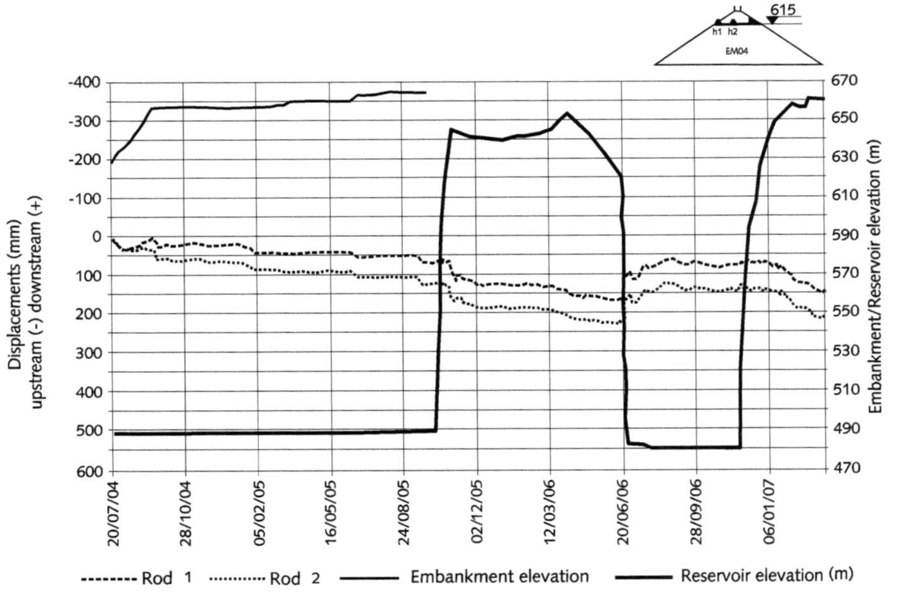

Figure 10.15 Horizontal displacements in Campos Novos CFRD (Cruz & Pereira, 2007), h = 615.

10.5 COMBINED MOVEMENTS

Combining the vertical and the horizontal displacements, it is possible to obtain the resultant in the transversal section of the rockfill mass as can be observed in Figure 10.16 for two sections of Itapebi Dam.

As part of the dam is founded in alluvium, the larger settlements are shifted to the downstream area as it can be seen in section A of Figure 10.16.

Comparing the vertical displacements with the horizontal displacements one can see that there is a marked predominance of the verticals in relation to the horizontals. As a general rule, this is the behavoir in any CFRD.

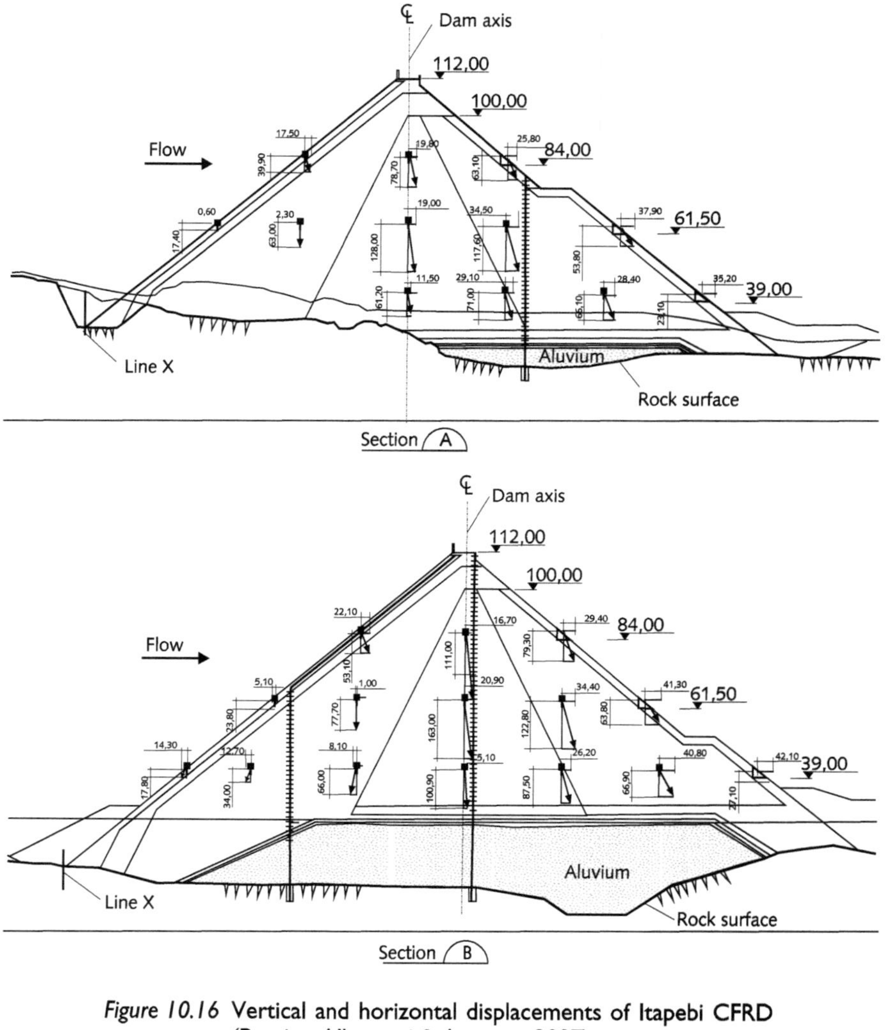

Figure 10.16 Vertical and horizontal displacements of Itapebi CFRD (Pereira, Albertoni & Antunes, 2007).

10.6 FACE DEFLECTION

The slab face deflections are related to the displacements of the rockfill mass, because the slab is a membrane simply supported by the rockfill mass without any lateral bond. The slab is placed near the plinth and connected by the perimetric joint only. The restrictions in movement toward the valley due to the parapet wall is virtually

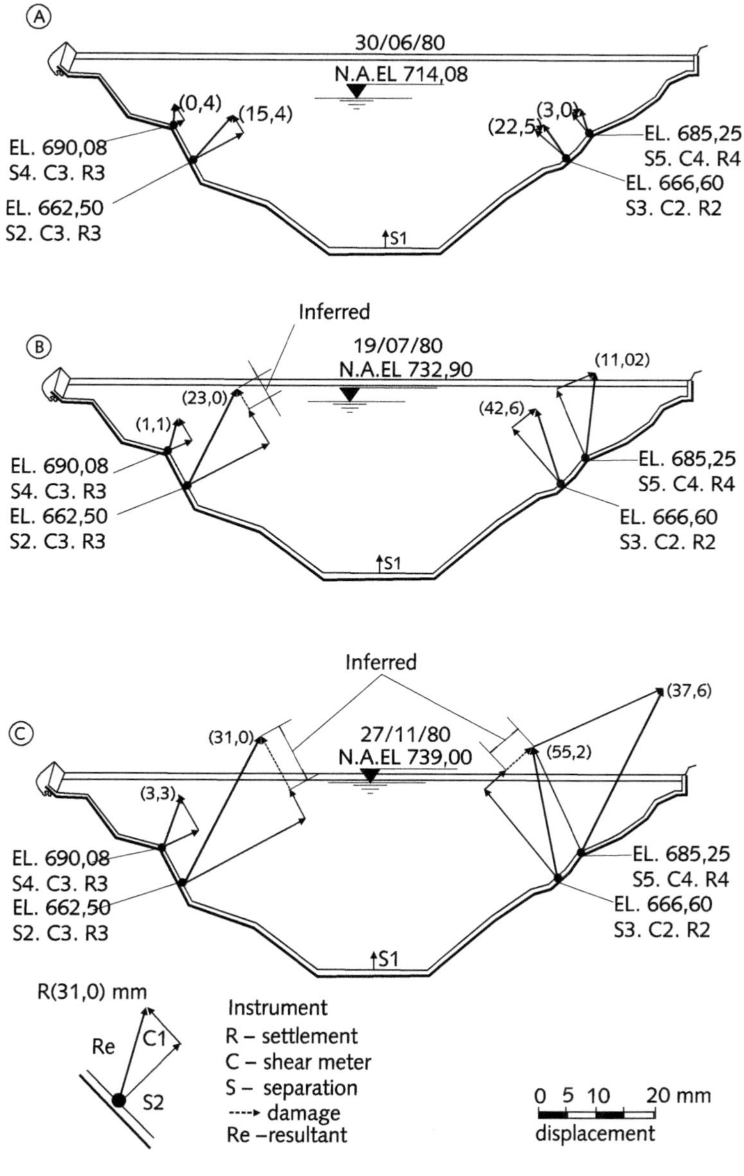

Figure 10.17 Foz do Areia Dam: perimetric joint displacement during impounding (Pinto, Materón & Marques Filho, 1982).

zero, since the parapet wall follows roughly the same displacement pattern as the slab. Once the slab bends in the central area and in the top following the displacement of the rockfill, the tendency is to open tension joints and close compressive joints in the central compressed area.

These displacements are quite clear in some measurements carried out on the perimetric joint of the concrete slab of Foz do Areia Dam (see Figure 10.17).

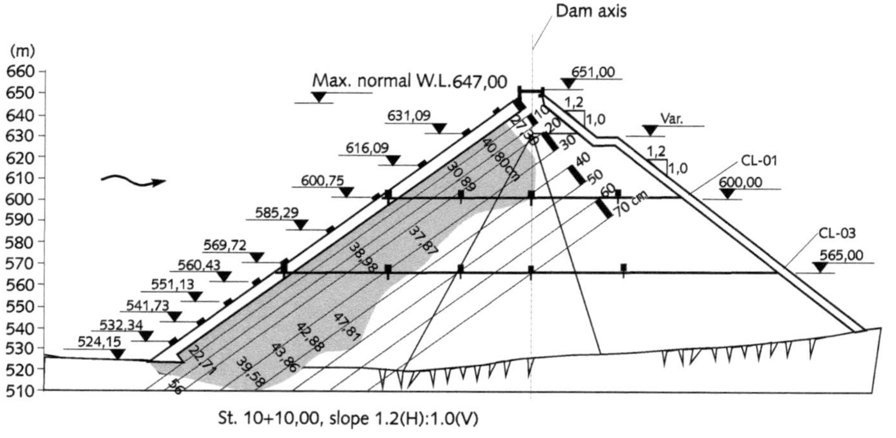

Figure 10.18 Deflection measured in the face of Barra Grande CFRD (right bank) (Freitas, 2008).

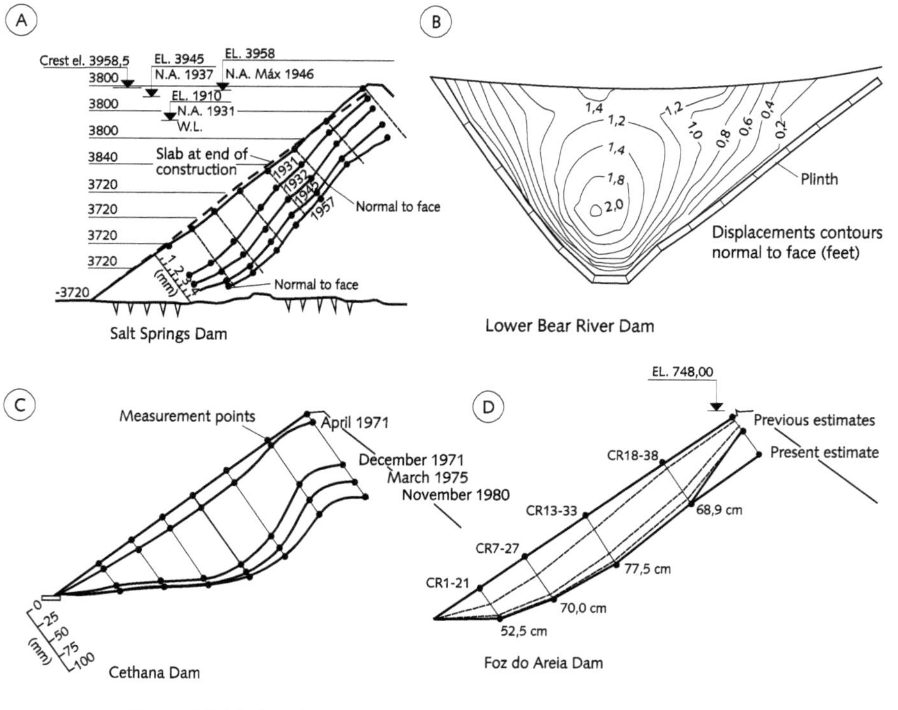

Figure 10.19 Salt Springs (A), Lower Bear River (B), Cethana (C) and Foz do Areia (D) CFRD slab displacements.

All instruments recorded separation, settlement and shear movements in a tangential upslope direction as shown in Figure 10.17B. The shear joint meters located on the steeper sections of the abutment (El. 662–666) were damaged during this period.

Figure 10.17C shows a situation when most of the readings were stabilized. According to Pinto, Materón and Marques Filho (1982), one can see that the shear movements corresponding to the damaged instruments are extrapolated values, starting from the date when the gauge was broken, based on existing records.

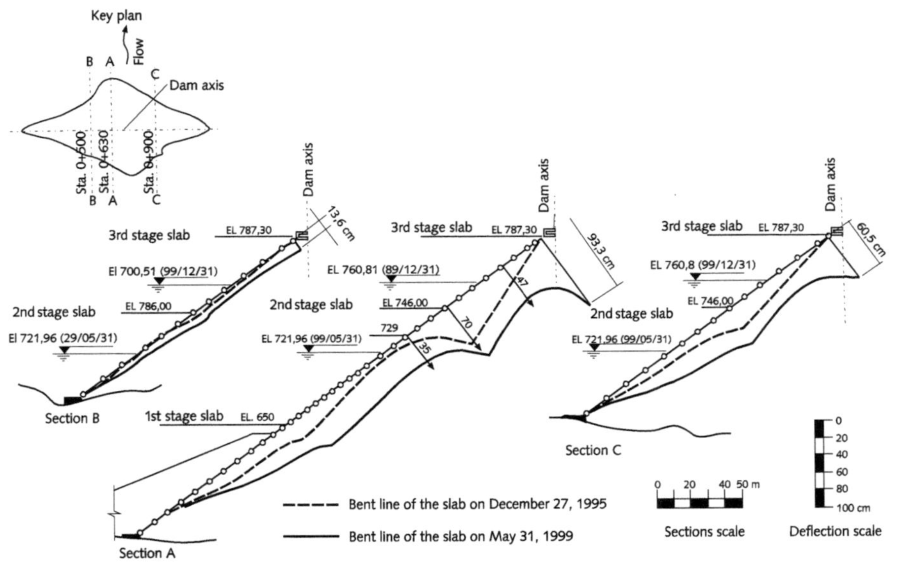

Figure 10.20 TSQ1 CFRD slab displacements (Penman & Rocha Filho, 2000).

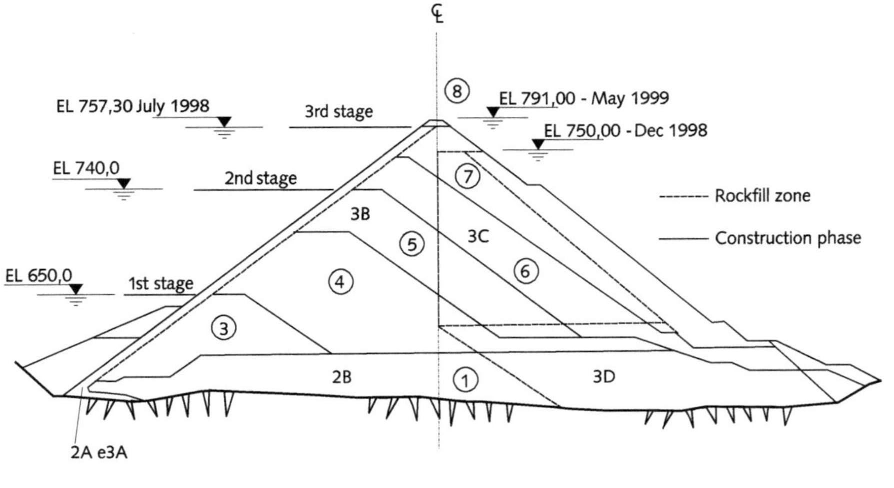

Figure 10.21 TSQ1 Dam construction phases (Guocheng & Keming, 2000).

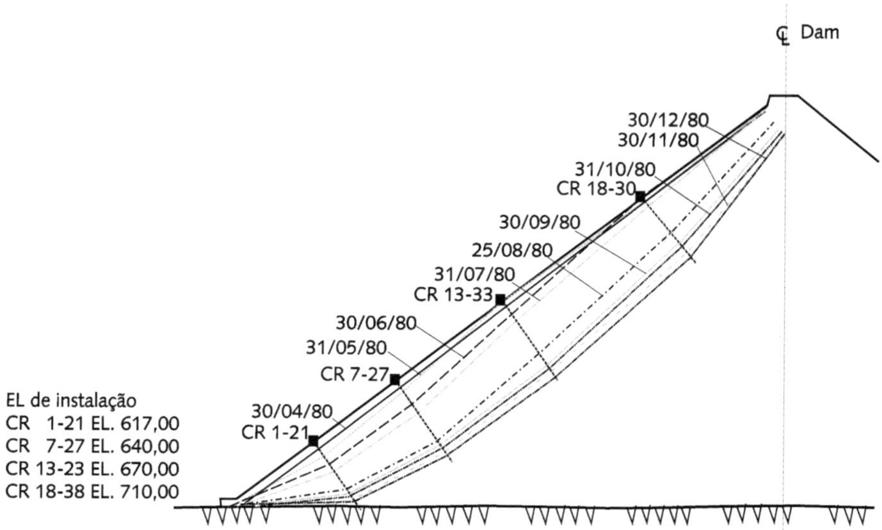

Figure 10.22 Foz do Areia CFRD: slab deformations after reservoir filling (Pinto, Materón & Marques Filho, 1982).

In the large majority of CFRDs the displacements of the slab are measured with electrolevels in the direction normal to the slab (Figures 10.18 and 10.19; see also Figure 4.4).

Figure 10.20 presents the displacements measured in the slab of TSQ1 Dam, raised in three stages as indicated in Figure 10.21, and with the reservoir partially filled up. The phenomenon of creep is more evident in case of slab deflections as can be seen during the heightening interruptions. In the case of Xingó CFRD, the deflections practically doubled in a six-year period (see Figure 4.4).

The evolution of the displacements during and after reservoir filling can be observed in Figure 10.22 for Foz do Areia CFRD.

In Foz do Areia, slab deflections due to creep were less intense. Cell 1–21 moved from 50 cm to 52 cm; cell 7–27 from 65 cm to 70 cm; cell 13–33 from 69 cm to 77 cm, and cell 18–36 from 56.5 cm to 69 cm (Figures 10.19 and 10.22). Slab deflections were computed from the settlement cells installed in zone 2B below the concrete slab.

Other references to deflection of the slab are included in chapter 2.

10.7 VERTICAL COMPRESSIBILITY MODULUS (E_V) AND TRANSVERSAL MODULUS (E_T)

In Machadinho Dam (125 m) magnetic cells to measure settlement were installed every 6 m and Swedish boxes every 12 m. Considering the settlement observed during the construction, it is possible to compute the vertical compressibility modulus E_v

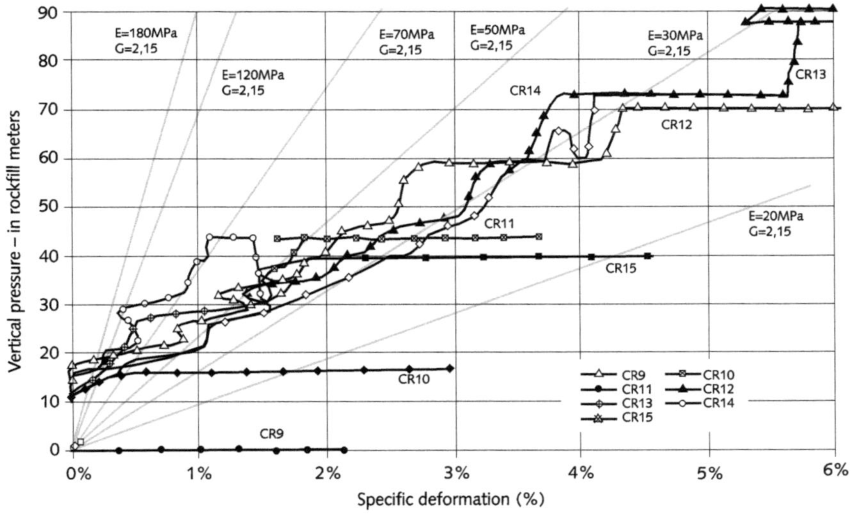

Figure 10.23 Machadinho CFRD: evaluation of the deformability modulus based on settlement measured by the Swedish boxes (Oliveira, 2002).

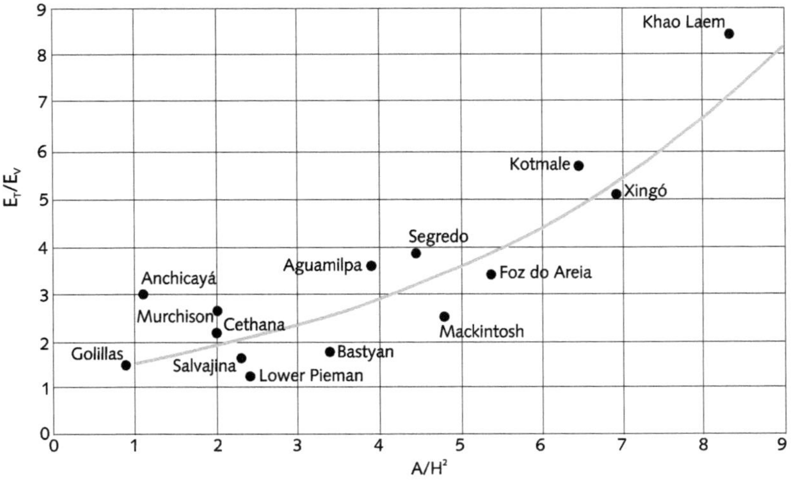

Figure 10.24 Relation between E_T and E_v modulus ratio versus A/H² (Pinto & Marques Filho, 1998).

(Figure 10.23). The vertical E_v modulus decreases as the specific deformation increase is higher, as already discussed in chapter 4.

Considering now the deflections of the slab in a similar way, one can compute the compressibility modulus normal to the slab. This is usually called the transversal deformability modulus.

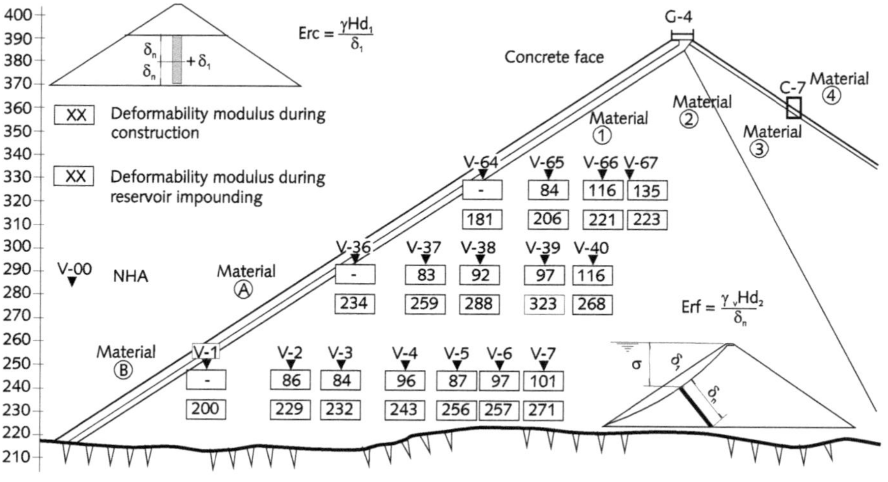

Figure 10.25 Deformability measurement at El Cajón CFRD (Mena Sandoval et al., 2007a).

When the water load is applied to the concrete face during the filling of the reservoir, there is a rotation of the principal stresses acting within the rockfill and even a reduction of the shear stresses in some areas of the upstream zone. Therefore, the new deformations will occur in a kind of "pre-compressed embankment" (pre-consolidated in soil mechanic language). Thus, it is not surprising to find that the transversal modulus value is higher that the vertical modulus value.

The ratio of measured values between E_T and E_V vary from 1.5 to 8.5 according to Figure 10.24 presented by Pinto and Marques Filho (1998).

Both 1.5 and 8.5 are extreme values, measured at the very narrow valley of Golillas and in an open valley (Khao Laem). In most of the dams the ratio E_T/E_V is between 2 and 4.

Figure 10.25 shows the modulus measured in El Cajón (Mexico). The ratio E_T/E_V tends to reduce as the measurement points move away from the slab, or move towards downstream.

10.8 TRI-DIMENSIONAL DISPLACEMENTS

Another instrument to measure displacements are the superficial marks normally installed on the crest and on the downstream slope of CFRDs.

The great advantage of this instrument, as opposed to others, is that it enables us to measure the displacement in full scale since all three dimensions are recorded. The disadvantage is that readings are possible only after construction is completed.

Figure 10.26 shows the displacements of the superficial marks measured at the Campos Novos Dam crest. It is clear that the displacements developed towards the valley. The downstream slope displacements have a more erratic behaivor.

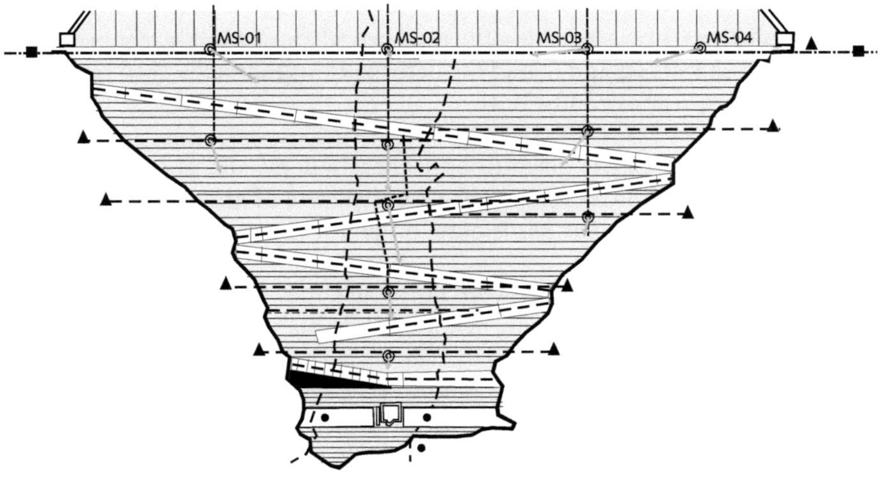

Figure 10.26 Campos Novos CFRD: horizontal displacements – surface marks (4/26/2006) (Cruz & Pereira, 2007).

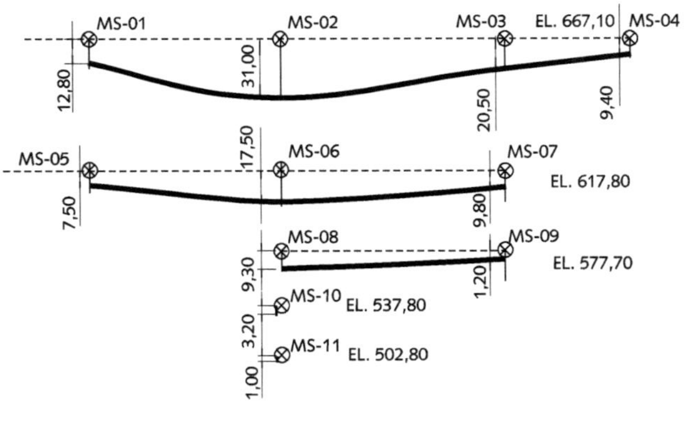

Figure 10.27 Campos Novos CFRD: settlement – surface marks (4/26/2006) (Cruz & Pereira, 2007).

Figure 10.27 shows the vertical displacements of the superficial marks on the crest and the downstream slope using the same marks of Figure 10.26. Combining these settlements with the displacements in the plane of Figure 10.26, the real displacements and their directions can be known.

Analyses of the magnitude of such displacements show that with exception of the surface mark on the right-hand side of the crest, the vertical displacement is predominant.

10.9 CONCLUSIONS

Considering chapter 3 case histories, the mechanical and hydraulic properties of rockfill described in chapters 4 and 6, and the deformations that have been reported in this chapter, some conclusions can be drawn.

- The generic name CFRD given to both rockfill dams and gravel dams should be altered to CFRD and CFGD since the performance of gravel dams with regard to displacements or deformations is quite different. Gravels are much less compressible than rockfills.
- The concrete slab of dams built with gravel in zones 3B and T and rockfill in zone 3C has performed satisfactorily, in spite of the high displacements that occurred on the upper part of the slab.
- The construction of CFRDs with consecutive heightenings and partial fillings of the reservoir has led to a path of displacements of the concrete face that are quite different from the slab displacements of dams in which the reservoir filling was done in one single stage.
- Two factors are dominant in CFRD performance: the height of the dam and the shape of the valley. The state of stresses that build up in the rockfill embankment is accountable for the displacements (both vertical and horizontal). And, because the deformability moduli of these materials decrease with increasing pressures, the higher the dams, the larger the deformations will be in percentage terms. The valley shape directly influences the arching effects that are more pronounced in narrow valleys as is well known.
- These findings have resulted in changes to the compaction requirements for all zones of the dam, because the displacements of the embankment are rapidly reflected in the slab stresses that tend to increase and can lead to failure.
- This behavior has also produced a revision to the vertical joint details in the compression zone of the dam.
- The construction sequence, the partial heightening, as well as the time to start the concrete slab, should get more attention in higher dams, because the slab behavior depends on the magnitude of the displacements that develop in the embankment. In high dams the moment to start the concrete slab should be postponed as much as possible, in order to reduce the effects of the displacements due to the dam construction on the face slab.
- Today's knowledge perhaps enables us to classify dams as far as their behavior is concerned according to their height:
  - CFRDs up to 50 m have not presented problems of any nature;
  - CFRDs between 50 m and 100 m high have displayed fissures and cracks on the slab that were easily repaired by dumping fines upstream;
  - CFRDs between 100 m and 150 m high have presented throughflow problems due to displacements on the concrete face, but in general have had good performance;
  - CFRDs between 150 m and 200 m high have required more repair work due to the opening of more cracks and, in some cases, due to failures of the concrete slabs;

  - CFRDs in the order of 200 m high are few and have required more attention and changes not only in the compaction requirements of the rockfill but also in the joint design.
- As far as leakage, or the flow through the rockfill, its magnitude is currently much lower than the one necessary to initiate a process of rock block instability on the downstream slope. But, seepage and leakage should be measured and observed because they may reflect existing problems in the slab, on top of the economical losses they impute.
- The phenomenon of creep, or slow deformation resulting from continuous fragmentation, crushing and accommodation of the rockfill, although decreasing with time, must be measured because the face slab may suffer damage resulting from its continuous displacement.

Chapter 11

Numerical analysis and its applications

11.1 INTRODUCTION

Friendship between Brazil and China in the field of CFRDs has lead the authors to invite Dr Xu Zeping to collaborate in the making of this book. We decided for his participation to be this chapter on numerical methods because the subject can be discussed in isolation from the topics analyzed in the other chapters. Results of numerical analyses carried out on the Brazilian dams of Itá, Itapebi, Machadinho, Barra Grande and Campos Novos are included at the end of the chapter.

CFRD is a kind of dam that concieves rockfill as the supporting structure and concrete slabs on the upstream slope as the impervious element. Modern CFRDs constructed by using thin layer compaction were first built in the 1970s. Within less than 30 years of development, rapid progress has been achieved on CFRD design and construction.

In the early days of CFRD development, the designs were empirical, i.e. mainly based on previous engineering experience and on engineers' judgment. Little systematic research was conducted. From 1980 to 1990, engineers in China, Brazil, Mexico and Australia developed a series of research work in the fields of rockfill testing, deformation analyses, waterstop elements, etc.

The design of a CFRD is gradually changing from fully relying on an engineer's judgment to an approach of being instructed by theoretical analyses and lab/field testing research.

Previous analyses of CFRD were performed mainly using linear elastic models and most were 2D analyses. In recent years, non-linear analysis has become widely accepted and the main analytical method is either the finite element method (FEM) or the finite difference method.

For concrete faced rockfill dams, the stress and deformation properties of the rockfill and the concrete face slab are the most important issues regarding dam safety and performance. In recent years, CFRDs have got higher and higher, and the topographic and geological conditions at the dam site have got more and more complicated, which presents an increasing number of challenges to be faced by both the theoretical modeling and the analytical method of CFRD numerical analysis.

How to accurately predict deformation tendency for high CFRDs and how to optimize the design and improve the stress status of the concrete face slab have become the key issues in CFRD design.

In CFRD design, numerical analysis is a powerful tool. It allows for the stress and deformation behavior of CFRDs in various conditions to be obtained and the possible deformation patterns or failure modes to be predicted. The results from numerical analyses can provide useful instruction for the structural design and for the construction of the dam.

11.2 ENGINEERING PROPERTIES OF ROCKFILL MATERIAL

CFRD construction materials come in a very wide range, and include sedimentary rock, igneous rock, and metamorphic rock. The main indices of the original rock of rockfill material are density, specific gravity, void ratio, compression strength, tensile strength, softening coefficient, etc. According to the saturated unconfined compression strength of the original rock, rockfill material can be classified into hard, medium, and soft. Rockfill with the saturated uniaxial compression strength of original rock larger than 80 MPa is classified as hard rockfill material; between 30~80 MPa, as medium strength rockfill material; and less than 30 MPa, as soft rockfill material.

The main engineering properties of rockfill materials include stress and deformation, gradation, compressibility and strength.

The exploitation of rockfill material is mainly done by blasting. So the gradation of rockfill material depends primarly on the method of blasting, the structure of the rock mass, and the development of fissures and joints within the rock mass. Usually, the gradation curve of rockfill material presents continuous distribution. When the uniformity coefficient (Cu) of rockfill material is larger than 15, it is considered a good gradation.

Another characteristic for the gradation of rockfill material is its variability. The most important impact on the variation of the gradation of rockfill is breakage during compaction, which mainly depends on the rock strength and the compaction energy. The variation of the gradation of rockfill material will directly relate to the change in the engineering properties of rockfill.

The particle shape of rockfill is polyhedral. The general compressibility of rockfill material is mainly controlled by the rearrangement of particles and it is also affected by the rock density, gradation, etc. Normally, compacted rockfill will have high density and a low void ratio. Its compressibility is relatively low. For high CFRDs, the high stress level in rockfill can lead to secondary breakage of rockfill particles and the rearrangement of the broken particles can cause creep deformation.

When rockfill is wetted, additional deformation results, caused by softening and breakage of the particle edges and by the moving and rearrangement of rockfill particles promoted by the lubricating action of water.

The wetting deformation is directly related to the properties of the rock. The deformation would be reduced if the rockfill density were high or the initial water content were large.

The compressibility of rockfill material is directly related to its gradation. The same rock with different gradation will result in totally different compressibility. The compression modulus of a rockfill will be increased significantly as its density is increased.

Rockfill is a kind of granular material which is composed of hard particles. The shear strength of rockfill includes the action of the sliding friction and the intercalation between particles. The interlocking action of particles is affected by the effect of shearing dilation and the influence of particle breakage. For the shear strength of rockfill materials, the expressions commonly used are both linear and non-linear.

The linear strength expression of rockfill is $\tau_f = c + \sigma_n \tan\varphi$, which is mainly used in slope stability analyses. However, for the numerical stress and deformation analyses, the non-linear strength expression will be applied. The widely used non-linear strength expression of rockfill is $\varphi = \varphi_0 - \varphi \log(\sigma_3/P_a)$. The non-linear strength parameters φ_0 and $\Delta\varphi$ can be obtained from large scale triaxial tests of rockfill material.

11.3 ROCKFILL MATERIAL CONSTITUTIVE MODELS

In numerical analyses of CFRDs, the constitutive model of the rockfill is the most important base of an analysis. The stress and deformation analyses of concrete faced rockfill dams need comprehensive consideration of different factors, which include the stress and deformation relationship, strength properties and deformation properties, etc.

Since the 1970s, the emergence of *FEM* and the rapid progress of computer science have promoted deeper research in the field of constitutive models of rockfill material. Furthermore, with the development of large scale testing machines for rockfill material and the progress of the related testing technologies, the engineering properties of rockfill materials have been deeply studied. As a consequence of the research, the constitutive model gradually developed from an empirical simple model to a theoretical complex model.

The constitutive relationship of rockfill materials possess complicated characteristics, including non-linear work hardening, volume change under shear stress, plastic yielding, anisotropy, creep, etc. It is also influenced by external factors, such as stress path, stress history, initial stress status, composition of particles, etc.

For such complicated material properties and influential factors, there is no comprehensive model representing all aspects and factors. In practice, the development of the constitutive model is mainly to determine the mathematic model through observation and analysis of laboratory test results that consider the main factors of the stress strain relations of the material.

Actual theories on the constitutive model include the non-linear elastic theory, the elasto-plastic, and the visco-elasto-plastic. In the numerical analysis of concrete faced rockfill dams, the commonly used constitutive models for rockfill material are the non-linear elastic model and the elasto-plastic model.

As the stress-strain relationship of rockfill material is highly non-linear, the constitutive model in the numerical analysis of CFRDs must correctly represent this non-linear relation. Moreover, the rockfill volume changes during shearing will also impact the stress distribution on the concrete face.

Furthermore, for the proper consideration of the volume changes of rockfill material under shear stress, which include volume reduction and volume dilation, the elasto-plastic models with the application of multiple yield surface and the non-associated flow rule are theoretically perfect.

But, at present time, these models are still facing some challenges in testing, parameter determination, and computational methods. Compared with the elasto-plastic model, a non-linear elastic model, such as Duncan-Chang's hyperbola model (E-B), is more practical and applicable.

11.3.1 Non-linear elastic model

Elastic material models based on the theory of continuum mechanics can be generally classified as linear elastic (generalized Hook's law), Cauchy elastic, hyperelastic, and hypoelastic model.

For Cauchy elastic material, the stress status of the material depends solely on the deformation (strain status), i.e. the stress and strain of the material have an unique non-linear relationship.

For the presently applied non-linear elastic models in CFRD engineering, most of the models are the Cauchy elastic model (variable elastic model). It employs the incremental form of the generalized Hook's law:

$$\{\Delta\varepsilon\} = [C]\{\Delta\sigma\} \tag{11.1}$$

and assumes that the elastic constants (E, v, K, G) in the flexibility matrix are only functions of the stress status, and not related to stress history.

The flexibility matrix can be expressed as:

$$[C] = \begin{bmatrix} C_1 & C_2 & C_2 & 0 & 0 & 0 \\ C_2 & C_1 & C_2 & 0 & 0 & 0 \\ C_2 & C_2 & C_1 & 0 & 0 & 0 \\ 0 & 0 & 0 & C_t & 0 & 0 \\ 0 & 0 & 0 & 0 & C_t & 0 \\ 0 & 0 & 0 & 0 & 0 & C_t \end{bmatrix} \tag{11.2}$$

After inversion, the stiffness matrix [D] is:

$$[D] = \begin{bmatrix} D_1 & D_2 & D_2 & 0 & 0 & 0 \\ D_2 & D_1 & D_2 & 0 & 0 & 0 \\ D_2 & D_2 & D_1 & 0 & 0 & 0 \\ 0 & 0 & 0 & G_t & 0 & 0 \\ 0 & 0 & 0 & 0 & G_t & 0 \\ 0 & 0 & 0 & 0 & 0 & G_t \end{bmatrix} \tag{11.3}$$

where:

$$C_t = 1/G_t;\ C_1 = 1/9K_t + 1/3G_t;\ C_2 = 1/9K_t - 1/6G_t;\ D_1 = K_t + 4G_t/3; \\ D_2 = K_t - 2G_t/3 \tag{11.4}$$

K_t and G_t are tangent bulk modulus and tangent shear modulus.

11.3.2 Duncan-Chang's hyperbola model

Under the condition of the conventional triaxial test, $\Delta\sigma_2 = \Delta\sigma_3 = 0$. One defines $E_t = \Delta\sigma_1/\Delta\varepsilon_1 = 1/C_1$, $v_t = -\Delta\varepsilon_3/\Delta\varepsilon_1 = -C_2/C_1$, and then the relationship between E_t, v_t, K_t, and G_t can be expressed as:

$$E_t = \frac{9K_tG_t}{3K_t + G_t} \quad v_t = \frac{3K_t - 2G_t}{6K_t + 2G_t} \tag{11.5}$$

Assuming that the curve of $(\sigma_1 - \sigma_3)$ and $(\varepsilon_3 - \varepsilon_1)$ is hyperbolic, the relationship of initial tangent modulus E_t and initial Poisson's ratio v_t to the stress can be expressed as:

$$E_t = KP_a\left(\frac{\sigma_3}{P_a}\right)^n \quad v_t = G - \text{Flog}\left(\frac{\sigma_3}{P_a}\right) \tag{11.6}$$

By derivation calculus, the tangent elastic modulus E_t and tangent Possion's ratio will be:

$$E_t = E_i(1 - R_fS_l)^2 \quad v_t = \frac{v_i}{\left[1 - D\frac{\sigma_1-\sigma_3}{E_i(1-R_fS_l)}\right]^2} \tag{11.7}$$

In 1980, Duncan suggested the E-B mode of the model, which uses the tangent bulk modulus to replace the tangent Possion's ratio v_t. The expression of tangent bulk modulus is:

$$B_t = K_bP_a\left(\frac{\sigma_3}{P_a}\right)^m \tag{11.8}$$

By establishing a certain loading and unloading principle, the Duncan's E-B model can also consider the stress history impact. The unloading modulus is defined as:

$$E_{ur} = K_{ur}p_a\left(\frac{\sigma_3}{P_a}\right)^n \tag{11.9}$$

In these expressions, P_a is atmospheric pressure; K, R_f, n, G, F, D and K_b are parameters of the model; K_{ur} is the unloading modulus number; S_l is stress level, defined as $\sigma_l = (\sigma_1 - \sigma_3)/(\sigma_1 - \sigma_3)_f$, $(\sigma_1 - \sigma_3)_f = 2\,(c\cos\varphi + \sigma_3\sin\varphi)/(1 - \sin\varphi)$.

11.3.3 Modified Naylor's K-G model

The modified Naylor's *K-G* model uses bulk modulus *K* and shear modulus *G* as the variable elastic parameters. The expression of the model is:

$$K_t = K_i + \alpha_k \sigma_m \tag{11.10}$$

$$G_t = G_i + \alpha_G \sigma_m + \beta_G \sigma \tag{11.11}$$

where K_i, G_i, α_k, α_G and β_G are parameters of the model. Under the condition of isotropic compression, $\Delta\varepsilon_v = \Delta\sigma_m/K_t$. Under the condition of shearing with σ_m constant, $\Delta\varepsilon_s = \Delta\sigma_s/2G_t$. After integration, one can obtain:

$$\varepsilon_v = \varepsilon_{v1} + \frac{1}{\alpha_k}\ln(K_i + \alpha_k\sigma_m) \tag{11.12}$$

$$\varepsilon_s = \varepsilon_{s1} + \frac{1}{2\beta_G}\ln(G_i + \alpha_G\sigma_m + \beta_G\sigma_s) \tag{11.13}$$

The modified Naylor's *K-G* model has relatively simple expressions, but its model parameters can only be determined by special isotropic compression tests and shear tests under the condition of σ_m constant.

11.3.4 Elasto-plastic model

In the elasto-plastic model, the total strain (increment) is decomposed into the elastic and the plastic components:

$$\{\Delta\varepsilon\} = \{\Delta\varepsilon^e\} + \{\Delta\varepsilon^p\} \tag{11.14}$$

Consequently, the elasto-plastic stress-strain relationship can be expressed as:

$$\{\Delta\sigma\} = [D](\{\Delta\varepsilon\} - \{\Delta\varepsilon p\}) \tag{11.15}$$

where the elastic strain $\{\Delta\varepsilon_e\}$ can be calculated by generalized Hook's law and the plastic strain $\{\Delta\varepsilon_p\}$, by the following formula:

$$\{\Delta\varepsilon_p\} = \Delta\lambda\{n\} \tag{11.16}$$

where $\Delta\lambda$ is a positive scalar of proportionality dependent on the state of stress and load history. It represents the magnitude of the incremental plastic strain and is determined by the hardening rule. $\{n\}$ represents the direction of the plastic strain increment vector and is determined by the flow rule. The boundary between the elastic strain and the plastic strain is defined by yield surface.

According to the latter formula, the general elasto-plastic matrix can be expressed as:

$$\{\Delta\varepsilon\} = \{\Delta\varepsilon^e\} + \sum_{i=1}^{l} A_i\{n_i\}\Delta f_i \tag{11.17}$$

where l is the number of yield surface. Thus, the flexibility matrix can be expressed as:

$$[C] = [C]_e + \sum_{i=1}^{l} [C_i]_p \tag{11.18}$$

where $|C_i|_p = A_i |n_i| \{\frac{\partial f}{\partial \sigma}\}^T$. Thus, when the stress increment $\{\Delta\sigma\}$ is known, the plastic strain increment can be determined by A_i (hardening rule), $\{n_i\}$ (flow rule) and $\frac{\partial f_i}{\partial \sigma}$ (yield surface).

When double yield surfaces are employed ($l = 2$), the elasto-plastic matrix can be expressed as:

$$\begin{aligned}[D_{ep}] = [D] - \frac{1}{D_{et}} \Bigg\{ & A_1[D]\{n_1\}\left\{\frac{\partial f_1}{\partial \sigma}\right\}^T + A_2[D]\{n_2\}\left\{\frac{\partial f_2}{\partial \sigma}\right\}^T \\ & + A_1A_2[D]\Bigg(\{n_1\}\left\{\frac{\partial f_2}{\partial \sigma}\right\}^T [D]\{n_2\}\left\{\frac{\partial f_1}{\partial \sigma}\right\}^T - \{n_1\}\left\{\frac{\partial f_1}{\partial \sigma}\right\}^T [D]\left\{\frac{\partial f_2}{\partial \sigma}\right\}^T \\ & + \{n_2\}\left\{\frac{\partial f_1}{\partial \sigma}\right\}^T [D]\{n_1\}\left\{\frac{\partial f_2}{\partial \sigma}\right\}^T - \{n_2\}\left\{\frac{\partial f_2}{\partial \sigma}\right\}^T [D]\{n_1\}\left\{\frac{\partial f_1}{\partial \sigma}\right\}^T \Bigg)\Bigg\}[D]\end{aligned} \tag{11.19}$$

where:

$$\begin{aligned}D_{et} = {} & 1 + A_1\left\{\frac{\partial f_1}{\partial \sigma}\right\}^T [D]\{n_1]+ A_2\left\{\frac{\partial f_2}{\partial \sigma}\right\}^T [D]\{n_2\} \\ & + A_1A_2\left(\left\{\frac{\partial f_1}{\partial \sigma}\right\}^T [D]\{n_1\}\left\{\frac{\partial f_2}{\partial \sigma}\right\}^T [D]\{n_2\} - \left\{\frac{\partial f_1}{\partial \sigma}\right\}^T [D]\{n_2\}\left\{\frac{\partial f_2}{\partial \sigma}\right\}^T [D]\{n_1\}\right)\end{aligned} \tag{11.20}$$

For a single yield surface, the elasto-plastic matrix can be expressed as:

$$[D]_{ep} = [D] - \frac{A[D]\{n\}\left\{\frac{\partial f}{\partial \sigma}\right\}^T [D]}{1 + A\left\{\frac{\partial f}{\partial \sigma}\right\}^T [D]\{n\}} \tag{11.21}$$

In geotechnical engineering, by considering the anisotropic hardening of the plastic deformation and the dependence of the incremental plastic strain on the direction of the incremental stress, the multiple yield surfaces model will be more applicable. Today, the most popular model used in numerical analysis is the double yield surface model which includes the Lade-Duncan's model and the Shen Zhujing's model (China).

The double yield surface suggested by Lade are:

$$f_1 = I_1^2 + 2I_2 \quad f_2 = \left(\frac{I_1^3}{I_3} - 27\right)\left(\frac{I_1}{P_a}\right)^m \tag{11.22}$$

where I_1, I_2 and I_3 are the first, the second and the third stress invariants, respectively; f_1 and f_2 are compression yield surface and shear yield surface, respectively.

The double yield surfaces suggested by Shen Zhujiang are:

$$f_1 = \sigma_m^2 + r^2\sigma_s^2 \quad f_2 = \frac{\sigma_s^2}{\sigma_m} \tag{11.23}$$

Under the condition of triaxial stress status, $\Delta\sigma_m = \Delta\sigma_1/3$, $\Delta\sigma_s = \Delta\sigma_1$. From the general expressions of the elasto-plastic matrix, the corresponding plastic coefficients of the model are:

$$\begin{aligned} A_1 &= \frac{\eta\left(\frac{9}{E_t} - \frac{3\mu_t}{E_t} - \frac{3}{G}\right) + 2s\left(\frac{3\mu_t}{E_t} - \frac{1}{K}\right)}{2(1+3r^2\eta)(s+r^2+\eta^2)} \\ A_2 &= \frac{\eta\left(\frac{9}{E_t} - \frac{3\mu_t}{E_t} - \frac{3}{G}\right) - 2r^2\eta\left(\frac{3\mu_t}{E_t} - \frac{1}{K}\right)}{2(3s-\eta)(s+r^2+\eta^2)} \end{aligned} \tag{11.24}$$

11.4 CFRD NUMERICAL ANALYSES METHODS

11.4.1 Simulation of surface contact and joints

The structure of CFRDs involves the contact interface between concrete face slabs and rockfill as well as the joints in between slabs and the joints between the slabs and the plinth. In CFRD numerical analyses, these interfaces and joints must be well simulated. As the material property of the concrete slab is much different from that of the rockfill, the sliding and separating deformation may occur at the interface by the action of external loads. Special elements should be used in order to simulate the interaction between different materials.

For the simulation of the interface of different materials, the commonly used interface element is the zero thickness Goodman's element. The element is configured by a pair of nodal points at both sides of the interface. The element is assumed to have no thickness at all and it is also assumed that the normal stress and shear stress have not coupled with the shear movement and normal displacement. The relationship between stress and the relative displacement of the nodal points of the element is:

$$\{\sigma\} = [\lambda]\{\omega\} \tag{11.25}$$

where:

$$[\lambda] = \begin{bmatrix} \lambda_s & 0 \\ 0 & \lambda_n \end{bmatrix} \tag{11.26}$$

When the interface is subject to compression action, the normal stiffness will have a large value all the while; when the element is subject to tensile action, the normal stiffness has a small value. The shear stress of the element will directly depend on the relative movement of the nodal point at both sides of the interface. For an interface of different materials, λ_s can be determined by a direct shear test. The commonly accepted hyperbolic relationship of shear stress and the relative displacement is:

$$\tau = K_s u \tag{11.27}$$

$$K_s = K_i \gamma_w \left(\frac{\sigma_n}{P_a}\right)^n \left(1 - \frac{R_f \tau}{\sigma_n \tan\varphi + c}\right)^2 \tag{11.28}$$

where R_f is the failure ratio; φ and c are the friction angle and the cohesion on the interface, respectively.

For the application of Goodman's element, the node points at interface may embed each other during the action of compression and the very big normal stiffness may also have an adverse effect on the computational results. Therefore, another element, called the thin layer interface element, was further developed.

From observation and laboratory tests, when the interface between two materials with large differences in properties is subject to shear stress, it can be found that a thin shear layer forms in the material with relatively weak properties. For that reason, the use of the thin layer interface element is more suitable to simulate the interaction of different materials.

For the thin layer interface element, the deformation at the interface is composed by elastic deformation and failure deformation. Under normal conditions, only elastic deformation $\{\varepsilon'\}$ occurs in the element. The material property is the same as the transitional rockfill and the element is treated as a normal element. When the shear stress of the element reaches its shear strength or tensile stress reaches its tensile strength, failure deformation $\{\varepsilon''\}$ occurs. The assumption for the patterns of the failure deformation is that there is no relative displacement at interface before failure and the relative displacement develops continuously after failure.

Failure deformation $\{\varepsilon''\}$ can be expressed as follows:

$$\begin{Bmatrix} \Delta\varepsilon_s'' \\ \Delta\varepsilon_n'' \\ \Delta\gamma_{sn}'' \end{Bmatrix} = \begin{bmatrix} 0 & 0 & 0 \\ 0 & \frac{1}{E''} & 0 \\ 0 & 0 & \frac{1}{G''} \end{bmatrix} \begin{Bmatrix} \Delta\sigma_s \\ \Delta\sigma_n \\ \Delta\tau_{sn} \end{Bmatrix} = [C]''\{\Delta\sigma\} \tag{11.29}$$

In the formula, E'' and G'' are the modulus parameters for tensile and shear failure. As the normal strain of the interface will not fail due to the restriction of concrete slabs, $\Delta\varepsilon_s'' = 0$ and the corresponding coefficient in the matrix can be set to zero.

The total deformation of the interface is the sum of elastic deformation and failure deformation.

$$\{\Delta\varepsilon\} = \{\Delta\varepsilon'\} + \{\Delta\varepsilon''\} = [C'][\Delta\sigma] + [C''][\Delta\sigma] = [C][\Delta\sigma] \tag{11.30}$$

In practical numerical analyses of CFRDs, different elements are applied to deal with the different interfaces of the structure. For the interface between the concrete face slab and the rockfill, the thin layer interface element is used. For the vertical joints between concrete face slabs, two nodes are set at the joint to simulate the different displacement at the joint of slabs. This joint element has zero thickness. Once the joint is open, the two nodes are separated; when the joint is closed, the two nodes are treated to have the same displacement. For the joints between face slabs and plinth (perimetric joint), a soft element is applied. When the element is subject to compression stress, it uses the material property of the concrete; when the element is subject to tensile or shear stress, it uses the material property of the soft material.

It must be noticed that, although the thin layer interface element has advantages in simulating the shear stress transfer and in avoiding material embedding, it still has limitations in simulating the separation of the interface. In future analyses, it will be important to develop an interface element able to consider discontinuous displacements.

11.4.2 Simulation of construction steps and reservoir impounding sequence

From the practice of numerical analysis, one can see that the construction steps and reservoir impounding procedure will have a significant effect on rockfill deformation and face slab stresses. Therefore, in numerical analyses, the real construction phases and the steps of reservoir impounding should be well simulated, especially the priority section construction.

For non-linear analysis, the incremental method is applied, where the total load is divided into several steps. These steps correspond to the rockfill construction steps and also to the procedures for reservoir impounding. In the analyses, every construction layer is dealt with like a load increment to perform an iteration calculation. The procedures of reservoir water impounding are also divided into several incremental load cases. By considering the real situation during the construction, the displacement of the nodal points on the top of each construction layer is set to zero at the end of the increment calculation.

During reservoir impounding, the directions of principal stresses in the upstream rockfill are deflected due to water load action. Compared to major principal stress values, the minor principal stress values increase more significantly. It could cause the decrease of $(\sigma_1 - \sigma_3)$ and an unloading of partial stress. For the computation during reservoir impounding, this unloading procedure should be considered by using unloading modules.

11.5 APPLICATION OF NUMERICAL ANALYSES ON CFRDs

11.5.1 The contribution of the numerical analyses for improving CFRDs designs

In the design of concrete faced rockfill dams, numerical analyses are the most important tool for understanding the stress and deformation of the dam and its face slab before hand.

By predicting the possible deformation of rockfill and the related stress distribution on the concrete face slab, different options can be compared in order to archieve an optimized design. In addition, the impact factors such as rockfill zoning, construction sequence, impounding procedure, valley shape, compaction density, etc. can also be studied via numerical analysis. It will be of further benefit as it improves design and construction methods.

11.5.2 Understanding the stress-strain status of the dam

By numerical analyses, the deformation status of different sections of the rockfill dam can be disclosed. The stress distribution of different parts of the rockfill dam can also be obtained. Figures 11.1 to 11.4 provide some examples of results.

11.5.3 Understanding the stress status of face slab

For the design of the concrete face slab, the deformation and stress of the slabs after reservoir impounding are of real concern, especially the maximum compression stress and the distribution of tensile stresses. Normally, the computational results of the stress are presented in the direction of dam axis and upstream slope.

Figures 11.5 to 11.7 provide some examples of the stress distribution and deformation vectors on a concrete face slab obtained from numerical analyses.

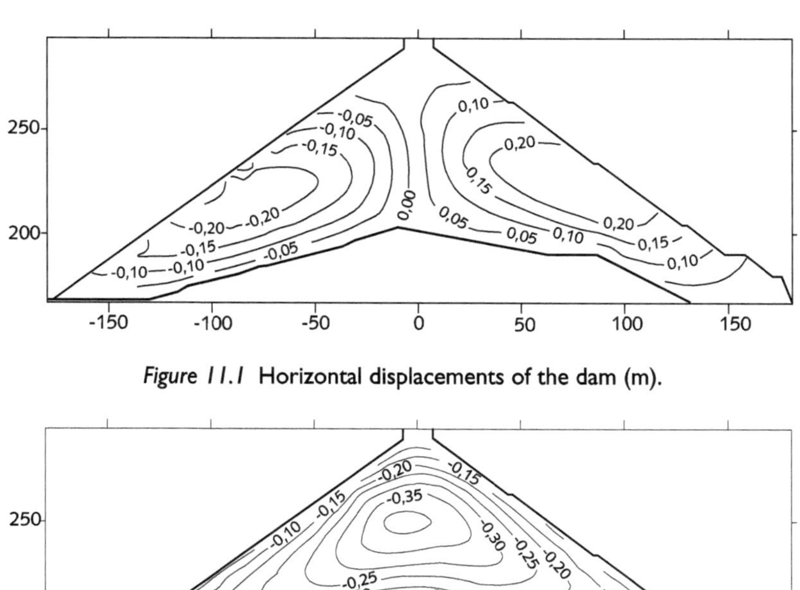

Figure 11.1 Horizontal displacements of the dam (m).

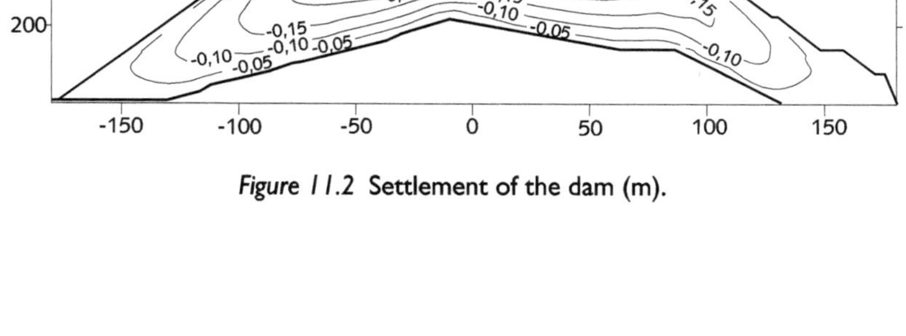

Figure 11.2 Settlement of the dam (m).

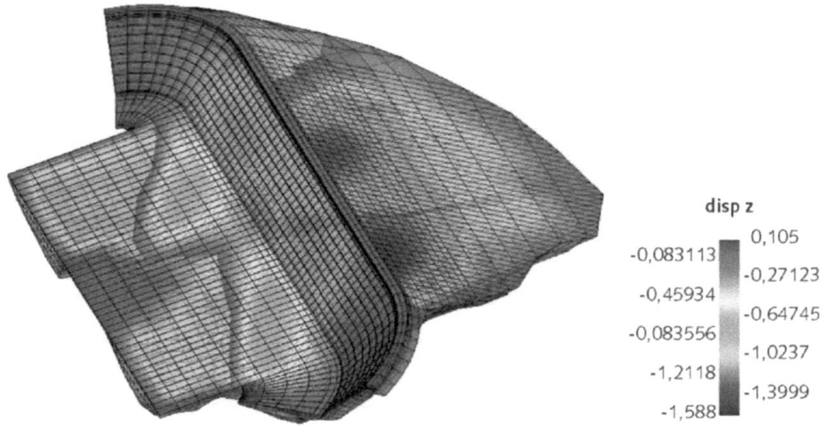

Figure 11.3 Displacement distribution of the dam. (See colour plate section).

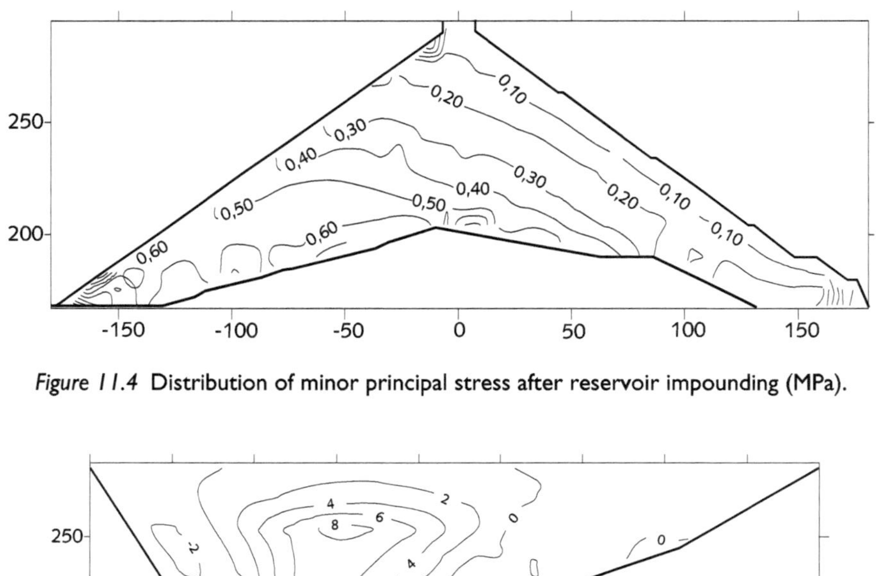

Figure 11.4 Distribution of minor principal stress after reservoir impounding (MPa).

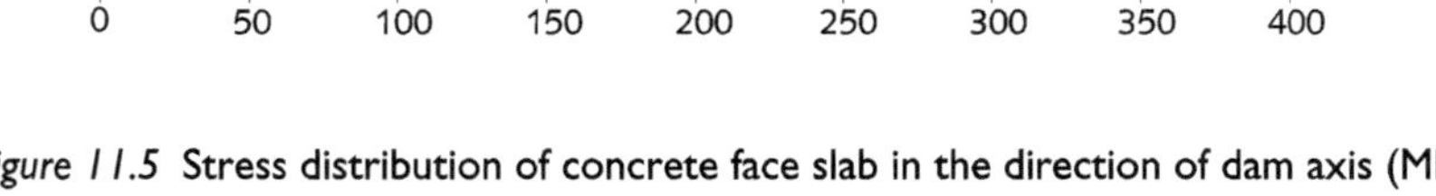

Figure 11.5 Stress distribution of concrete face slab in the direction of dam axis (MPa).

11.5.4 Predicting the displacement of joints

For the design of the joint waterstop devices, the predicted joint displacement will be an important reference. With numerical analyses, the movement of vertical joint and

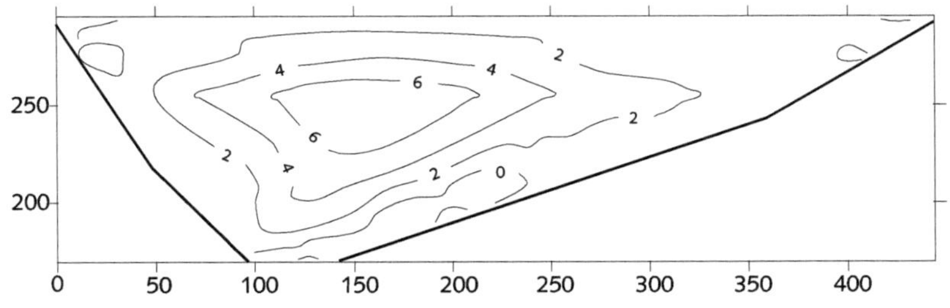

Figure 11.6 Stress distribution of concrete face slab in the direction of dam slop (MPa).

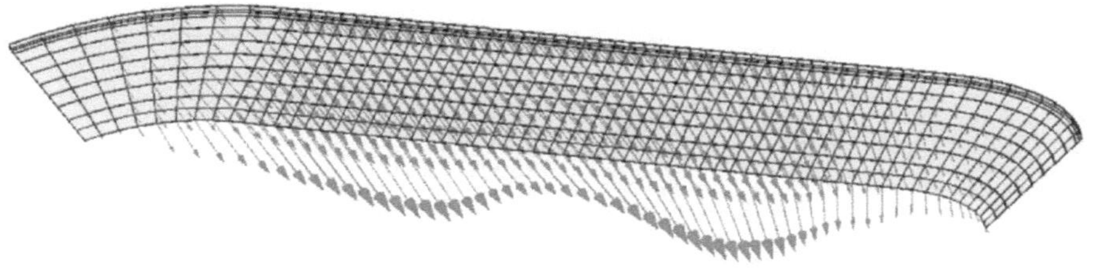

Figure 11.7 Displacement vectors of concrete face slab. (See colour plate section).

the displacements of perimetric joint in three directions (opening, shearing and settlement) can be presented. Figure 11.8 shows an example of the presented displacements of peripheral joints.

11.5.5 Case studies

Case 1: *Analysis on a CFRD built in a narrow valley (Hongjiadu Dam, China).*

Introduction to the project

Hongjiadu CFRD is located in the Guizhou Province, southwest China. The maximum dam height is 179.5 m and the crest length is 427.79 m. The factor of valley shape is 2.32. The width of the dam crest is 11 m and the crest elevation is 1148.00 m. The upstream and downstream slopes are 1.4(H):1.0(V). The direction of river flow at the dam site turns from S45°W to S45°E, forming a rectangular river bend. The river valley has a very unsymmetrical *V* shape.

The abutment mountain height is over 300 m. From upstream to downstream, the left abutment has two cliffs with a height of about 100 m. In between the cliffs, there is mudstone gentle slope, 80 m to 120 m wide. The slope of the right abutment is about 25° to 40°. Figure 11.9 shows the river valley at the dam site and Figure 11.10 is a typical design section of the dam.

Constitutive model of rockfill and analysis parameters

The accepted constitutive model of the rockfill material is Duncan-Chang's hyperbola, non-linear elastic model. The non-linear increment method is used in the analysis.

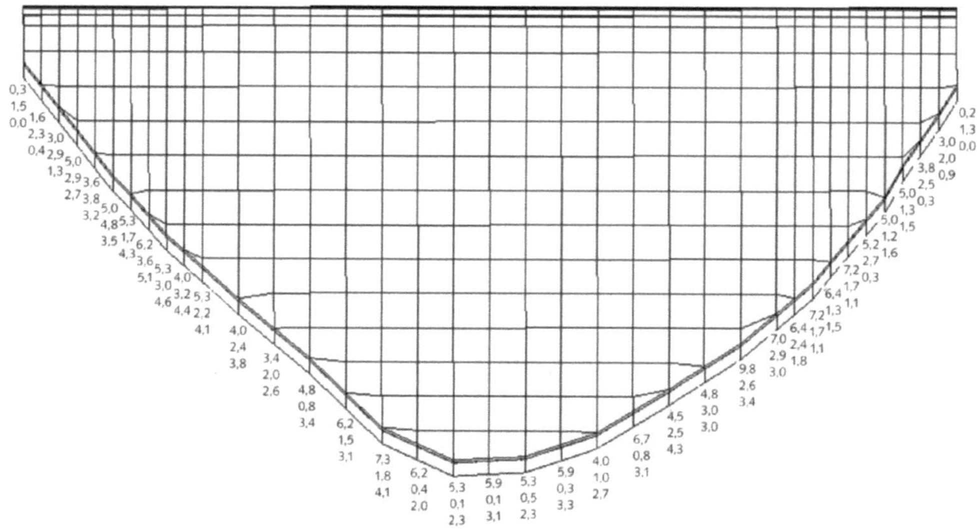

Figure 11.8 Displacements of perimetric joint.

Figure 11.9 River valley at Hongjiadu Dam site. (See colour plate section).

For each increment, a two step iteration process is adopted. The parameters used in the analysis come from large scale laboratory triaxial tests, as shown in Table 11.1.

Computation results – Rockfill dam stress and deformation

The stresses and deformations of the river valley section are shown in Figures 11.11 to 11.15 by contour lines. The deformations of the longitudinal section (along the dam axis) are shown in Figures 11.12 and 11.13.

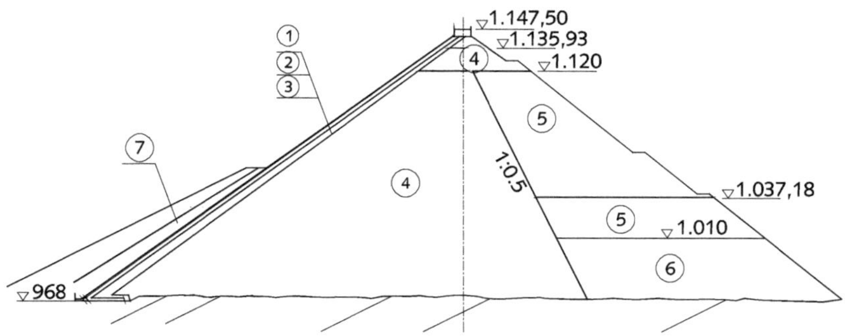

Figure 11.10 The zoning of Hongjiadu CFRD.

Table 11.1 Computation parameters for case 1 (E-B model).

Material	*γd kN/m³*	*K*	K_{ur}	*n*	R_f	K_b	*m*	*φ (°)*	*Δφ (°)*
2B	22.05	1.100	2.250	0.40	0.865	680	0.21	52	10
3A	21.90	1.050	2.150	0.43	0.867	620	0.24	53	9
3B	21.81	1.000	2.050	0.47	0.87	600	0.40	53	9
3C	21.20	850	1.750	0.36	0.29	580	0.30	52	10

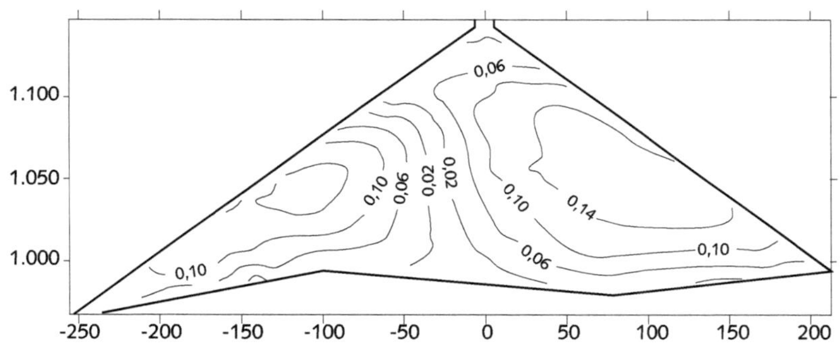

Figure 11.11 Hongjiadu CFRD: horizontal displacements after dam construction (m).

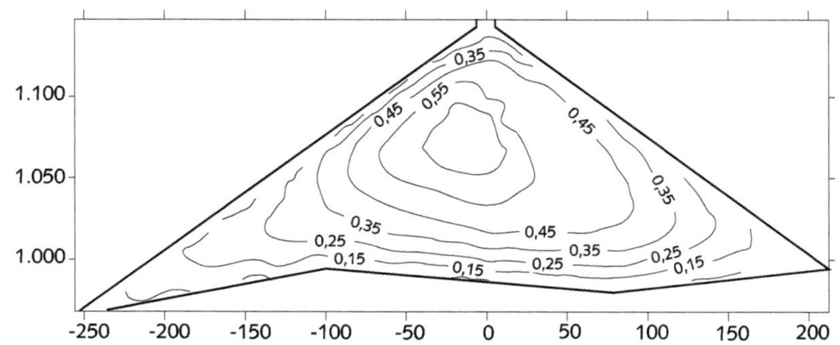

Figure 11.12 Hongjiadu CFRD: vertical displacements after dam construction (m).

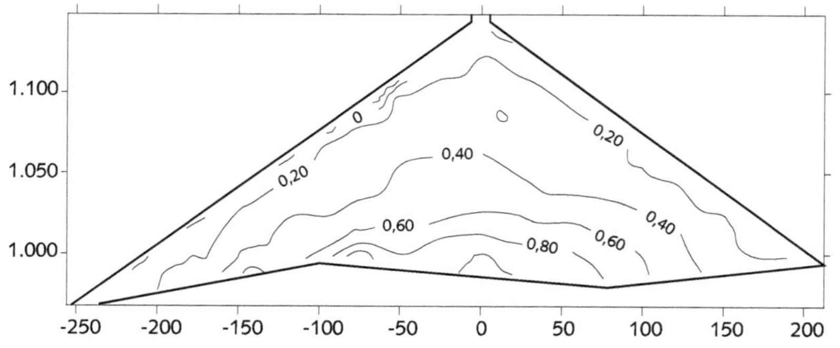

Figure 11.13 Hongjiadu CFRD: minor principal stress after dam construction (MPa).

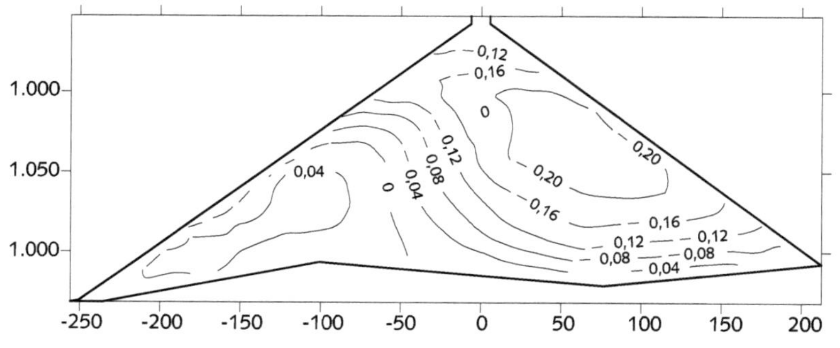

Figure 11.14 Hongjiadu CFRD: horizontal displacement during reservoir impounding (m).

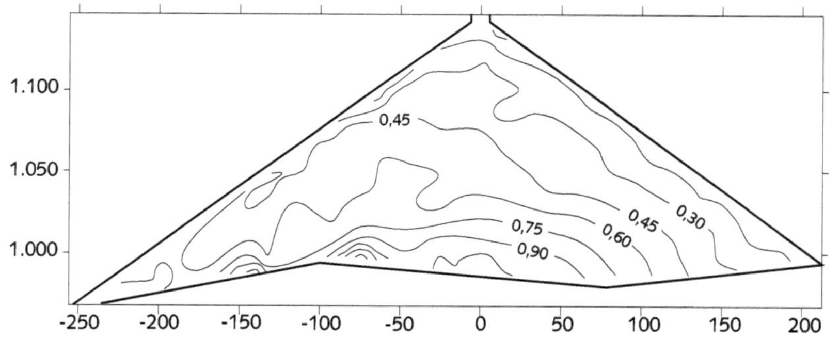

Figure 11.15 Hongjiadu CFRD: minor principal stress of reservoir impounding (MPa).

From the results, the maximum settlement of the dam body after completion of dam construction is 0.78 m, about 0.44% of the dam height. The position of maximum settlement lies in the riverbed section, at about the middle height of the dam. The maximum horizontal displacements on the upstream and downstream sides are 0.14 and 0.16 m respectively. After reservoir impounding, the settlement of the dam body increases slightly – the maximum settlement reaches 0.81 m. The distribution of horizontal displacements on the transverse section has an obvious change after reservoir impounding.

The horizontal displacements decrease on the upstream part and increase on the downstream part. The maximum horizontal displacement towards downstream is 0.24 m.

Along the dam axis, the rockfill at the abutment moves toward the river center and the displacements at the right abutment are relatively large. As for the gradient of settlement from the abutments towards the riverbed, the left bank gradient is larger than the one on the right bank. The maximum stress in the rockfill lies at the bottom of the dam, with a major principal stress of 4.7 MPa and a minor principal stress of 1.6 MPa.

The deformation and stress distribution of the concrete slab during reservoir impounding is shown in Figures 11.16 to 11.19.

The maximum normal displacement of the concrete slab lies at the upper part of the slab on the riverbed section. From this position to the abutment, normal displacements gradually reduce. On the left bank, the gradient of this displacement is relatively large. Along the dam axis, horizontal displacements are unsymmetrical to the river center. The displacements of the slab are basically pointing to the river center. The displacement on the right bank is relatively larger than the one on the left bank.

By analysing the contour of slab stress distribution during reservoir impounding, the larger position of the concrete slab is subjected to compression. The tensile stresses are mainly developed along the perimeter and crest. The maximum compression stress along the slope is 8.2 MPa and the maximum compression stress along the dam axis is 8.9 MPa.

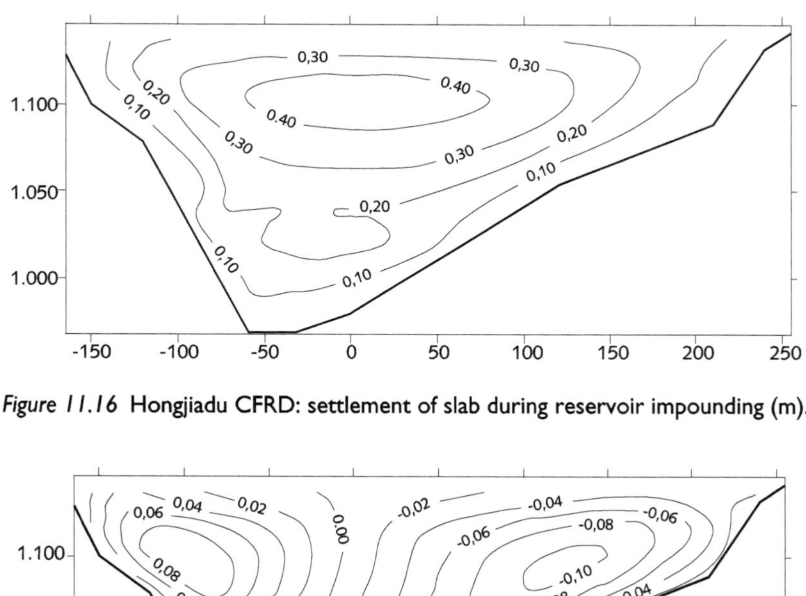

Figure 11.16 Hongjiadu CFRD: settlement of slab during reservoir impounding (m).

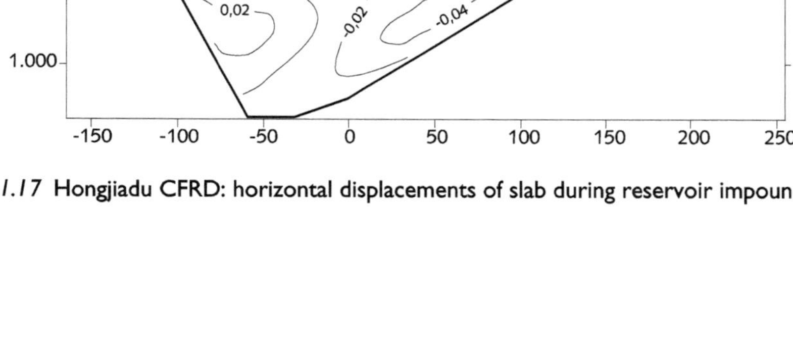

Figure 11.17 Hongjiadu CFRD: horizontal displacements of slab during reservoir impounding (m).

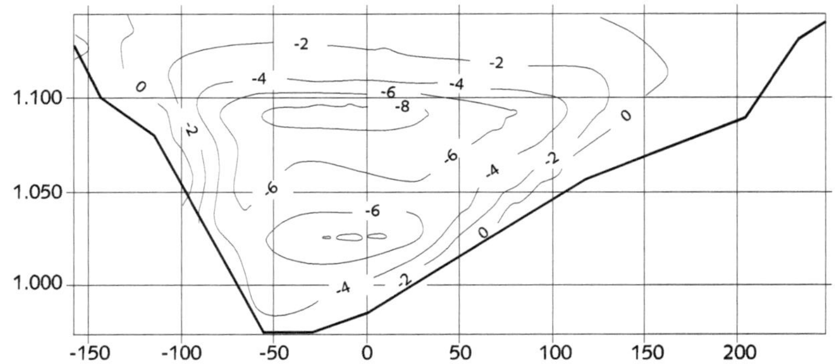

Figure 11.18 Hongjiadu CFRD: stresses along dam axis during reservoir impounding (MPa).

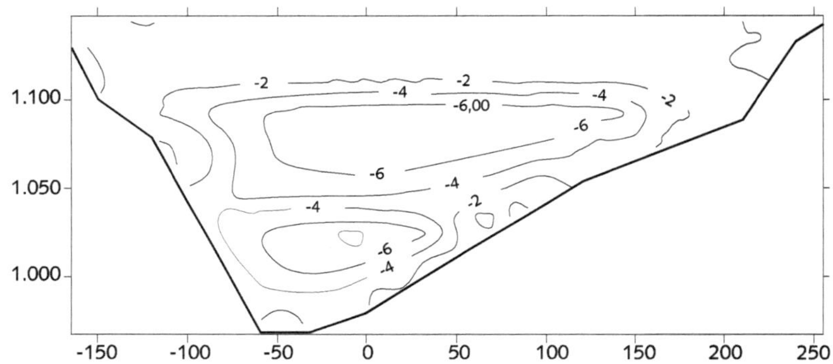

Figure 11.19 Hongjiadu CFRD: stresses along slope during reservoir impounding (MPa).

As the valley is narrow and the abutments slopes are quite steep, the area of tensile stress along the dam axis occurs on the slabs of the abutment. For the steeper abutments, the tensile stress of the slab is fairly high, even though the area is relatively small. For the gentle slope abutment, the tensile stress of the slab is fairly low, and the area is relatively large. Along the upstream slope, tensile stresses have occurred on the top of the slab.

Numerical research indicates that the most effective method of deformation control of CFRDs is rockfill zoning and compaction. Rational arrangement of rockfill zoning and suitable control of rockfill compaction specifications will play an important role in improving the working conditions of the face slab and the safety of the dam.

Numerical analyses show that the compaction density of rockfill material, the sequence of rockfill construction, and the material properties of the downstream 3C zone will have a significant impact on the stress and deformation of the dam and face slab. For high CFRDs, the rockfill material in upstream and downstream zones should not have large difference in modulus and densities. Considering the special topographical conditions of the Hongjiadu CFRD and in order to eliminate adverse impacts of the differential deformation between upstream and downstream rockfills, the same compaction density of upstream and downstream rockfills was required in the construction. In addition, near the steep abutment a special compaction zone with higher density was demanded with the purpose of reducing the differential deformation of rockfill along the dam axis direction.

Case 2: *Analysis of a CFRD built on deep alluvium (Chahanwusu CFRD, China).*

Introduction to the project

The Chahanwusu CFRD is located in the Chinese province of Xinjiang, and is designed to be constructed on deep alluvium layers of sandy gravel and coarse sand. The alluvium foundation is 34 m to 47 m in depth. The maximum height of the dam is 110 m with the crest elevation at 1654.0 m. The slopes inclination are 1.5(H):1.0(V) for the upstream slope and on average 1.8(H):1.0(V) for the downstream slope.

The dam section is shown in Figure 11.20. A layout combining a diaphragm, connecting slabs, plinth, and face slab has been adopted to form the impervious defense. The concrete diaphragm with a maximum depth of 40.8 m is 1.2 m thick and it is connected to the plinth by two concrete slabs, as shown in Figure 11.21. Each connecting slab is 3.0 m long and 0.8 m thick, with a thin layer of asphalt concrete underneath.

The alluvium – with an average depth of 40 m – can be simplified into three layers: the upper sandy gravel layer with a depth of 19.24 m, the coarse sand layer with a depth of 5.92 m in the middle, and a sandy gravel layer 11.18 m thick at the bottom.

Constitutive model of rockfill and the finite element mesh

In numerical analyses, Duncan's *E-B* model has been applied for the earth and rockfill materials. Bedrock, concrete slab and diaphragm behavior are being simulated by a linear elastic model.

The finite element mesh for the numerical analysis is shown in Figure 11.22, which has 6,189 elements and 8,163 nodal points.

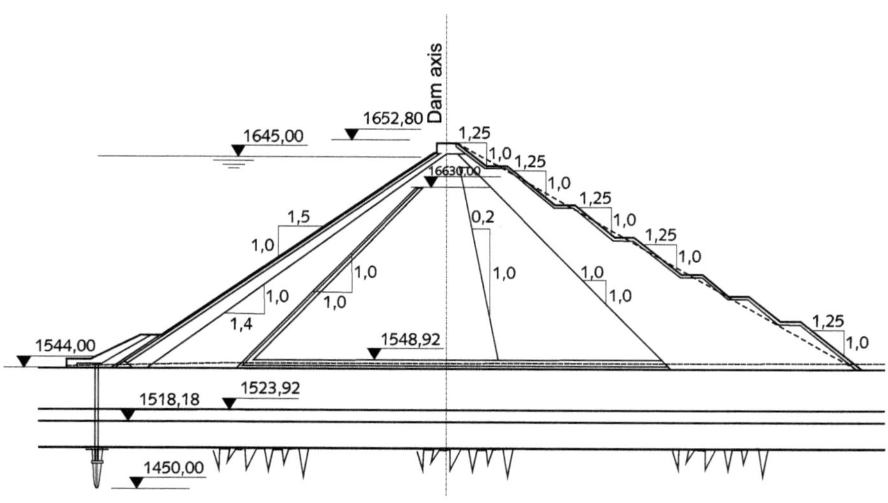

Figure 11.20 Typical section of Chahanwusu Dam.

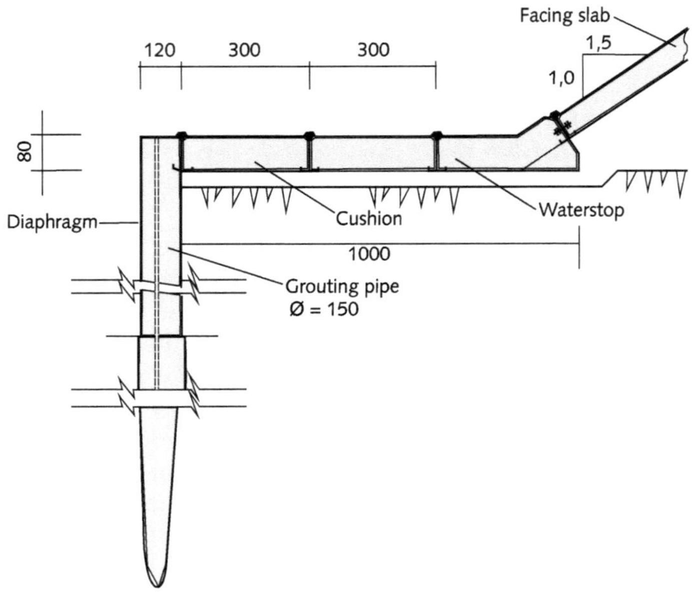

Figure 11.21 Chahanwusu CFRD: detailed design of seepage control structures.

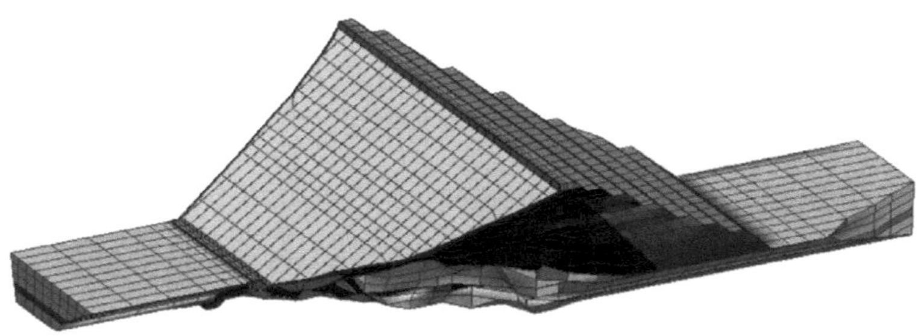

Figure 11.22 Chahanwusu CFRD: finite element mesh for the analysis. (See colour plate section).

The numerical analysis results, the stresses, and deformations of the dam, facing slab and the diaphragm are shown in Figures 11.23 to 11.30.

Computation results analysis

From the figures of the dam and its foundation deformations, it can be seen that the foundation compressibility has a significant impact on the deformations of the dam. CFRDs constructed on rock foundation will display very small settlements. The main deformation of the dam is caused by dead weight and water load. For CFRDs built on deep alluvium, the compressible foundation produces a profound deformation under the load of the dam. Thus, the position of the dam maximum settlement moves towards its lowest part and its crest sinks toward the center.

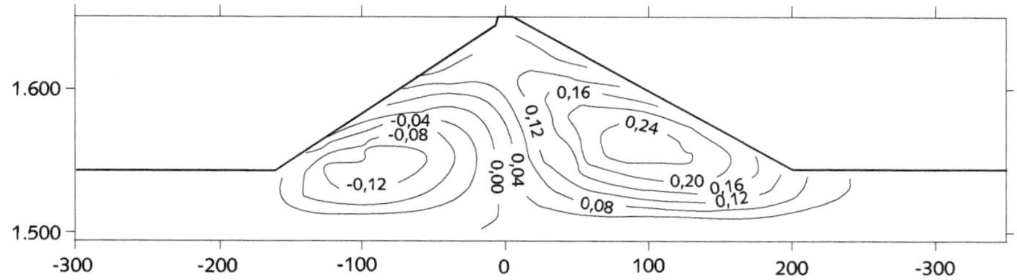

Figure 11.23 Chahanwusu CFRD: horizontal displacements of the dam and its foundation (m).

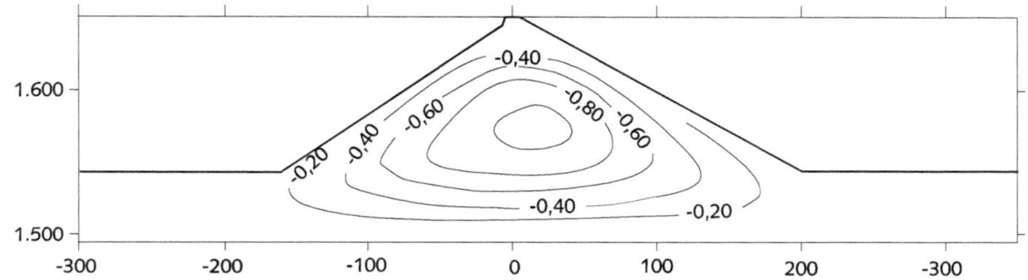

Figure 11.24 Chahanwusu CFRD: vertical displacements of the dam and its foundation (m).

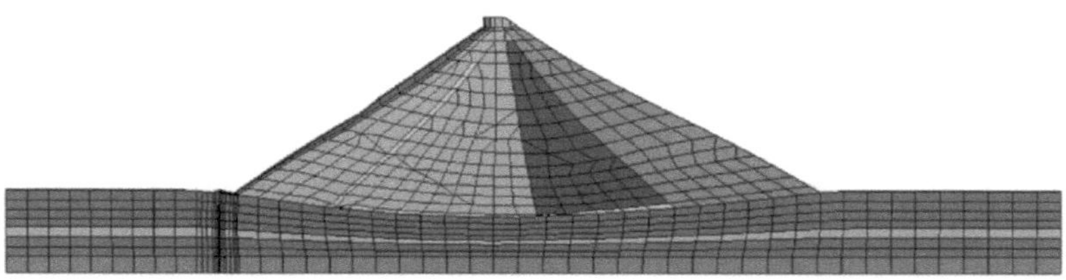

Figure 11.25 Chahanwusu CFRD: deformation of the dam and its foundation. (See colour plate section).

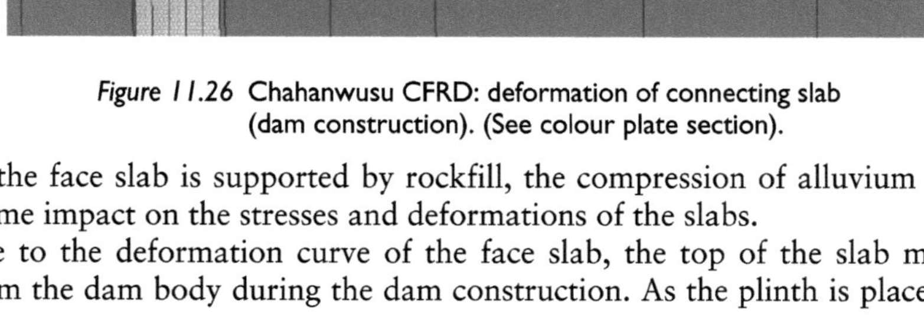

Figure 11.26 Chahanwusu CFRD: deformation of connecting slab (dam construction). (See colour plate section).

As the face slab is supported by rockfill, the compression of alluvium will also have some impact on the stresses and deformations of the slabs.

Due to the deformation curve of the face slab, the top of the slab may separate from the dam body during the dam construction. As the plinth is placed on the

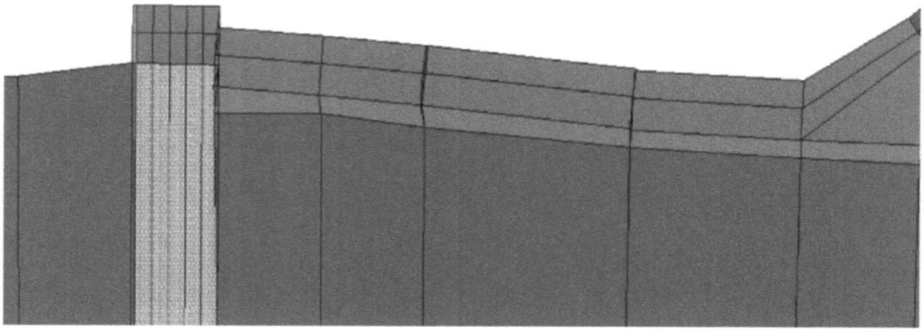

Figure 11.27 Chahanwusu CFRD: deformation of connecting slab (reservoir impounding). (See colour plate section).

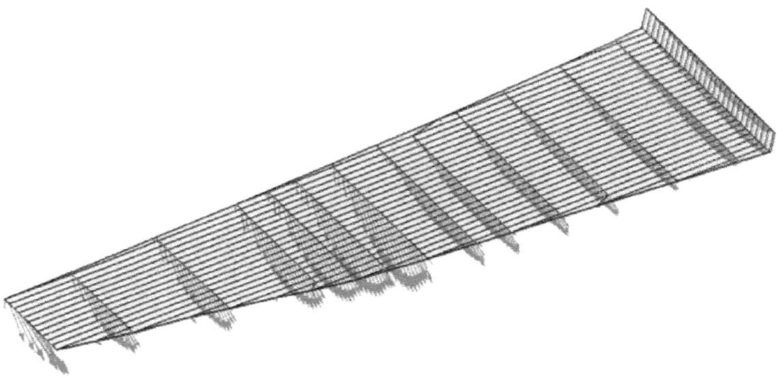

Figure 11.28 Chahanwusu CFRD: deformation of face slab. (See colour plate section).

compressible alluvium, along with the face slab it will have a certain displacement during dam construction and reservoir impounding. The main tendency of the displacement is towards settlement.

As the alluvium is sunk by the weight load of the dam and the bottom of the alluvium is confined by bedrock, horizontal displacement both upstream and downstream will occur in the alluvium during the construction period. Therefore, the diaphragm will move towards the upstream side. After reservoir impounding, the diaphragm will be pushed back by water load.

As the bottom of the diaphragm is restricted by bedrock, the maximum displacement will occur at the top of the diaphragm. Tensile stresses appear near the bottom and the top of the diaphragm, and the tensile stress value at the top is relatively large. Except for the bottom and top, most of the diaphragm is subjected to compression stresses. During dam construction, the diaphragm is mainly subjected to friction load caused by alluvium settlement. During reservoir impounding, the diaphragm is mainly subjected to water pressure load at its top and sides.

From the deformation distribution of the connecting slabs during the dam construction, tensile displacement will occur at the joint between the diaphragm and the connecting slab. No obvious differential settlement occurs. After reservoir impounding, a relatively large differential settlement appears at the joint between the diaphragm

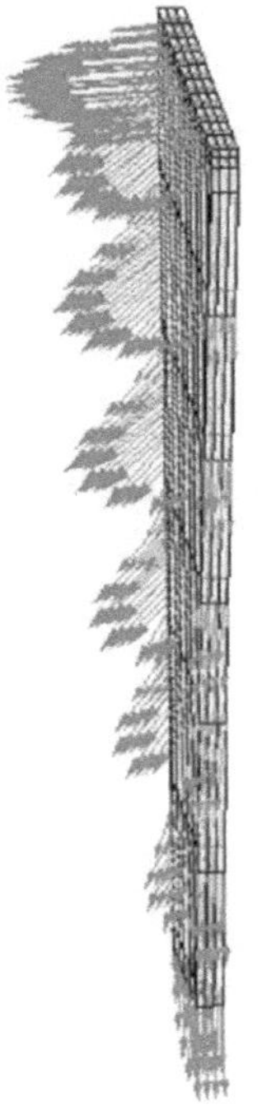

Figure 11.29 Chahanwusu CFRD: diaphragm deformation (after construction). (See colour plate section).

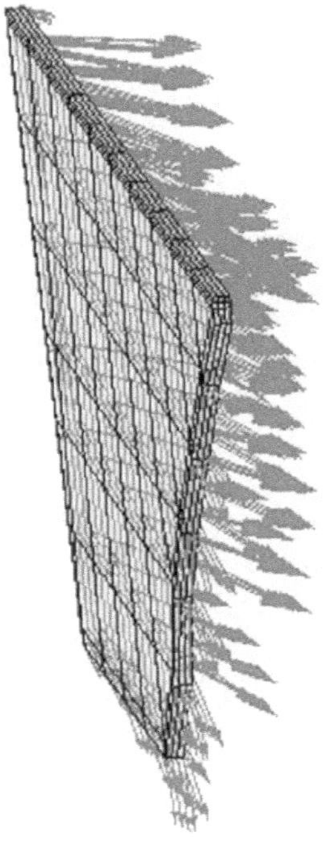

Figure 11.30 Chahanwusu CFRD: diaphragm deformation (reservoir impounding). (See colour plate section).

and the connecting slabs. Thus, one can see that the differential settlement between diaphragm and connecting slabs is mainly caused by reservoir impounding.

For CFRDs constructed on deep alluvium, the connection between the diaphragm and the plinth is very important. In practice, some projects directly connect the diaphragm to the plinth, other projects use connecting slabs to join the diaphragm and the plinth together. In order to accommodate the differential settlement of the alluvium foundation, a connecting slab is more appropriate; however, this may increase the number of joints, reducing the reliability of the whole connecting system. Either way, a certain distance should be kept between the diaphragm and the plinth to allow for the interaction of the plinth and the diaphragm.

From the displacement tendency of the alluvium during dam construction, the impact of foundation deformation will depend on the distance from the dam axis. The larger the distance from the dam axis, the smaller the horizontal displacement will be. After reservoir impounding, the connecting slabs will be subjected to reservoir water load.

A long slab has a relatively large area for water loading. Therefore, there is a relation between diaphragm displacement and slab length. The longer the connecting slab, the smaller the horizontal displacement of the diaphragm towards upstream during dam construction. However, there will be larger horizontal displacement of the diaphragm towards downstream during reservoir impounding. Conversely, the shorter the connecting slab length, the larger the horizontal displacement of the diaphragm towards upstream during dam construction. However, there is less horizontal displacement of the diaphragm towards downstream during reservoir impounding. Thus, one can conclude that the diaphragm displacement has a close relation to the length of the connecting slab (or plinth). For different projects, there should be an ideal length optimized by numerical analyses.

Regarding stress status, the diaphragm will be subjected to a pushing action by the alluvium from its downstream during the dam construction. After reservoir impounding, the water load from upstream and its top will be the dominating effort. Also, the traction produced by the settlement of the alluvium foundation is an important external load acting on the diaphragm. Thus, the stress status of the diaphragm is very complex. Generally, the compression stress is dominant in the broader part of the diaphragm. Near the top and the bottom, the diaphragm will be subjected to tensile stresses.

11.6 CLOSING REMARKS

In general, numerical analyses are useful tools for understanding the stress and deformation status of a CFRD under different load conditions. It can also provide the changing dam stress and deformation trends in different design alternatives. Although we cannot fully rely on the calculated values from numerical analyses for designs, they definitely can provide a good reference point for designers to improve on their designs.

Good numerical analyses results depend on good modeling. For CFRDs, the correct simulation of the interaction between the concrete face slab and the rockfill is very important. Moreover, the constitutive model of rockfill material, simulation of various joints, and correct consideration of construction steps and water load play a fundamental role in the analyses. Although, the present modeling falls short of perfection, it can really represent the main characteristics of CFRDs during construction and operation.

With the development of CFRD technology, more challenges will be faced in future design and construction. Formal experience and design criteria may need reconsideration for the very high CFRDs or those built in more complicated conditions. For future CFRDs under complex engineering conditions, numerical analyses methods will be developed continuously by meeting such challenges.

11.7 NUMERICAL ANALYSES APPLIED TO BRAZILIAN CFRDs

In Foz do Areia, the first CFRD built in Brazil, finite element analyses were developed early on to support the design of the dam.

Among the first published works on numerical analyses applied to CFRDs in Brazil, a paper by Peixoto, Saboya Jr. and Karan (1999) discusses the relative displacement between the slab and the rockfill at Xingó Dam during its construction. Figures 11.31 and 11.32 show the predicted relative displacements in the horizontal and vertical directions between the concrete slab and the rockfill.

A paper by Saboya Jr. (1999) presents the predicted and the observed vertical and horizontal displacements in Segredo CFRD (see Figures 11.33 and 11.34).

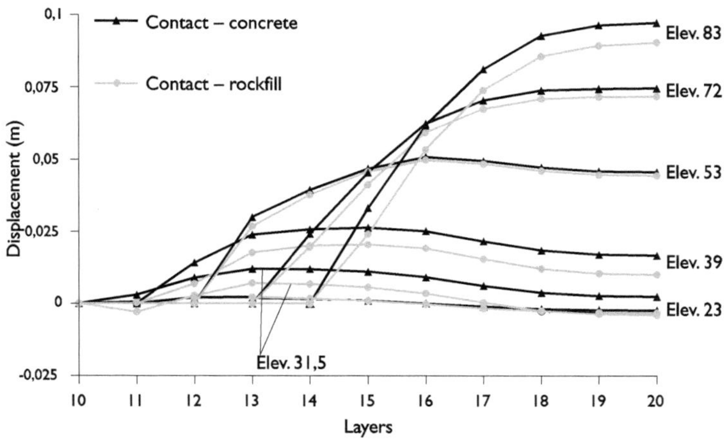

Figure 11.31 Xingó CFRD: horizontal displacements – slab-rockfill (Peixoto, Saboya Jr. & Karan, 1999).

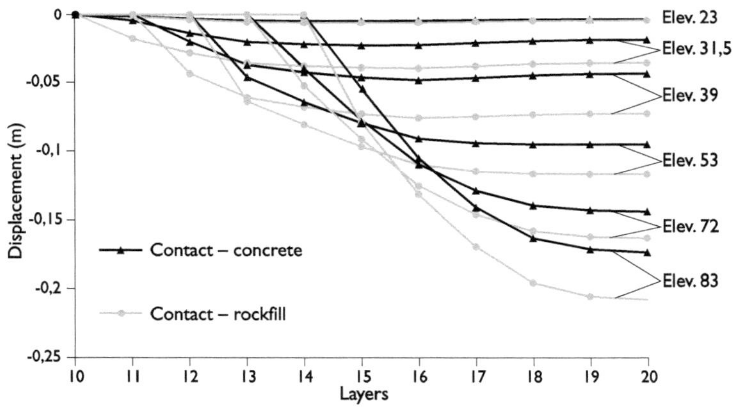

Figure 11.32 Xingó CFRD: vertical displacements – slab-rockfill (Peixoto, Saboya Jr. & Karan, 1999).

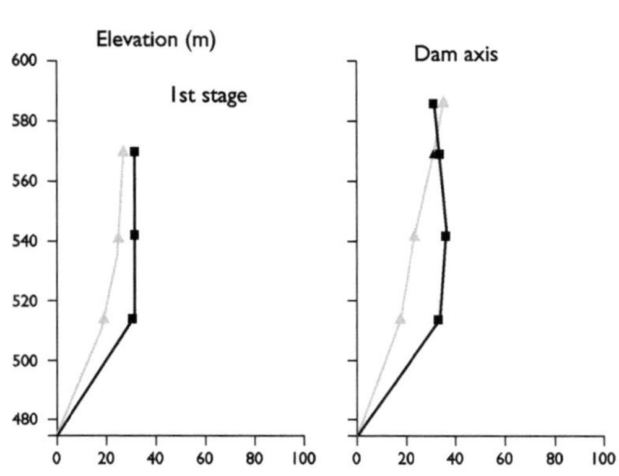

Figure 11.33 Segredo CFRD: predicted and observed vertical displacements (Saboya Jr., 1999).

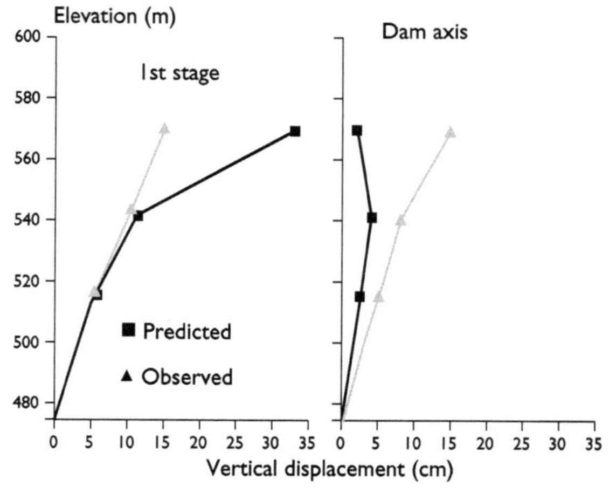

Figure 11.34 Segredo CFRD: predicted and observed horizontal displacements (Saboya Jr., 1999).

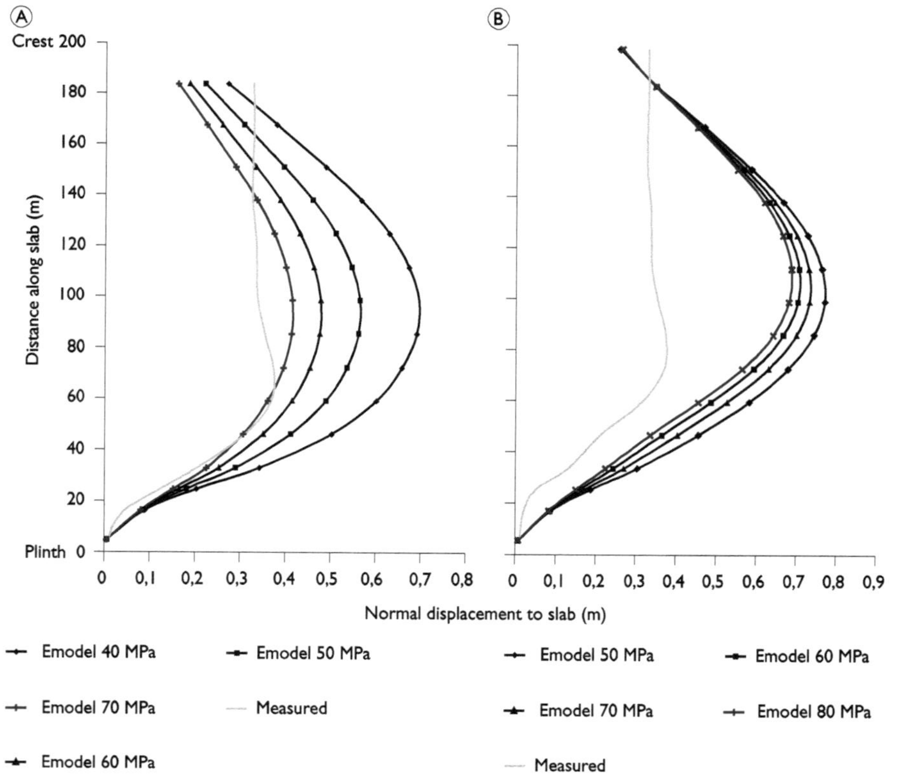

Figure 11.35 Machadinho CFRD: (A) modeling for E from 40 MPa to 70 MPa in different dam zones; (B) modeling for E from 50 MPa to 80 MPa in different dam zones (Oliveira, 2002).

Predicted and observed displacements of the concrete face of Machadinho Dam are discussed by Oliveira (2002). Predictions were made for different combinations of the compression modulus. One can observe some agreement between predictions and measurements for the lower third of the slab. From that point on the observed values deviate significantly from the predictions. Displacements of Itá and Xingó slabs some years after the reservoir impounding show the same tendency of the Machadinho Dam (Fig. 11.35).

Basso (2007) and Basso and Cruz (2007) show that from the construction period to the reservoir filling, rotations of the principal stresses occur on the upstream shell and even an initial reduction of the deviator stress during the beginning of the impounding

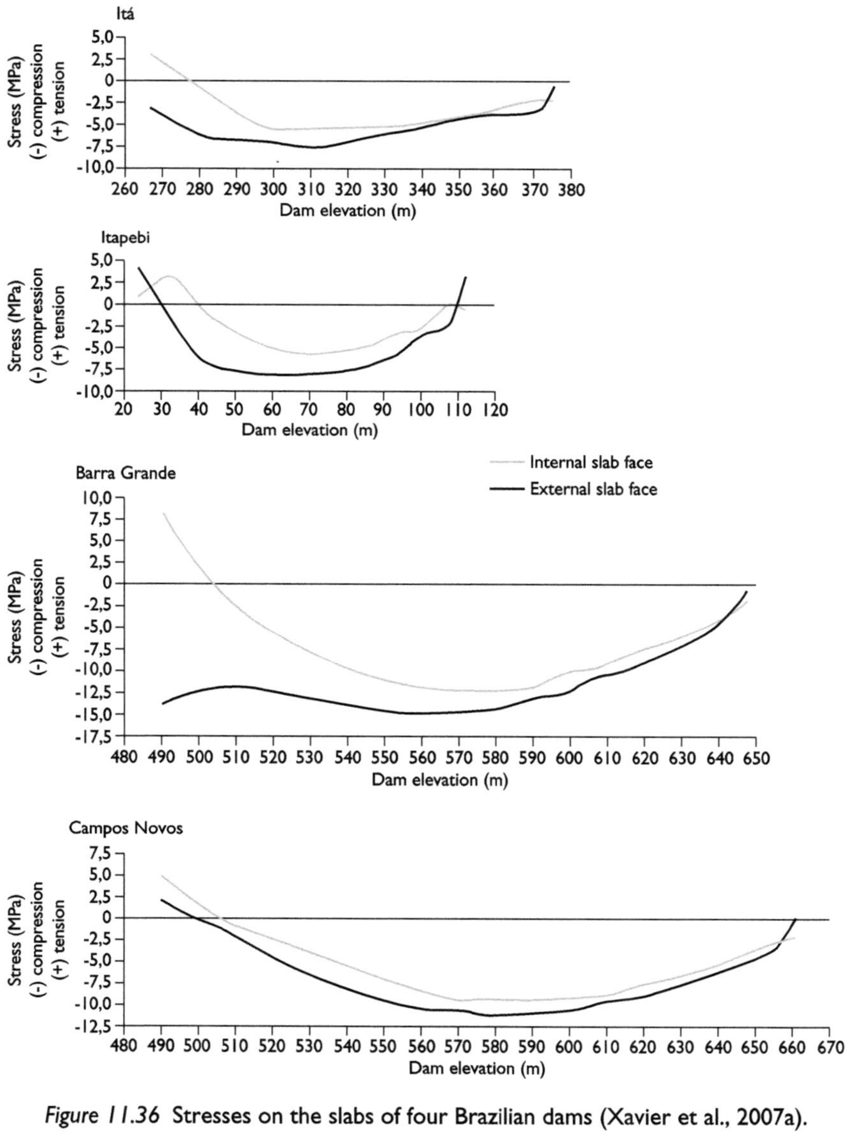

Figure 11.36 Stresses on the slabs of four Brazilian dams (Xavier et al., 2007a).

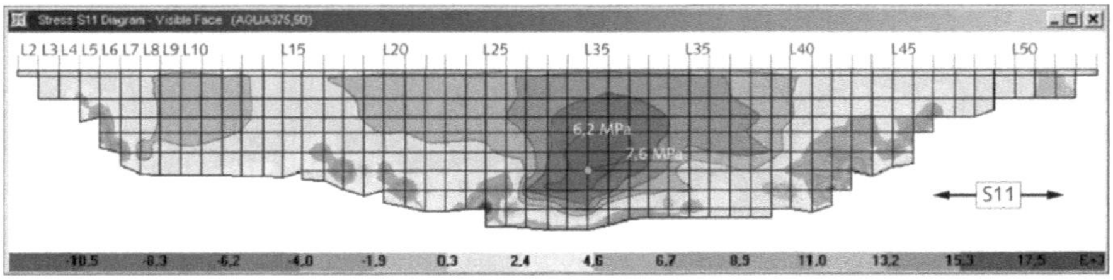

Figure 11.37 Itá CFRD: longitudinal stress diagram in the concrete face (Xavier et al., 2007a). (See colour plate section).

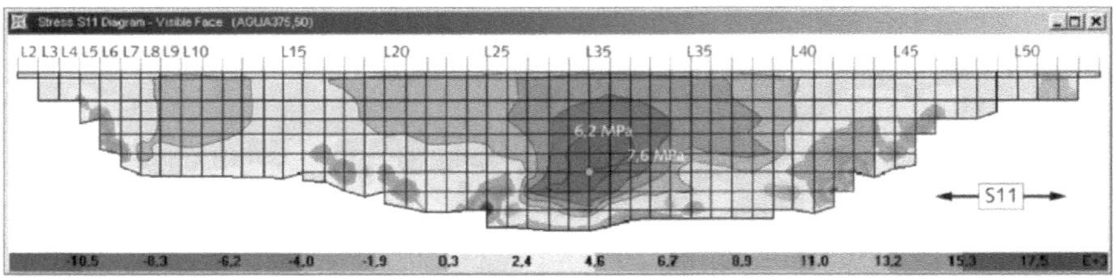

Figure 11.38 Itapebi CFRD: longitudinal stress diagram in the concrete face (Xavier et al., 2007a). (See colour plate section).

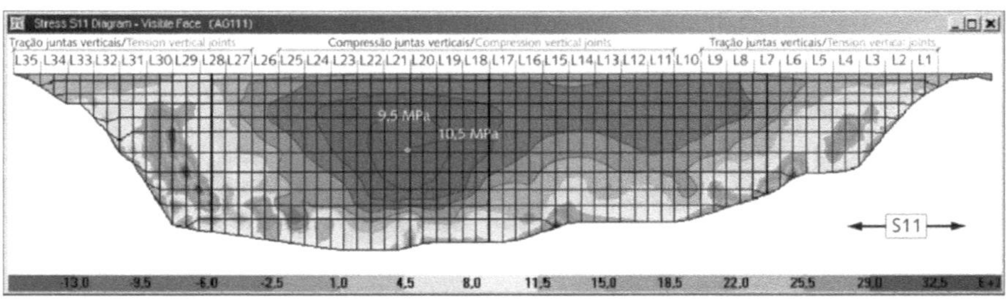

Figure 11.39 Campos Novos CFRD: longitudinal stress diagram in the concrete face (Xavier et al., 2007a). (See colour plate section).

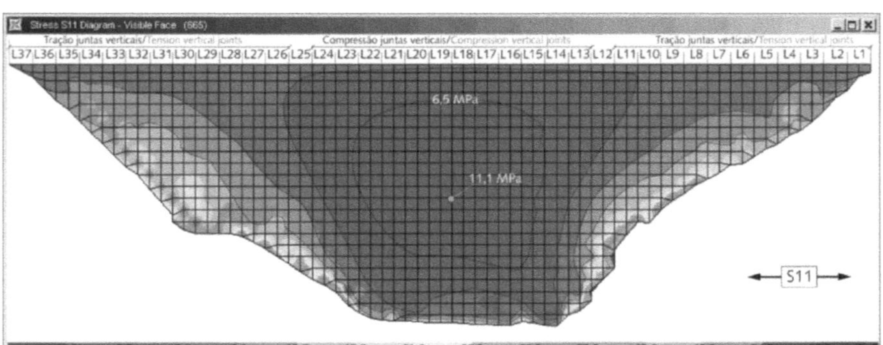

Figure 11.40 Barra Grande CFRD: longitudinal stress diagram in the concrete face (Xavier et al., 2007a). (See colour plate section).

Table 11.2 Predicted compressive stresses on slab.

Characteristics	*Itá*	*Itapebi*	*Campos Novos*	*Barra Grande*	*Hongjiadu*
Height (m)	125	121	202	185	179.5
Length (m)	880	583	590	665	428
Face area (m^2)	110,000	70,400	10,4600	10,6000	74,751
A/H^2	7.04	4.81	2.56	3.15	2.32
Max. compressive stress (MPa)	7.5	11	11	14.80	8.90

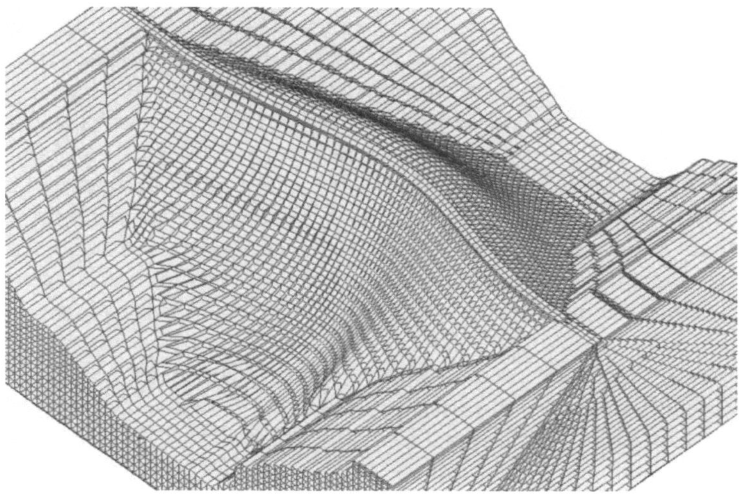

Figure 11.41 Campos Novos Dam: mathematical model 3D (Xavier et al., 2007). (See colour plate section).

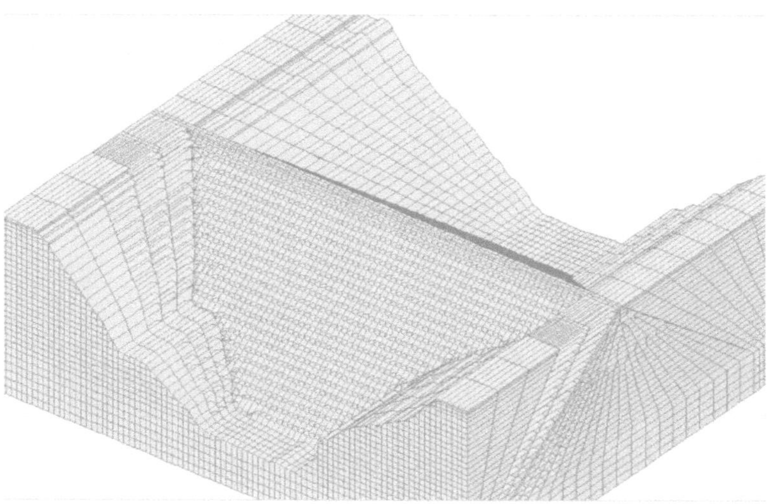

Figure 11.42 Campos Novos Dam: deformed mathematical model (deplacements after reservoir filling) (Xavier et al., 2007). (See colour plate section).

can be noticed. A similar conclusion is drawn by Zeping (see section 11.5.5) in his analysis of the stresses in the dam rockfill.

Both bi- and tridimensional models were developed by Xavier et al. (2007a) for Itá, Itapebi, Barra Grande and Campos Novos CFRDs. Those models were focused on the determination of the stresses acting on the concrete face in the inner and outer side of the slab (see Figure 11.36).

Diagrams with the stress distribution in the slabs are shown in Figures 11.37 to 11.40.

The maximum compressive stresses are recorded near the center of the slab. Values varied from 7.5 to 15 MPa. Some characteristics of these dams are summed up in Table 11.2.

Both in Campos Novos and Barra Grande CFRDs, slabs failures occurred due to excessive compression between slabs. Such compressive stresses must have been higher than 21 MPa (the compressive strength of the concrete). These stresses were not foreseen by the numerical analysis. Table 11.2 also suggests that there are no correlations between the predicted compressive stresses and the dam height (e.g. Itapebi and Campos Novos), and the valley shape A/H^2. Even in the Hongjiadu case, reported by Zeping (see section 11.5.5), the maximum compressive stress is lower than it is in other dams, such as in Campos Novos and Barra Grande. In these cases, the actual analyses were not able to predict the failure of the slabs. It seems that improvement in the analyses must be made before further correlations are proposed.

Tridimensional models were also developed by Xavier et al. (2007a) to predict the displacements of the rockfill during construction and reservoir impounding for Campos Novos CFRD (see Figures 11.41 and 11.42).

Chapter 12

Construction features

12.1 INTRODUCTION

This chapter gives a general review of the construction aspects, commented in the *J. Barry Cooke Volume* (Materón & Mori, 2000), for concrete face rockfill dams based on the experience of the authors on Brazilian dams and on observations of international work on this type of dam.

The construction technology for concrete face rockfill dams has developed very fast due to the simplicity, economic procedures, and inherent safety of this type of structure.

The demand for rapid construction as imposed by the new type of contracts (EPC – Engineering Procurement and Construction) has motivated designers and contractors to develop new design techniques and construction methodologies when applicable. All the while, the development of heavier smooth vibratory rollers since 1960 has allowed for the design and construction of high CFRDs as long as there is good compatibility between the compacted rockfill compressibility modulus and the face slab deformations.

This chapter discusses the construction techniques of the different elements that constitute a CFRD, while presenting modern trends in construction and construction outputs for the largest projects in operation or under construction.

Table 12.1 presents a list of dams, in a chronological order, which have contributed to the development of CFRD design and construction over the past 40 years. In most of these dams the authors have participated in the design and in the construction.

12.2 GENERAL ASPECTS

Design concepts and construction methods have been presented and discussed in the following techinical meetings:

a A symposium sponsored by the Geotechnical Engineering Division of the ASCE (American Society of Civil Engineers), Detroit, USA. In that meeting the "Green Book": *Concrete Face Rockfill Dams – Design, Construction and Performance*, was published, as well as the post-conference volumes 112 and 113 in the *Journal of the Geotechnical Engineering Division* of the ASCE, edited by J. Barry Cooke and James L. Sherard.

Table 12.1 High dams in chronological order.

Dam	*Year of completion*	*Country*	*Height (m)*	*Area slab (m^2)*	*Type of rockfill*
Cethana	1971	Australia	110	30.000	Quartzite
Alto Anchicayá	1974	Colombia	140	22.000	Hornfels-schists
Foz do Areia	1980	Brazil	160	139.000	Basalt
Salvajina	1983	Colombia	148	57.500	Gravels, siltstones/ sandstones
Aguamilpa	1993	Mexico	187	137.000	Gravels ignimbrite
Xingó	1994	Brazil	150	135.000	Granite gneiss
Santa Juana	1995	Chile	113	39.000	Cascalho Gravels
Tianshengqiao I	1999	China	178	180.000	Limestone, mudstone
Itá	1999	Brazil	125	110.000	Basalt
Puclaro	2000	Chile	83	68.000	Gravels
Antamina	2002	Peru	109	67.000	Limestone
Machadinho	2002	Brazil	125	77.000	Basalt
Itapebi	2003	Brazil	120	67.000	Gneiss, mica schist
Mohale	2003	Lesotho	145	87.000	Basalt
Barra Grande	2006	Brazil	185	108.000	Basalt
Campos Novos	2006	Brazil	202	106.000	Basalt
El Cajón	2007	Mexico	189	99.000	Ignimbrite
Kárahnjúkar	2007	Iceland	196	93.000	Basalt
Shuibuya	2008	China	233	120.000	Limestone
Merowe	2008	Sudan	53	135.000	Gneiss
Bakún	2008	Malaysia	205	127.000	Greywacke/shale

b A symposium sponsored by the Chinese Society for Hydro-electric Engineering and ICOLD, held in Beijing, China, in 1993 resulting in the publication of three volumes: *Proceedings of International Symposium on High Earth-Rockfill Dams (Especially CFRD)*, edited by Jian Guocheng et al.
c The II Symposium on CFRD sponsored by the Brazilian Committee of Dams (CBDB), Engevix and Copel, held in Florianópolis, Brazil, in 1999 resulting in the book *Concrete Face Rockfill Dams Proceedings*.
d Proceedings of the "Symposium on 20 years for Chinese CFRD Construction", September 2005, Yichang, China, edited by Chen Qian.
e Proceedings of the III Symposium on CFRD, in honor of J. Barry Cooke, October 2007, Florianópolis, Brazil.

Designations and international nomenclature for the CFRD zoning suggested by these references are in chapter 3 and they are strongly recommended.

12.3 PLINTH CONSTRUCTION

The construction of the plinth depends on the dam site's topography. Logistically, the plinth has to be excavated earlier, during the excavation of the diversion tunnels in order to enable construction continuity after the river diversion. However, the shape of the valley and the amount of excavation sometimes restrict the sequence of works.

In wide valleys, as Foz do Areia (160 m), Xingó (150 m) or Itá (125 m), it was a general practice to develop access roads to strategic points of the plinth alignment where construction equipment could be mobilized. It is very common to provide bulldozers, loaders, heavy trucks, backhoes and complementary drilling equipment to carry out the excavations.

In narrow valleys, where the ratio crest/length is less than 3, sometimes it is convenient to build the plinth and the rockfill placing simultaneously. The fill provides access to execute the excavation and concrete operations.

During the construction of the Alto Anchicayá (140 m) and Golillas (125 m) dams in Colombia, where the steep slopes of the abutments limited the construction of access roads, the excavation of the plinth foundation and the fill placing were coordinated simultaneously.

12.4 EXCAVATION

The excavation of the plinth varies according to the type of foundation. Although the general concept is to found the plinth on sound non-erodible groutable rock, there are some examples of plinths founded on lower quality rock – and they will be discussed further on.

12.4.1 Excavation on sound rock

The general practice is that excavations are planned to avoid excessive over break by pre-splitting the rock with holes spaced every 0.60 m.

Generally, the drilling equipment is comprised of air tracks with hydraulic drills or in places with restricted access, jack legs or manual drillers. The material blasted is pushed away by bulldozers or loaded into trucks by either hydraulic backhoes or front loaders.

The general concept is to get long alignments and avoid placing the plinth over high walls. A generalized method is to excavate to the sound foundation and then build a dental concrete to level up the foundation allowing for the installation of anchors and the required reinforcement of the plinth.

Some designs specify that anchors be installed in the excavated rock before placing the dental concrete, but this practice complicates construction. Relaxation of rock in places requires treatment with rock bolts or tendons for adequate stability to the plinth. The dental concrete to level the foundation is of lower strength; although, some designers require the same quality for the dental concrete as they do for the plinth structural concrete.

12.4.2 Excavation in weathered rock

Although the main objective is to place the plinth on sound rock foundation, there are cases where it is necessary to place the plinth over weathered zones. In the places it is important to evaluate the geo mechanical characteristics of the foundation and to adapt the plinth dimensions to the foundation by establishing a conservative gradient, avoiding potential erosion of the foundation.

During the construction of the Machadinho Dam (125 m, Brazil) a stretch of the plinth foundation was founded on weathered acid basalt with open and clay filled joints. The area was duly treated and high walls were built to support the plinth.

The foundation of the plinth of the Berg River Dam (60 m, South Africa) was mostly on weathered rock. The rock was classified by Bieniawski's RMR and adequate dimensions of the plinth were specified to prevent weathered rock erosion. The excavation was carried out with bulldozers, loaders and small trucks.

12.4.3 Excavation in saprolite

There are cases where the plinth design has to overcome the presence of hydrothermal weathering or weak layers or shear zones – avoiding large excavations.

This was the case at Salvajina (148 m, Colombia), where the upper plinth was placed on top of highly decomposed rock containing cohesive soil. In Itapebi (125 m, Brazil) stretches of the plinth were found with mica-schist layers within the foundation gneiss.

There are more cases in the technical literature where local treatments were carried out to avoid excessive excavations. The plinth has been placed on weathered rock and the foundation treated with reinforced shotcrete and covered with fillers to prevent foundation erosion by the expected high hydraulic gradients.

12.4.4 On alluvium

The very well-known experience of locating the plinth of asphalt face dams over alluvium foundations has been successfully applied to CFRDs.

The articulated plinth is placed on compact alluvial materials, duly protected by filters to prevent migration of fines or gravels into the rockfill.

The excavation of the foundation is done after lowering the water table by draining pumps or well points, removing the loose material by loaders and dozers to be loaded and transported away by trucks.

There are many examples of this type of solution in Europe, China and recently in South America. Santa Juana (110 m) and Puclaro (85 m) in Chile, and Potrerillos (116 m) and Caracoles (130 m) in Argentina, are projects built with articulated plinths tied to diaphragms walls to assure an impervious foundation.

Puclaro and El Bato (55 m) in Chile allowed some water seepage underneath the dams since the construction of a diaphragm down to the rock was very expensive.

12.5 CONCRETE CONSTRUCTION

The concrete construction of the plinth varies according to the geometric solution and the possibility of developing adequate access. Figure 12.1 shows different types of plinth.

The conventional plinth has a slab and a head with a face perpendicular to the face slab – as adopted for Foz do Areia (Fig. 12.2), Cethana and other dams on sound rock.

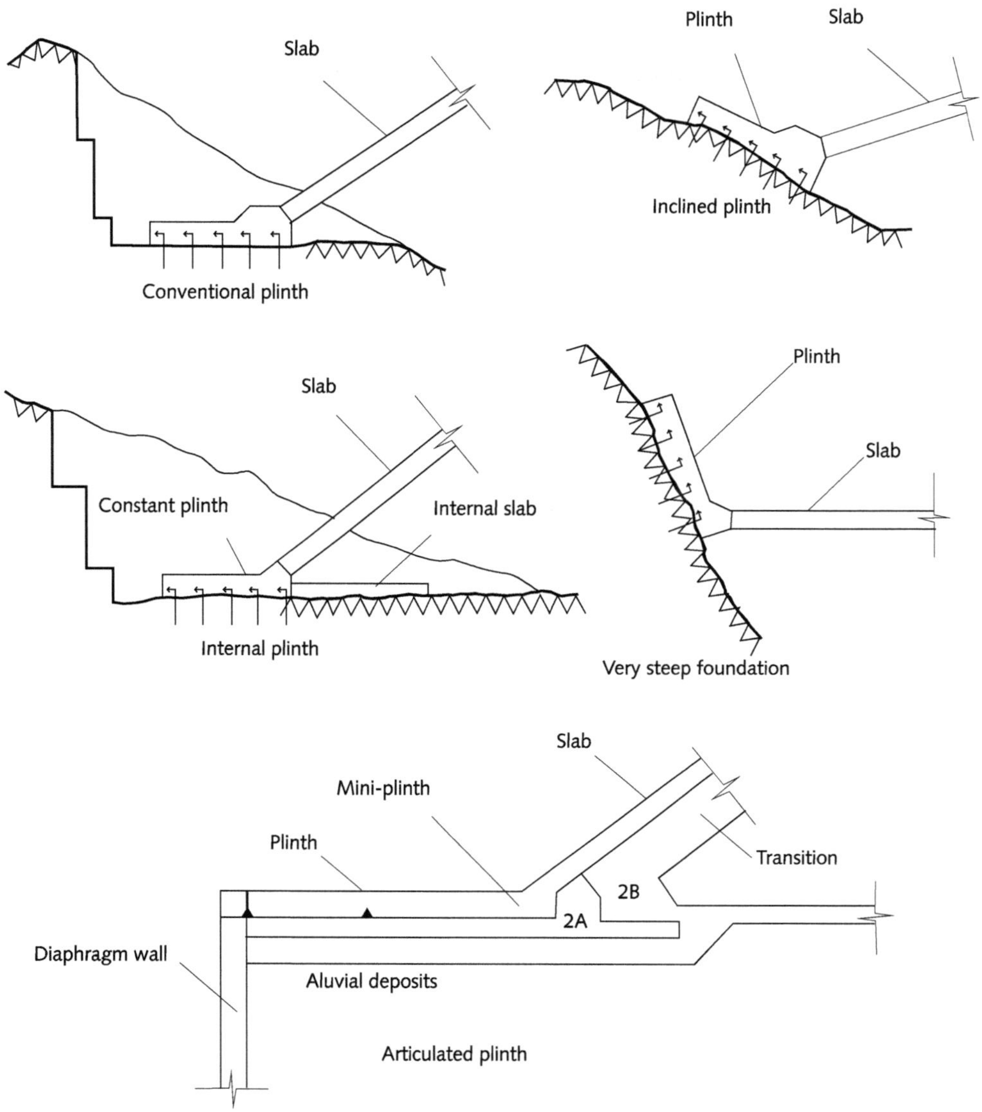

Figure 12.1 Types of plinth.

In some abutments, it is advisable to design the plinth with an external slab of 3 m–4 m and an internal slab to complete the necessary distance to meet the gradient requirement as in Itá and Itapebi (Fig. 12.3), Brazil.

On very steep abutments in narrow valleys, it is common to build a plinth as if it were a wall anchored to the rock, as was the case in Alto Anchicayá and Golillas, in Colombia.

Figure 12.2 Foz do Areia Dam: plinth construction. (See colour plate section).

Figure 12.3 Itapebi Dam: plinth with internal slab. (See colour plate section).

12.5.1 Concrete type

The specified concrete is generally fabricated using pozzolans or pozzolanic cement with strength of 21 MPa at 28 days. However, depending on the operation schedule sometimes the strength is specified for 60 or 90 days.

Concrete is fabricated in conventional batching plants, transported by mixers or truck of 5–6 m^3 capacity, to the place where the forms are. The placing into the form is done by cranes with buckets, or by concrete pumps when there is no direct access.

12.5.2 Forms type

Timber forms are generally preferred for the construction of most plinths. However, in places where the plinth has been designed in long streches, it is convenient to use slip forms (Fig. 12.4) operated by hydraulic jacks on rails, or by winches with cables to climb up the form. In Xingó and Itapebi dams, Brazil, the plinth was built by slipforming.

Figure 12.4 Plinth slipforming. (See colour plate section).

In Merowe, Sudan, where the plinth was many kilometers long, the practical solution adopted was a metallic form with winches inside the form recovering a cable anchored ahead of it.

12.5.3 Articulated plinth

The articulated plinth is designed over alluvial deposits. Usually, depending on the thickness of the river bed alluvium, the articulation is done by two or three slabs as adopted at Santa Juana (Fig. 12.5) and Puclaro dams, in Chile.

The construction sequence is:

- Leveling out the foundation after removing all loose material and preparing the foundation by compacting the alluvium with many passes of a vibratory roller;
- Placing the transition 2A and 2B materials, and compacting and leveling the area;
- Building first the mini-plinth (once the mini-plinth is complete, construction of gravel dam can be continued);
- Simultaneously, placing the guide walls for construction of the diaphragm wall;
- Pouring of the diaphragm wall in panels as an independent job;
- Finally, when the gravel fill is complete, building the intermediate slabs of the articulated plinth.

The construction of the mini-plinth and intermediate slabs will require the same equipment as a conventional plinth.

12.5.4 Diaphragm wall

The construction of the diaphragm wall requires special equipment for the excavation and for the placing of concrete. Depending on the width of the wall (0,80 – 1 m) the construction of guide walls in reinforced concrete (21 MPa) is required. These guide

Figure 12.5 Santa Juana Dam: articulated plinth on alluvium. (See colour plate section).

walls have a thickness of 0.20–0.25 m and a height of 1–1.50 m. They will help to keep the alignment of the diaphragm and to control the plastic concrete level. Furthermore, these walls allow for fixing the tremie pipes during the construction of the diaphragm (Fig. 12.6).

The gap between the guide walls is always a little larger than the theoretical thickness of the diaphragm. The excavation is done with a backhoe.

The excavated material can be re-utilized to give the walls stability before introducing bentonite.

The diaphragm excavation is done with Kelly type clamshells with dimensions to suit the primary panels; generally if the panels are 6.0 m long, the clamshell is 2.50 m.

Therefore, the panel excavation may be carried out in three stages, two of 2.50 m and one central of 1.0 m overlapping the two side panels.

Stabilization of the excavation is done by placing a bentonite cake and removal of material simultaneously.

For the bentonite cake preparation it is a general practice to use mixers and pumps to deposit it in ponds close to the diaphragm excavation. Once the excavation is done in the bottom, penetration into the rock is carried out using "chisels" or special steel equipment as shown in Figure 12.7.

The diaphragm may be built using conventional concrete, reinforcing the upper first 6–10 meters, or by plastic concrete, whichever suits better the alluvium deformation during the fill construction and during reservoir filling.

Figure 12.6 Typical guide walls for diaphragm construction. (See colour plate section).

Figure 12.7 Chisels for rock penetration. (See colour plate section).

Pouring of concrete in the excavated panel is performed with equipment as follows:

- 1 Batching plant mixer trucks as required;
- 2 Tremie concrete pipes;
- 4 Funnels for unloading the concrete;
- 1 Crane;
- 1 Pump for circulating the bentonite cake;
- 2 Laboratory as required;
- 1 Foreman;
- 8–10 Workers as required.

An alternative method of excavation is provided by a heavy hydrofraise, as at Merowe (Sudan). This equipment is able to excavate into the rock, penetrating 5 m down and encasting the diaphragm into the rock foundation (Fig. 12.8).

Figure 12.8 Hydrofraise. (See colour plate section).

12.5.5 Grouting

One of the main advantages of a CFRD is that grouting is an operation that does not interfere with dam construction, as it is done outside the dam. The operation should not affect the overall construction schedule.

Once the plinth is built, the grouting operation may start by using the plinth slab as a grout cap. Drilling for grouting is normally carried out with percussion equipment. When the plinth is horizontal, various drilling equipment may operate simultaneously as illustrated in Figure 12.9.

When the plinth is inclined, the drilling equipment moves over the plinth by means of mechanical tirfors or electrical winches operated from the higher point of the plinth alignment. During the construction of Messochora, 150 m, in Greece, the drilling equipment was mounted on rails and mobilized by winches thanks to the straight alignment of the plinth. Figure 12.10 shows the alignment of the plinth and drilling equipment.

In Itá (Brazil), the drilling equipment was mounted on platforms which kept the drilling machine level. The equipment was pulled up by a system of cables and pulleys. Figure 12.11 shows the equipment for grouting used in Foz do Areia.

In narrow valleys, where the plinth is steep or nearly vertical, drilling is done by jack legs or jack hammers using the fill as a close platform.

Recently, the GIN method for grouting has been adopted for some concrete faced dams, such as Aguamilpa, Pichi Picún Leufú, and Corrales in Chile, and Mohale (145 m) in Lesotho. Grout is mixed in containers and pumped into the drill holes by using Moyno or displacement pumps.

Some of these pumps may be supplemented with automatic devices that stop the pump when the product of the grouting flow and the pressure reaches the hyperbola of the selected GIN number. In Pichi Picún Leufú, Jean Lutz equipment was installed to automatically control grouting injection.

Figure 12.9 Merowe Dam: drilling machines for plinth grouting. (See colour plate section).

Figure 12.10 Messochora Dam: drilling equipment for grouting. (See colour plate section).

Grouting has been carried out by two methods: grouting from the top downwards and re-drilling the grouted zone to continue the next stage; and grouting from the bottom up, using calibrated packers to control grouting into the selected length, of approximately 3 m.

Figure 12.11 Foz do Areia Dam: drilling equipment on timber platform.

12.6 RIVER DIVERSION

Diversion of the river can be carried out by two classical methods:

a by tunnels;
b by encroaching the river to one abutment so a concrete structure (spillway) can be built, through which the river will be later diverted.

For the first case (tunnels), the diversion of the river can be optimized in the dry season, by taking calculated risks about the timing of the next flood season.

Rockfill embankments allow overtopping, if adequate protection measures are provided. On smaller rivers, overtopping protection has been designed by means of reinforced rockfill (Cethana, Australia) and gabions placed downstream, as described in Australian projects.

On larger rivers, the diversion is designed by defining a strategy to reduce the size of the diversion tunnels. It requires some rockfill stages within the dam to prevent any overtopping during construction.

At Xingó (Brazil) it was decided to build the dam with an internal cofferdam that could be overtopped, so that the river would flow over the dam in the next wet season after diversion. A downstream protective layer of roller compacted concrete was provided for occasional flows larger than 10,500 m^3/s. This flow value corresponded to a turnover period of 180 years. Then, a diversion strategy was planned for future stages of the dam construction to give a protection against a 500 years recurrent flood.

In TSQ1, the rockfill embankment was flooded several times during the first wet season as mentioned in chapters 3 and 6. Protection measures were adopted to ensure safe overtopping, such as reinforced rockfill on side walls and slopes, and prior flooding of the space between the downstream cofferdam and the dam itself. Downstream and upstream cofferdams had very flat downstream slopes that were covered with concrete slabs. The overtopping was successful with only minor damage to the downstream cofferdam. Figures 12.12 and 12.13 show aspects of this overtopping.

A very interesting aspect of the Xingó project was the river closure by using three instead of the conventional two rockfill cofferdams to diverte the river into the tunnels. The very high water head loss between the upstream and the downstream cofferdams required a third dike, located at the dam axis. The dumped underwater

Figure 12.12 TSQ1 Dam: gabion protection for overtopping. (See colour plate section).

Figure 12.13 TSQ1 Dam: rockfill overtopping. (See colour plate section).

rockfill of this dike was allowed to be incorporated into the embankment due to its adequate location.

12.6.1 Diversion strategy

Construction of the dam has to be adjusted to the selected diversion strategy to avoid fill overtopping. The river diversion is designed by reducing the size of the diversion tunnels and height of cofferdams to allow a priority section to be built inside the dam, so that during the next rainy period after diversion, the dam section can face floods of turnover periods of 300 to 500 years.

12.6.2 Priority sections

The priority section can be built by raising up the upstream portion of the dam as planned for Aguamilpa (Mexico), Foz do Areia and Segredo (Brazil), and TSQ1 (China), or by building up an internal portion of the dam as carried out at the Itá and Machadinho projects (Brasil).

The main cofferdam controls overtopping risk during a dry period. For the first rainy period a fuse channel, or any other device, is built to prevent losing the main cofferdam to overtopping. The priority section inside the dam will prevent any overtopping over the main rockfill, which could be catastrophic.

To avoid high floods passing directly through the main rockfill, a semi-pervious transition material is placed on the upstream slope of the priority section when it is built downstream. This transition will provide a restriction on the hydraulic gradient to safely control any flow through the rockfill. Later, this transition cover should be removed in order to avoid any undesirable water impoundment inside the rockfill situated upstream of the priority section. If this is not done, the water may induce additional post-construction differential settlement.

Logistically, it is always better and more economical to first build the plinth and then build an upstream priority section using only the transition material (2B) which, being semi-pervious, provides a gradient restriction to effectively control flows through the main rockfill. In today's dams, this 2B material is protected by the extruded curb. However, in some dams, poor construction access or local topography prevents the initiation of the plinth ahead of diversion. These projects adopt an internal priority section. Figure 12.14 presents diversion stages in four dams, indicating the selected priority sections.

12.6.3 Stages

According to the shape of the valley, it is always possible to create a strategy to divide the dam construction into stages. Adequate protection to the fill and economical rockfill placing is provided by this type of dam construction schedule, as shown in Figure 12.15.

This is one of the most significant CFRD issues. High dams, such as Foz do Areia, Segredo, Mohale, Itá, and TSQ1, have been built in stages during their construction.

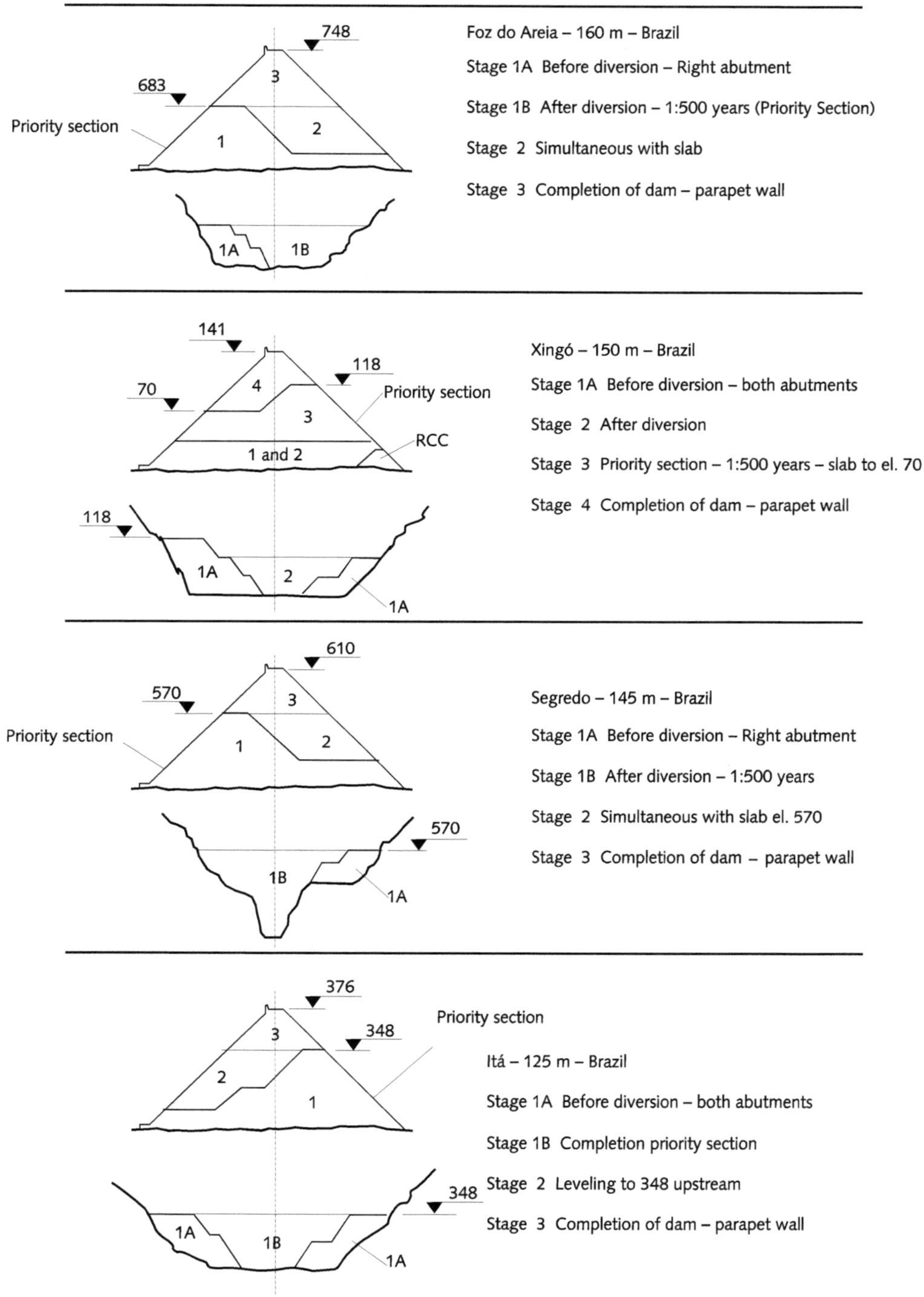

Figure 12.14 **Diversion stages of four dams, indicating the selected priority sections.**

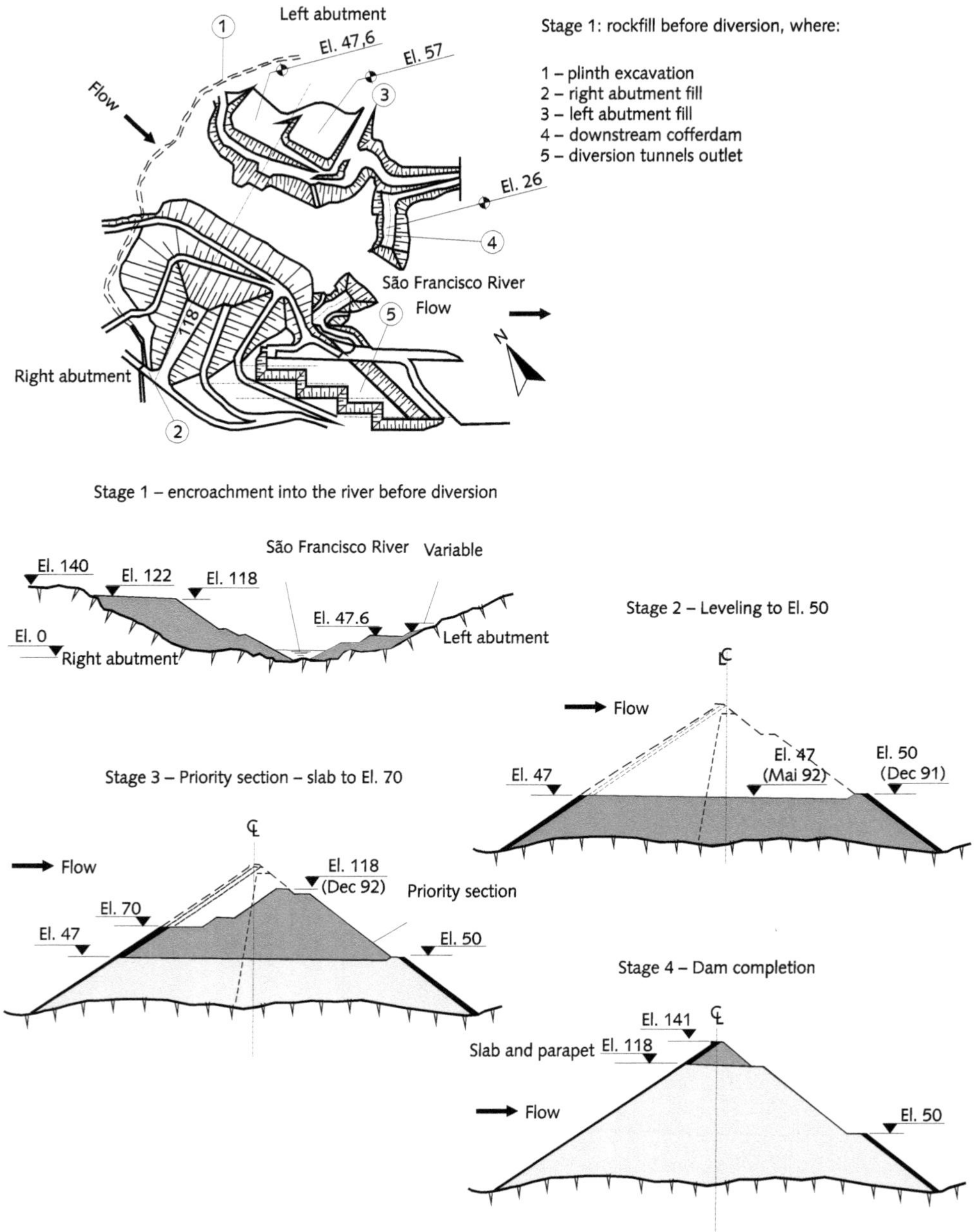

Figure 12.15 Xingó CFRD: construction strategy in different stages.

Figure 12.15 depicts the stage strategy adopted in Xingó (150 m, Brazil), where it was possible to build the dam as follows.

Stage I – A substantial volume of rockfill was placed on both abutments which had been obtained from the excavations for the power intake, powerhouse and diversion structures before diversion. This stage and the construction of the four diversion tunnels were built simultaneously.

Stage II – After river diversion, the dam was raised up to the same level of the cofferdam; a RCC protection was built in case any flood overtopped the cofferdam and the embankment during the first rainy period after diversion.

Stage III – The priority section was heightened to El. 118 m. The upstream rockfill was raised to El. 70 m in order to allow for the simultaneous construction of the face slab and the rockfill placing, aiming to complete the priority section to withstand a 500-year recurrent flood event.

Stage IV – Completion of the dam up to the crest.

In Aguamilpa (187 m, Mexico), the main cofferdam was designed for a frequency of 1:100 years. After diversion, a river flood (10,800 m^3/s) overtopped the main cofferdam. A lateral fuse was broken to avoid losing the cofferdam. Since the main fill was high enough to force the river floods into the diversion tunnels, the dam was not overtopped and construction continued. Aguamilpa did not have extruded curb and the 2B material was somewhat damaged (Castro et al., 1993).

12.6.4 Scheduling

The construction schedule for a high CFRD is related to the specific site and to the construction stages needed to complete the structure.

For conventional layouts, as shown in Figure 12.16, where the river is diverted through diversion tunnels and it is feasible to place some rockfill before diversion, the dam construction may be divided in the following phases:

A – Mobilization, accesses, abutment stripping, diversion construction, plinth construction above river water level, and fill before diversion.

B – Diversion of the river, cofferdams, riverbed excavation, dewatering, rockfill placing and plinth construction on the river bed, initiation of priority section, grouting.

C – Completion of priority section, first stage slab construction, downstream rockfill placing, completion of grouting.

D – Completion of the fill up to crest, final stage slab construction.

E – Parapet and complementary fill.

In phase *A*, it is extremely important to develop full access for dam construction and to start the excavation of the plinth above river level, and its concrete placing before the diversion.

Delays in plinth construction always affect the general schedule of the dam. The plinth is built above the maximum water level of the river. In wide valleys, as the

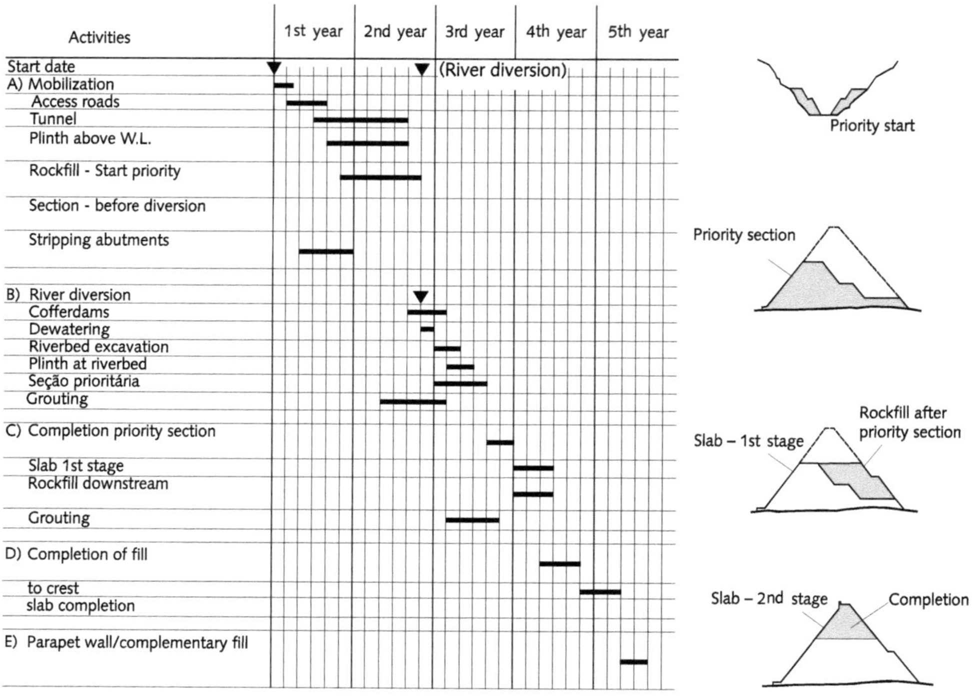

Figure 12.16 CFRD typical construction schedule.

Brazilian sites, it is advisable and economical to initiate placing rockfill over the abutments before diversion.

In phase *B*, the river is diverted through the tunnels by building corresponding cofferdams. After river bed dewatering it is possible to initiate river bed excavation in order to complete the plinth over it. Simultaneously, the priority section is initiated by placing rockfill 20 m–30 m downstream of the plinth. Grouting may start as standalone activity.

In phase *C*, the priority section is completed. The first stage of the slab construction starts along with placing the downstream rockfill in a simultaneous action to leveling the dam to the same elevation of the slab. Grouting then proceeds.

In phase *D*, completion of the fill is carried out and it becomes the parapet foundation, and then the remaining slab is built. It is important to bear in mind that during phases *B* and *D* the transition materials are to be placed. These materials are processed and it is strongly recommended to have them stockpiled to avoid limiting the progress of the dam. There are many cases in CFRD construction where the lack of transition material has delayed the rockfill placing in the dam.

Finally, during phase *E*, the parapet and complementary rockfill are executed. Construction of high parapets are sometimes time-consuming operations. Premolded parapets reduce the completion time for the dam, as experienced in Pichi Picún Leufú (Argentina) and Itapebi (Brazil). (see Figure 12.17).

Figure 12.17 Itapebi Dam: precast parapet. (See colour plate section).

12.7 EMBANKMENT CONSTRUCTION

Generally, all kinds of granular fills have been used in the construction of CFRD by adequate zoning within the dam body.

12.7.1 Types of fill

Types of fill vary from fresh rockfill coming from the excavation for other structures to gravel deposits or combinations of material. Gravels and rockfill have been placed in alternate layers in Santa Juana (Chile) and Caracoles (Argentina).

12.7.1.1 *Rockfill*

Rockfill may come from quarries or from excavation of the structures. This might be fresh well-graded sound rock, or quite uniform hard rocks lacking fines, such as the basalt used in Brazilian dams or in Mohale (Lesotho).

Rockfills from soft rock have been used at the downstream portion of the dam (zone 3C) or at the central zone (*T*) where the rock is placed far away from the concrete face slab.

Rockfill use has been widely discussed in the technical literature. Theoretically, all types of rock can be used in a CFRD, by optimizing their location in the dam section, with the less compressible material at the upstream shoulder.

12.7.1.2 *Gravel fill*

Experience using compacted alluvial gravels in high dams, as in Aguamilpa (Mexico) or Salvajina (Colombia), has opened up the possibility to extend its use in new dams, such as Santa Juana and Puclaro in Chile and Pichi Picún Leufú, Los Caracoles and Punta Negra in Argentina.

Gravel compressibility, in general, is much lower than rockfill compressibility.

Other non-cohesive materials – as fluvial deposits – are also excellent for the construction of internal zones of this type of dam.

12.7.1.3 *Combined fill*

A combination of rockfill and gravels have been utilized, for financial purposes, in the construction of various projects. Usually, the gravels are placed in the upstream zone and the rockfill in the downstream zone. Since the compressibility of the rockfill may be five to six times greater than that of the compacted gravels, special care is necessary in the higher parts of the dam where its width decreases.

In gravel dams, where the available volume of rock excavation is not significant to justify a separated zone of rockfill from gravels, a combination of both materials have been used.

As commented earlier, during the construction of the Santa Juana Dam, material from the spillway excavation was alternated with gravels in the downstream shoulder. The best compacted gravels were placed upstream. A similar solution has been applied to Los Caracoles and Punta Negra dams in Argentina.

Combination of materials requires an analysis of the differential settlements between the two zones, which may induce cracking of the cushion zone or of the concrete slabs.

12.7.2 Embankment zoning

Construction equipment and construction techniques are defined according to the zoning of the dam. Since each zone is placed with a layer thickness and a number of passes of the compaction equipment, it is important to recall each zone in this chapter. The international nomenclature for the different zones of this type of dam is given in chapter 3.

Zone I

Consisting of random material covering a deposit of silty fine sand, this is located over the perimetric joint upstream of the face slab. The silty fine sand is a migrating element, and in a case of a waterstop failure it clogs the open joint. The former practice of using clay material has been abandoned, as this cohesive material keeps any gap between the earth fill and the concrete face slab open.

The granular material volume is small, because it is not meant to be an upstream blanket, but just a deposit for clogging an open joint or incidental fissures of the slabs.

This zone is divided in two sub zones.

- Zone 1A – This is located over the perimetric joint. It is generally alluvial silty fine sand, coal ash or fly ash. Its volume is small, as such material is needed only for filling the open joint or adjacent slab fissures.
- Zone 1B – This is random material to confine and protect the material of zone 1A against washing by runoff water during construction. A dumped random fill at the river bed and sand bags along the abutments are satisfactory ways of placing zone 1B.

Recently, in very high dams, this material has been placed to protect the lower zone of the face slab.

Zone 2

Consisting of filters located under the face slab, this is divided in two subzones.

- Zone 2A – This is located underneath the perimetric joint. It is made of processed material with a filter gradation in order to retain migrating silty fine sand through the joint, clogging it and restraining any leakage.
- Zone 2B – This is the cushion zone for the face slabs. It is also made with a processed material located under the main slab with a maximum size of 3 to 4 inches. The function of this material is to provide support for the face slab and control seepage in case of occasional leakage through the face slab joints, or during construction when accidental overtopping of the cofferdam may occur, as in Aguamilpa (Mexico). With the development of the extruded curb this cushion transition is well protected against erosion by rain.

During construction, deformation is continuously taking place. There is no need to replace the design offset line for the 2B zone. TSQ1 CFRD might have been the single case where the offset line was replaced by manually spreading and compacting additional material.

Zone 3

This is a rockfill zone that embodies the main portion of the dam.

Zone 3A generally is a transition zone located between zones 2B and 3B. The material sometimes is processed but usually is the finer rockfill selected at the quarry and especially stockpiled for this purpose. It is spread and compacted in layers of the same thickness as the zone 2B material.

Zone 3B constitutes the main rockfill, compacted in layers of a maximum thickness of 0.80–1.00 m depending on the vibratory roller selected. It is located immediately downstream of zone 3A with a variable downstream boundary depending on the rockfill grading, height of the dam, and design specifications. It is common practice to place zone 3B material at least over a third of the width of the cross section. Recently this zone has been widened in a downstream direction beyond the dam axis.

Zone 3C is built with the same material or with a more weathered rock, compacted in layers up to 1.2 m high. Softer rock and larger rock blocks are accepted. It is located downstream the dam axis; or, in more recent dams, downstream of zone 3B or *T*.

There is occasionally a "dead zone" between zones 3B and 3C, called the "*T*" zone where materials of poorer quality are placed, improving the rock balance and providing greater economy for the project. This was beneficial the case of the Xingó and TSQ1 projects.

Although in some high dams there have been layers built up to 2 m thick, the authors recommend a lower maximum thickness within the values mentioned earlier.

Zone 4

Zone 4 is a rockfill of over sizes that it is placed on the downstream slope. Generally, the material is pushed to the slope and arranged to give a pleasant appearance to the slope. Good examples are Foz do Areia and Xingó dams, and recently, in Bakún Dam (Fig. 12.18).

Figure 12.18 Bakún Dam: downstream slope. (See colour plate section).

Drainage zone

When the rockfill has plenty of fines and thus lower permeability, provision of a drainage zone is required. The most suitable location for the drainage layer is approximately the dam axis.

Due to construction loading, the center of the rockfill section settles more, and water from rainfall tends to penetrate the dam and seep into the center section, even after the face slabs are built and the reservoir impounded.

Each compaction layer has more fines at the top, which can force the water path towards the center. This can cause additional undesirable settlement. Additionally, the downstream foundation surface should be covered by a sounder uniform and pervious rockfill.

Porce III, built with deformable schists with fines, has a drainage zone at the dam axis connected to a processed horizontal filter.

12.8 FILL CONSTRUCTION

Construction techniques are briefly reported here, and complemented with comments on new practices observed during the execution of recent dams.

12.8.1 Placing layers

Placing layer thickness varies between 20 cm to 50 cm for zones 2 and 3A, and from 0.80 m to 1.20 m for remaining rockfill zones. Very high dams, with uniform grading, should be compacted with 0.80 m thick layers.

12.8.2 Compaction

The silty material located in zone 1A is simply dumped or slightly compacted in layers of 0.20 m to 0.30 m by means of vibratory plates and confined against the dam slab by the *1B* material, a random material spread and compacted by the construction equipment (20 ton trucks and bulldozers).

Zone 2A, located under the perimetric joint, is placed in layers of 0.20 m and compacted by manual vibratory compactors. On the slope face it is also compacted by vibratory plates mounted on a hydraulic backhoe. In some projects, an additional 3% to 4% of cement gives some cohesion and stability to the material during the removal of the waterstop protection.

Zone 2B material is generally compacted in layers of 0.30 m (gravels) and up to 50 cm for processed rockfill. With the development of the extruded curb, commonly called the "Itá Method", the need for compaction on the slope was eliminated.

The typical construction equipment used for transportation, placement, and compaction of zone 2B materials in high dams must be able to handle large volumes. In general, the transportation of processed materials to the dam is accomplished in 20 to 25 ton articulated trucks that are loaded by front-end loaders with a capacity compatible with the transportation units. It is useful to create a stockpile of material containing 20% to 30% of the total volume to be placed, since the demand for this material increases at higher elevations. Graders and/or bulldozers are also used to spread materials, as well as 6 ton vibratory rollers for compaction.

When the material is processed from a quarry lacking fines, as the basalt found in southern Brazil or South Africa, the gradation may not be exactly as defined by Sherard. In recent dams, such as Itá, Machadinho, Mohale, Antamina, Bakún and Kárahnjúkar, an extruded curb is built to eliminate face compaction. This method is explained below.

Placing of 2B material (cushion material) has lead to segregation in some dams, because of the available type of material and the construction method adopted. A maximum size reduction and the addition of sand in rates between 35% and 55% has improved placing conditions. Nevertheless, there is still a tendency to segregation during unloading and spreading operations.

Segregation normally occurs when:

- gravel is used, for its particles tend to roll more easily because of their roundness;
- the material is unloaded from trucks near the edge of the upstream slope;
- the material is unloaded from trucks in a position perpendicular to the axis of zone 2B (cushion material).

In Alto Anchicayá, Salvajina, Cirata and other dams, erosion of 2B material was occurring frequently. The problems observed in the dams motivated Itá's contractor to develop a method for the transition 2B placing, assuring the required protection in order to avoid raveling and to minimize segregation and erosion of the upstream face.

This method, known as the "Itá Method", consists of building an extruded curb before the construction of a layer of the 2B material (Fig. 12.19). The method has been described in technical literature. Its use, with complementary innovations, has been applied to recent new dams with construction advantages such as:

- improved segregation, raveling and erosion control, since the protected face is more resistant to heavy rains;
- reduced activities on the face, resulting in economy and safety;

Figure 12.19 Extruded curb construction. (See colour plate section).

- production is not affected as demonstrated in Itá, Itapebi, Machadinho and other modern high dams;
- reduced amount of equipment to stabilize the face;
- easier application of the reinforcement and construction of face slab.

Details of mix design and equipment is given in some technical references and explained in this chapter in the section on slab construction.

Zones 3 using rockfill are built by unloading the trucks near the edge of the layer under construction, pushing the material with a tractor D8 or similar, and allowing segregation of the material in such a way that the coarser particles move to the bottom and the finer are kept on the surface. This method provides a favorable horizontal internal permeability and provides a surface that does not lead to excessive wear of the tires of the trucks transporting the loads. Layer thickness may vary between 0.80 m for zone 3B and 1.20 m for zone 3C. Zone 3A is compacted in layers similar to 2B (0.20-0.50 m).

Zone 3B using compacted gravel is built in a similar way, but the layers are less thick (0.60 m). The segregation along the layer is not noticiable as much as it is in rockfill, although the resulting horizontal permeability is also greater than the vertical permeability.

The equipment used for rockfill production depends on the source of material, transportation distances and volume of the dam.

A proper rock balance is essential to optimize the costs of the project, reducing quarry demands and controlling material waste.

In dams that have required large volumes of material, such as Foz do Areia and Segredo (Brazil), and Bakún (Malaysia), in which the material has originated from the excavation of several structures, it was convenient to use mobile loading units (front end loaders), D8 bulldozers and 25 t to 35 t transportation units. The modern trend is to use larger stationary crawler hydraulic loaders when the volumes to be excavated are high.

A typical fleet for high rockfill production consists of the following:

- stationary or eletromechanical cranes;
- hydraulic crawler loaders (≥3 m^3 bucket capacity);
- 35 t to 50 t trucks, eventually 75 t;
- bulldozers (320 HP);
- light wheel loaders (CAT 966 or similar) for segregation correction;
- graders;
- vibratory compactors 6 t, 12 t, 18 t (5 t/m roller at minimum);
- vibratory plate mounted on hydraulic backhoe;
- extruding machine or 2B equipment as defined earlier.

In projects that specify the application of water to the rockfill, it is important to calculate the number of water monitors according to demand peak. The ideal scheme is to use elevated natural water streams that by gravity may supply the volume of water required. For projects lacking such natural resource, pumps are installed to store water in tanks and thus provide the required water volume.

The specified volume of water depends on the type of rock, and will vary between 10% and 30% of rockfill volume. Water is not specified for gravels, but it is convenient to compact gravels that are slightly wet.

In Merowe, where the plinth is far from the river, showers were used for the application of water.

In mining projects where the industrial process places high demands on the transportation, loading equipment and trucks can became very heavy. Special access for the rock placing into the dam can be needed, as in Antamina (Peru), which will be 235 m high in the second stage.

12.8.3 Ramping

A very convenient aspect of CFRDs is the ease with which ramps can be placed in any direction, therefore reducing the number of access roads from the banks. The dimensions of the ramps depend mostly on the equipment selected, but these are a few rules of thumb.

- The ramps may be built within the fill, with up to 15% slope in any direction. Permanent ramps located on the downstream slope may be up to 12% inclined.
- It is desirable that the changes in direction should be executed with leveled platforms (turning platforms).
- The slopes between ramps may be up to 1.2(H):1.0(V) for rockfill. For gravel, the intermediate slopes should be flatter 1.3–1.35(H):1.0(V) to avoid loosening round particles.

12.8.4 Dumping under water

Rockfill can be dumped into the water in the middle of the dam, since it will result in a minimum displacement during the impounding. In Xingó, rockfill was dumped in 20 m deep water at the lower part of the dam lengthwise, to provide and increase the abutments' placing area (Fig. 12.15). Dumping was carried out before the diversion of the river. During river closure additional rockfill was dumped under water for a third dike. Dumped rockfill was also used at Itá Dam.

The authors of this book believe that it is possible, in high dams, to build a moderate rockfill string (±20 m) by dumping rockfill into the water at the central area to ease diversion, as accomplished in Xingó with success.

This saturated rock will support the highest loads during the building of the rockfill; and, because of its low relative depth compared to the height of the dam, it will be sufficiently compacted. Furthermore, it will result in negligible post-construction settlement before the impounding of the reservoir.

12.8.5 Stage construction

The fill construction flexibility used in CFRDs enables practical and economical construction planning.

- In wide valleys, considerable volume of rockfill can be placed with incorporated ramps, which will help to reduce transportation distances within the dam site. This will reduce of the volume to be placed after the diversion and will help the achievement of optimum dam heights for the economical handling of the river without overtopping (Fig. 12.20).
- CFRDs allow easy placing of material while other activities are being simultaneously conducted.

Figure 12.20 Slopes incorporated in the rockfill. (See colour plate section).

In Xingó, Machadinho, Segredo, Itá, Campos Novos and Barra Grande (Brazil) it was possible to place rockfill in both banks before diverting the river. Figure 12.15 shows the schematic sequence of stages for Xingó Dam construction, typical of some high Brazilian dams.

12.9 SLAB CONSTRUCTION

12.9.1 Surface preparation

In recent years, the construction of the face slab has improved from the traditional design of square panels separated by compressive joints to a continuous placing of the panels by using slip forms. Table 12.1 gives the area and height of some of the highest CFRDs built or under construction.

12.9.2 Conventional slope protection

Preparation of the upstream slope surface depends on the type of 2B material selected and on whether an extruded curb is used. In general, all modern dams have an extruded curb, but some designers still keep to the old method.

In a conventional dam without a curb, protection during construction can be made by either using asphalt, or shotcrete or mortar coating. The surface lining is important to prevent erosion during heavy rainfall and to provide a firm base for assembling the reinforcement and forms of the slab.

When using asphalt protection, the following rules are generally adopted.

- When the material follows the Sherard gradation with high percentage of sand (35–55%) and fines passing Nr. 200 sieve in order of 2–8%, the asphalt treatment is important to minimize the eroding action of the rain. A cutback agent is selected due to its higher penetrating capacity. It is normal to apply a second priming coat before the construction of mortar pads and reinforcement placing.
- When the cushion material is coarser and well graded, the asphalt treatment is carried out using asphalt emulsions. This method was first used in the Brazilian dams, Foz do Areia and Segredo.

The difference in gradation eliminates the need for backhoe excavator or telescopic equipment for cutting the surface. Normally, the material is spread at about 1.5 m to 2.0 m from the edge of the slope, and this zone is later filled with finer materials, as in Foz do Areia and Segredo.

In Messochora (Greece), zone 2B material was stabilized with shotcrete applied by a robot, after the face of the slope had been compacted. Similar methods have been also employed in other dams (Golillas, Salvajina). Shotcrete cracking and buckling demand continuous repair treatment of the face as observed in Messochora.

There are technical references reporting that in China, at Guanmenshan Dam in Lianoning province, compacted cement mortar was employed with satisfactory results.

In staged constructed face slab, open gaps can occur in high dams between the previous slab and zone 2B due to the deformation of the embankment which cannot be followed by the concrete slabs. An open gap was clearly observed in Xingó and TSQ1.

These were 10 cm wide on average and 15 cm at their maximum, reaching a depth of 7 m, at the river bed. In TSQ1 the gaps observed at the top of the first and second stage slabs were filled with a fluid mixture of 90% fly ash and 10% cement. This grout penetrates satisfactorily. In Mohale and Bakún dams gaps were encountered between the extruded curb and the slab, and between the extruded curb and the fill.

12.9.3 Concrete extruded curb

When the extruded curb method is used, preparation of the slope surface is simplified. An extruding machine is employed using a low cement mix with composition as shown in Table 12.2.

The mold of the machine is set to give the same inclination as of the upstream face (1.3(H):1.0(V) or 1.4(H):1.0(V)). The height of the extruded curb varies between 30 cm and 50 cm as explained earlier.

The construction of the curb follows these steps.

- Leveling the compacted layer of zone 2B to have a horizontal surface for moving the extruding machine (Fig. 12.21).
- Building the extruded curb by using a metallic mold with the design height of the layer (usually 0.40 m) and the upstream slope of the face (1.4(H):1.0(V)).
- Using a dry mix for concrete as indicated in Table 12.2.
- Controlling the alignment of the machine by laser equipment mounted on a fixed position on the plinth, or otherwise by topographic survey crew.
- After one hour of initiation, zone 2B material can be spread. The material may be placed by using an open steel dispenser or unloading the material directly from trucks.
- Leveling of the zone 2B material with a grader and compacting with 4–6 passes of the 6–10 ton vibratory roller. At Antamina CFRD (Peru) the curb is 0.50 m high and zone 2B material was spread and compacted in two 0.25 m high layers.

The benefits of this method are:

- little segregation;
- lower losses from material spilling upstream;
- immediate protection against erosion and raveling;
- less construction equipment needed;
- safer method of construction avoiding people working on the upstream face;
- high production rates (two layers per day are built in dams with crest length of 500 m);
- construction equipment is simplified (the extruding machine is low-cost equipment);
- clean work (the face is prepared for rebar placement and construction of the slab reducing excess of concrete).

This method was used for the first time at Itá Dam and has been applied on Antamina, Machadinho, Mohale, Campos Novos, Barra Grande, Kárahnjúkar, Bakún, Porce III, and El Cercado dams, with excellent results from a construction point of view.

Table 12.2 Curb concrete mix.

Cement	70–75 kg/m³
Aggregate ¾′	1.173 kg/m³
Sand	1.173 kg/m³
Water	125 liters

Figure 12.21 Extruded curb machine. (See colour plate section).

12.9.4 Mortar pads

Mortar pads are built to be used as a base for the bottom copper waterstop and the alignment of the slip forms. The distance between mortar pads is defined by the width of the concrete lane.

Several construction methods have been adopted. In Foz do Areia, mortar pads were manually built. Other dams employed special trolleys mounted on wheels and operated by winches. More recently, in Xingó, mortar pads were built with conventional gunite and finished manually. It has also been necessary to reinforce the upper part of the mortar pads with a steel mesh to prevent cracks caused by the load repetitions during reinforcement placing. The mortar ratio used was 1:3 cement/sand.

In Itá, where the extruded curb was used, the construction of the mortar pads was restricted to the tension joints with the bottom copper waterstops. Mortar pads should be placed outside of the theoretical thickness of the slab.

12.9.5 Waterstops

The most recent dams have eliminated the central PVC waterstop, but have kept the bottom copper waterstop in the perimeter joint and main slab joints. The tendency is to place a self-healing fine sandy material on top of the perimeter joint and to install a neoprene or EPDM mushroom shaped waterstop or the self-healing soil on top of these vertical joints, which tend to open with reservoir impounding.

Copper waterstops are industrially fabricated by means of rolling mills, similar to the one employed for the first time in Salvajina (Colombia).

At Xingó, a machine operated by the same hydraulic jacks that operate the plinth slip forms was used to produce the copper waterstops. Measuring from 12 m to 14 m long, the waterstops are transported to the dam site for welding. In Aguamilpa and Pichi Picún Leufú, machines with disks were used to gradually mold the waterstop until the final shape was achieved. At TSQ1 and Antamina, similar manufacturing mills were adopted.

At Itá, copper waterstops were used only in the perimeter joint and in the vertical joints close to the abutments, which work in tension. In the compression joints located at the highest sections of the face slabs, Jeene mushroom shaped neoprene joints were placed on top of the joints. As this was being used for the first time, exhaustive testing in laboratory was successfully carried out to pressures equivalent and even higher than the expected hydrostatic pressure from the reservoir impounding. The samples were stretched to the same magnitude of the predicted deformations.

Figure 12.22 illustrates the external waterstops used in Merowe Dam (Sudan).

It is important to protect the copper waterstop with metal or wood boxes to prevent damage during construction of the fill. The Messochora project adopted a metallic protection which is easy to install and remove. Figure 12.23 shows a wood protection.

A new type of corrugated waterstop is used in China. Similar waterstops were used in the Bakún (Malaysia), Mazar (Ecuador) and Merowe (Sudan) dams. Figure 12.24 shows an application of this waterstop.

12.9.6 Mastic

Since Alto Anchicayá, most of the high dams have opted to place mastic over the perimetric joints and tension joints (Fig. 12.25). Foz do Areia, Xingó, and Segredo are among the dams where mastic was placed and covered with a rubber membrane. In Aguamilpa and TSQ1, the highest dam ever constructed, the mastic was replaced by fly ash and coal ash over these joints, with very satisfactory results.

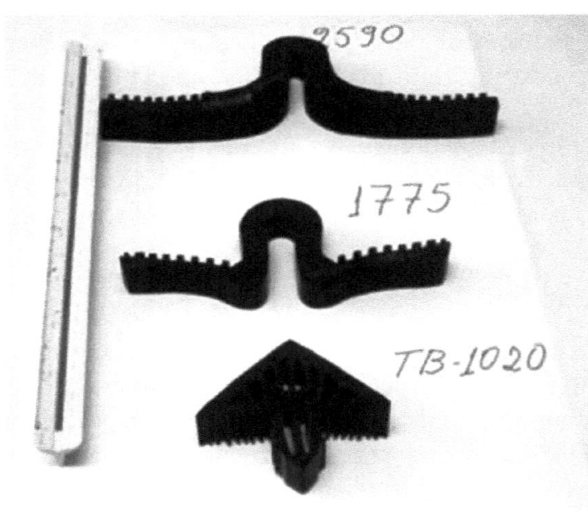

Figure 12.22 External waterstops. (See colour plate section).

Leakage was very small. There is a tendency to use this feature in the new dams under construction.

The Chinese have developed a mastic, called *GB*, with adequate performance. This mastic is placed over the joints and protected with an EPDM cover.

12.9.7 Concrete

The type of concrete used for the construction of the main slab is generally a workable mix of pozzolanic cement and water-reducing agents with air entrained ranging between 4–6%.

Concrete used in high CFRDs has 2″ to 4″ slump and a cement factor averaging 250 kg/m^3.

Figure 12.23 Waterstop protection with timber boxes. (See colour plate section).

Figure 12.24 Corrugated waterstops as used in Merowe (Sudan). (See colour plate section).

Figure 12.25 Mastic over the perimetral joint. (See colour plate section).

The specified strength is 20 MPa at 28 days; although in some dams where the slab is built in two or three stages, the required strength is specified for 60 or 90 days.

Forms

The construction of the main slab usually requires two types of forms: timber forms for the execution of the starter slabs, and slip forms for the construction of the main lanes. Some projects are using the slip form to build the starter slabs.

Starter slabs

The starter slabs are the segments of the slab in contact with the perimeter joint. They are built first in order to help to lift up the slip form.

The starter slabs and the rockfill placing are built simultaneously, well ahead of the main slab construction. Temporary forms made of timber boxes with dimensions of 2.00 × 0.50 m are used. They are moved up between pipe guides. The concrete is confined by these boxes. Then, after the box is moved the concrete is given a surface finishing by a trowel. Special care should be given to the construction of these fillets since the perimetric joint is there and the adequate execution of the concrete around the copper waterstop is important to avoid incidental leakage.

Starter slabs can be built by using the slip form with the help of cables or hydraulic jacks as already commented on.

Main slab

The construction of the main slab is carried out by slip forming after installing the steel reinforcing. The size of the face defines the number of slip forms. In high dams, the use of two light-weight slip forms can be accomplished by either electromechanical winches or hydraulic jacks. Both systems are efficient, although the use of hydraulic jacks presents economic advantages.

The jacking system is used by placing a rail mounted on the lateral form. This rail allows a controlled advance of two lateral jacks of 15 ton capacity each to raise up the form. The same system can lower the form from the crest to the bottom to start a new lane. Recently, some slip forms are being operated by cables as guides for the jacking system. The type of jack is similar to the one used for post-stressed cables.

The advantage of using light slip forms is that in large projects the same form can be utilized for the construction of the spillway slab. This was done in Messochora (Greece) and Xingó (Brazil).

TSQ1 CFRD had a unique experience of using two different types of slip forms: a railed one, as explained above, and a non-railed one, as usually used in China. The railed slip forms operated by hydraulic jacks were used in the first stage slabs, and the non-railed ones operated by winches were used in the second and third stage slabs. Both methods were satisfactory in terms of productivity and concrete quality.

The design of a light slip form is simple and normally executed for permanent loads of about 300 kg/m, including the weight of the elements comprising the form.

Live loads are about 70 to 100 kg/m, and additional loads such as vibration, wind effect, contingencies, etc, are assumed to be 20% of this value.

These slip forms are not designed to resist uplift caused by floating. However, since this is a very important aspect that may occur during the construction process, the following rules should be considered:

- avoid vibration near the form;
- concrete pumping pipes or metallic chutes should be located at 1.00 to 1.50 m from the work front;
- the surface skin plate must be tapered to prevent the development of total pressures on the sliding area;
- surface skin should introduce counter deflections to neutralize floating effects, without affecting the slab thickness;
- friction forces are considered at approximately 0.025 kg/cm^2.

Handling of the slip forms has been simplified by using raising platforms that can be transported laterally to the next lane. Figure 12.26 shows the one used in Kannaviou (Cyprus).

Control

Construction of the face slab requires continuous monitoring of the rate of placement, excess of concrete, slump, air entrainment and expected strength to record a statistical analysis of each pour. This control has allowed for a reduction of excessive concrete losses to values close to 3% of the theoretical volume.

Transporting and delivery

Concrete from the batching plant is generally transported to the crest of the slab stage by concrete-mixers of 6 m^3 capacity discharging into hoppers. Then, the concrete is discharged from the hoppers into metal chutes directly placed on the reinforcing steel. Two or three lines of metal chutes are required for good concrete distribution (Fig. 12.27).

Figure 12.26 Structure for slipform transportation from one panel to the next. (See colour plate section).

Figure 12.27 Concrete placing by metallic chute. (See colour plate section).

In Messochora (Greece) and Kannaviou (Cyprus) one line of metal chutes was located laterally to feed a conveyor which distributed the concrete into the form (Fig. 12.28).

The construction of slabs in different stages is convenient for the filling of the reservoir and for starting the construction of the face slab when rockfill is simultaneously built downstream. In higher dams, there has been no restriction on the construction stages of the slab. It has been proved convenient to discuss with the designer or owner the best sequence of construction in relation to concrete production histograms.

Steel reinforcing

The trend in the last years has been to reduce the percentage of reinforcement in the slab; however, the experience of slab cracking in high dams have forced designers to return to old practices. It is important to evaluate the cost of placing the steel reinforcing directly on the site, as compared to the productivity gained when mechanized methods are applied.

Figure 12.28 Concrete placing by conveyor. (See colour plate section).

Figure 12.29 Reinforcing detail.

In Xingó, because of the tight schedule, the dam elevation itself was required to produce an effective hydrologic protection. A mechanized scheme was applied to fabricate mats in advance and lower them in place by means of electric winches. The results were excellent.

The method of reinforcing placement depends on local labor wages and the required production needed to meet the schedule. Placing of reinforcement determines the schedule for the construction of the concrete.

Manual methods are efficient in countries with low to moderate labor wages. Semi-mechanical systems have also been used, depending on demand for face construction. Weldable reinforcing steel with maximum carbon percentage of about 0.20% to 0.24% and low percentages of manganese, as the steels fabricated in Brazil, allows for fusion welding, which prevents waste caused by overlapping.

All pre-fabricated meshes used in Xingó were fusion welded, resulting in savings in the use of commercial rebars. Normally, the bar size is 11 m long and the mesh 16×11 m (Fig. 12.29).

12.10 OUTPUTS

Peak productions of 60,000 m^3/month in cushion 2B have been registered in high CFRDs. It is feasible to build 2–3 layers of extruded curb per day in dams with crest length larger than 500 m. It is normal to place 30,000 m^3/month of 2B material.

Placing of rockfill has been more than 1,200,000 m^3/month as registered in Barra Grande (Brazil). In a high dam it is normal to place 500,000 m^3/month.

Extruded curb production varies between 40–60 m/hour for a high dam and with the use of one machine.

Average slip form production ranges between 2–4 m/hour. The maximum rate in narrow sections close to the top of the dam is about 6 m/hour. In Xingó it was possible to pour 20,500 m^2/month during the second stage construction of the face slab. Normal output is 13–22 m^3/hour. Peak values can reach 60 m^3/hour.

Reinforcement placement has ranged between 500–920 ton/month for peak production. Average values are in order of 300 ton/month.

References

Amaya, F.; Marulanda, A. Golillas Dam – Design, construction and performance. In: Cooke, J.B.; Sherard, J. L. (Eds.). *Concrete Face Rockfill Dams – Design, construction and performance* – Proceedings ASCE Symposium, Detroit, USA, 1985.

Anguita, P.; Alvarez, L.; Vidal, L. Two Chilean CFRD designed on riverbed alluviums. In: *Proceedings International Symposium on High Earth-Rockfill Dams (Especially CFRD)*, Beijing, China, October 1993.

Antunes Sobrinho, J. et al. Performance and concrete face repair at Campos Novos. *Intern. Journal on Hydropower & Dams*, n. 2, 2007.

Antunes Sobrinho, J. et al. Development aspects of CFRD in Brazil. In: *J. Barry Cooke Volume – Concrete Face Rockfill Dams*, Beijing, 2000.

Barbarez, Milijove. *Olmos dam project CFRD design and construction challenges.* Workshop on High Dam Know-how, Yichang, China, 2007.

Basso, R.V.; Cruz, P.T. Deformability study of a granular material submitted to different stress paths aiming CFRDs. In: III Symposium on CFRD-Dams Honoring J. Barry Cooke, 2007, Florianópolis. *Proceedings*, Florianópolis, 2007.

Borges, J.M.V.; Pereira, R.F.; Antunes, J. Design, construction and performance of Barra Grande Dam. In: III Symposium on CFRD-Dams Honoring J. Barry Cooke, 2007, Florianópolis. *Proceedings*, Florianópolis, 2007.

Boughton, N.O. Elastic analysis for the behavior of rockfill. ASCE, *Journal of the Soil Mechanics and Foundations Division*, v. 96, n. 5, pp. 1715–1733, 1970.

Casinader, R.; Rome, G. Estimation of leakage through upstream concrete facings of rockfill dams. In: *Transactions of the 16th International Congress on Large Dams.* San Francisco: ICOLD, 1988. v. 2, Q. 61, R. 17.

Castro, J. et al. Behavior of Aguamilpa and diversion works during January 1992 floods. In: *Proceedings International Symposium on High Earth-Rockfill Dams (Especially CFRD)*, Beijing, China, October 1993.

Charles, J.A.; Skinner, H.D. Compressibility of foundation fills. Proceedings of ICE, *Geotechnical Engineering*, v. 149, n. 3, pp.145–157, 2001.

Charles, J.A.; Soares, M.M. Stability of compacted rockfill slopes. *Géotechnique*, v. 34, n. 1, 1984.

Clements, R.P. *The deformation of rockfill: inter-particle behaviour, bulk properties and behaviour in dams.* PhD Thesis, Faculty of Engineering, King's College, London University, 1981.

Cooke, J.B. New Exchequer. In: Davis, V.C.; Sorensen, K. *Handbook of Applied Hydraulics.* 3. ed. New York: McGraw-Hill, 1969.

Cooke, J.B. *Notes on Aguamilpa face crack.* Memo n. 165, 1999.

Cooke, J.B. Progress in rockfill dams. 18th Rankine Lecture. *Journal of Geotechnical Engineering*, ASCE, v. 110, n. 10, 1984.

Cooke, J.B.; Sherard, J.L. Concrete Face Rockfill Dam: I. Assessment; II. Design. *Journal of Geotechnical Engineering*, ASCE, v. 113, n. 10, 1987.

Cooke, J.B.; Sherard, J.L. (Eds.). *Concrete Face Rockfill Dams – Design, Construction and Performance*. Proceedings ASCE Symposium, Detroit, USA, 1985.

Cooke, J.B. The high CFRD Dam. In: International Symposium on Concrete Faced Rockfill Dams, 2000, Beijing, China. *Invited lecture*, Beijing: ICOLD, 2000a.

Cooke, J.B. The plinth of the CFRD Dam. In: International Symposium on Concrete Faced Rockfill Dams, 2000, Beijing, China. *Proceedings*, Beijing: ICOLD, 2000b.

Cruz, P.T. *Leakage on concrete face rockfill dams*. Proceedings of the International Conference on Hydropower, Yichang, China, May 2005a.

Cruz, P.T. *Stability and instability of rockfills during throughflow*. Dam Engineering, n. 3, serial n. 59, August 2005b.

Cruz, P.T.; Freitas, M.S. Jr. Cracks and Flows in Concrete Face Rockfill Dams. In: 5th International Conference on Dam Engineering, Lisboa, Portugal, 2007.

Cruz, P.T.; Materón, B.; Freitas, M.S. Jr. Design criteria for CFRD – An actual review of 1987 papers by J. Barry Cooke and James L. Sherard – Concrete Face Rockfill Dams: I – Assesment; II – Design. *Journal of Geotechnical Engineering*, v. 113, October 10, 1987.

Cruz, P.T.; Nieble, C.M. *Engineering properties of residual soils and granular rocks originated from basalts – Capivara Dam – Brazil*. São Paulo: IPT, 1970.

Cruz, P.T.; Pereira, F.R. *The Rockfill of Campos Novos* CFRD. In: III Symposium on CFRD-Dams Honoring J. Barry Cooke, 2007, Florianópolis. *Proceedings*, Florianópolis, 2007.

Cruz, P.T.; Silva, R.F. Uplift pressure at the base and in the rock basaltic foundation of gravity concrete dams. In: International Symposium on Rock Mechanics Related to Dam Foudations, 1978, Rio de Janeiro. *Proceedings*, Rio de Janeiro: ISRM/ABMS, 1978. v. 1.

Dakoulas, P.; Thanopoulos, Y.; Anastassopoulos, K. Non linear 3D simulation of the construction and impounding of a CFRD. Intern. *Journal on Hydropower & Dams*, v. 15, n. 2, 2008.

De Mello, V.F.B. "Reflection on design decisions of pratical significance to embankment dams". 17th Rankine Lecture. *Géotechnique*, v. 27, n. 3, pp. 279–355, 1977.

Eigenheer, L.P.Q.T.; Mori, R.T. Xingó Rockfill Dam. In: *Proceedings International Symposium on High Earth-Rockfill Dams (Especially CFRD)*, Beijing, China, October 1993.

Escande, L. *Experiments concerning the infiltration of water through a rock mass*. Proceedings Minnesota Int. Hydro Convention, 1953.

Fernandez, R. et al. Instrumentation and auscultation of Itapebi CFRD during and after filling the reservoir. In: III Symposium on CFRD-Dams Honoring J. Barry Cooke, 2007, Florianópolis. *Proceedings*, Florianópolis, 2007.

Fitzpatrick, M.D. et al. Design of Concrete Faced Rockfill Dams. In: Cooke, J.B.; Sherard, J.L. (Eds.). *Concrete Face Rockfill Dams – Design, Construction and Performance*. Proceedings ASCE Symposium, Detroit, USA, 1985.

Fitzpatrick, M.D. et al. Instrumentation and performance of Cethana Dam. In: xi International Congress on Large Dams. *Proceedings*, Madrid: ICOLD, 1973.

Freitas, M.S. Jr. Contribution on designing and construction of high CFRDs. In: *Proceedings of the International Conference of Hydropower*, v. 1, Yichang, China, May 2004.

Guocheng, J.; Keming, K. The Concrete Face Rockfill Dams in China. In: *J. Barry Cooke Volume – Concrete Face Rockfill Dams*, Beijing, 2000.

Hartung, F.; Scheuerlein, H. *Design of overflow rockfill dam*. In: X ICOLD Congress, Montreal, 1970, Q. 36, R. 35, v. 1, pp. 587–598.

Hellström, B. Compaction of a rockfill dam. In: *Transactions of the 5th International Congress on Large Dams*. Paris: ICOLD, 1955. v. 3. pp. 331–337.

Hoek, E.; Bray, J. *Rock slope engineering*. London: The Institution of Mining and Metallurgy, 1974.

Hong Tao, Li. Enlargement of Hengshan Dam. In: *Proceedings International Symposium on High Earth-Rockfill Dams (Especially CFRD)*, Beijing, China, October, 1993.

International Commission on Large Dams (ICOLD). Rockfill Dams with Concrete Facing – State of the Art. *ICOLD Bulletin* 70, 1989.

International water power & dam construction. *International Water Power & Dam Construction Yearbook 2008*. 456p.

Johannesson, P. Design improvements of high CFRDs constructed of low modulus rock. In: III Symposium on CFRD-Dams Honoring J. Barry Cooke, 2007, Florianópolis. *Proceedings*, Florianópolis, 2007.

Johannesson, P.; Perez, H.J.; Stefansson, B. Updated behavior of the Kárahnjúkar Concrete Face Rockfill Dam in Iceland. In: 23rd ICOLD CONGRESS, Brazil 2009.

Johannesson, P.; Tohlang, S.L. Lessons learned from Mohale. *International Water Power & Dam Construction*, August, pp. 16–25, 2007a.

Johannesson, P.; Tohlang, S.L. Updated assessment of Mohale Dam behavior, including of slab cracking and seepage evolution. In: III Symposium on CFRD-Dams Honoring J. Barry Cooke, 2007, Florianópolis. *Proceedings*, Florianópolis, 2007b.

Koch, O.G. et al. Concrete Face Rockfill Dam of Xingó – construction control. In: *Proceedings International Symposium on High Earth-Rockfill Dams (Especially CFRD)*, Beijing, China, October 1993.

Kulasingle, A.N.S.; Tandon, G.N. Technical and behavior aspects of Kotmale Dam. In: *Proceedings International Symposium on High Earth-Rockfill Dams (Especially CFRD)*, Beijing, China, October 1993.

Larson, E.; Kelly, R. Geomembrane installation at Salt Springs CFRD Dam. Pacific Gas & Electric Co, 2005.

Leps, T.M. Flow through rockfill. In: Hirschfeld, R.C.; Poulos, S.J. (Eds). *Embankment Dam Engineering; Casagrande Volume*. NY: John Wiley & Sons, 1973.

Leps, T.M. Review of shearing strenght of rockfill. *Journal of the Soil Mechanics and Foundations Division*, v. 96, n. SM4, 1970.

Leps, T.M.; Cashatt, C.A.; Janopaul, R.N. *New Exchequer Dam, California*. Concrete Face Rockfill Dams, ASCE Convention, Detroit, Michigan, October 1985. pp. 15–26.

Lombardi, G. Selecting the grouting intensity. *Intern. Journal on Hydropower & Dams*, v. 3, n. 4, pp. 62–66, 1996.

Long, Tan Yong et al. Bakun Dam – Some design and construction challenges. In: *Proceedings of the Symposium on 20 Years for Chinese CFRD Construction*, Yichang, China, September 2005.

Macedo, G.G.; Castro, A.J.; Montañez, L.C. Behavior of Aguamilpa Dam. In: *J. Barry Cooke Volume – Concrete Face Rockfill Dams*, Beijing, 2000.

Machado, B.P. et al. Pichi Picun Leufu – The first modern CFRD in Argentina. *Proceedings International Symposium on High Earth-Rockfill Dams* (*Especially CFRD*), Beijing, China, October, 1993.

Marachi, D. et al. *Strength and deformation characteristic of rockfill materials*. Report n. TE-69-5, University of California, Berkeley, 1969.

Maranha Das Neves, E. Fills and embankments. International Conference on Geotechnical Engineering of Hard Soils and Soft Rocks. *General Report*, Athens, v. 3, pp. 2023–2037, 1993.

Marengo-Mogollón, H.; Aguirre-Tello, S. Instrumentation and Behavior of El Cajón Dam. In: III Symposium on CFRD-Dams Honoring J. Barry Cooke, 2007, Florianópolis. *Proceedings*, Florianópolis, 2007.

Marsal, R.J. Mechanical properties of rockfill. In: Hirschfeld, R.C.; Poulos, S.J. (Eds). *Embankment Dam Engineering; Casagrande Volume*. NY: John Wiley & Sons, 1973. pp. 109–200.

Marsal, R.J. Large scale testing of rockfill materials. *Journal of the Soil Mechanics and Foundations Division*, ASCE, v. 93, n. SM2, pp. 27–43, 1967.

Marulanda, A.; Pinto, N.L.S. Recent experience on design, construction and performance of CFRD dams. In: *J. Barry Cooke Volume – Concrete Face Rockfill Dams*, Beijing, 2000.

Marulanda, A.; Amaya, F.; Millan, M. Quimbo Dam. In: International Symposium on Concrete Faced Rockfill Dams, 2000, Beijing, China. *Proceedings*, Beijing: ICOLD, 2000.

Materón, B. Design, construction and behavior of slabs of the highest concrete-faced dams. CFRD *World*, v. 2, n. 1, CFRD International Society and HydrOu China, March 2008.

Materón, B. Alto Anchicaya Dam – Ten Years Performance. In: Cooke, J.B.; Sherard, J.L. (Eds.). Concrete Face Rockfill Dams – Design, construction and performance – *Proceedings* ASCE Symposium, Detroit, USA, 1985.

Materón, B. Hengshan Dam. *Internal report for the raising of Carén dam.* Codelco, Chile, 2007.

Materón, B. *Innovative design and construction methods for CFRDs.* In: xxii International Congress on Large Dams. *Proceedings*, Barcelona: ICOLD, 2006.

Materón, B. Responding to the demands of EPC contracts. *International Water Power & Dam Construction*, August 2002.

Materón, B. Transition material in the highest CFRDs. *Int. Journal on Hydropower & Dams*, England, v. 5, n. 6, 1998.

Materón, B.; Mori, R.T. Construction features of CFRD dams. In: *J. Barry Cooke Volume – Concrete Face Rockfill Dams*, Beijing, 2000.

Materón, B.; Resende, F. Construction innovations for the Itapebi CFRD. *Int. Journal on Hydropower & Dams*, v. 8, n. 5, pp. 66–70, 2001.

Materón, B. et al. Alto Anchicayá Concrete Face Rockfill Dam – Behavior of the concrete face membrane. In: International Congress on large dams, 14., 1982, Rio de Janeiro. Proceedings, Rio de Janeiro: CBGB-ICOLD, 1982.

Mauro, V. et al. The use of high gravity walls for plinth foundation on Machadinho CFRD. In: III Symposium on CFRD-Dams Honoring J. Barry Cooke, 2007, Florianópolis. *Proceedings*, Florianópolis, 2007.

McHenry, D. Discussion on stress conditions for the failure of saturated concrete and rock. *Proceedings of American Society of Civil Engineers*, v. 45, 1945.

Mena Sandoval, J.E. et al. Behavior observed at the El Cajón Dam (CFRD) Mexico during the first filling and one year of operation. In: III Symposium on CFRD-Dams Honoring J. Barry Cooke, 2007, Florianópolis. *Proceedings*, Florianópolis, 2007a.

Mena Sandoval, J.E. et al. El Cajón Hydroelectric Project (CFRD), Mexico – Dam auscultation system and behavior observed during construction. In: III Symposium on CFRD-Dams Honoring J. Barry Cooke, 2007, Florianópolis. *Proceedings*, Florianópolis, 2007b.

Mendez, F. Rapid construction of the El Cajón CFRD. *Int. Journal on Hydropower & Dams*, v. 12, n. 1, 2005.

Mendez, F. et al. The behavior of a very high CFRD under first reservoir filling. *Int. Journal on Hydropower & Dams*, v. 14, n. 2, 2007.

Montañez-Cartaxo, L.E; Hacelas, J.E.; Castro-Abone, J. Design of Aguamilpa. *Proceedings International Symposium on High Earth-Rockfill Dams.* Beijing, China: ICOLD, 1993.

Noguera, G.; Pinilla, L.; San Martin, L. CFRD constructed on deep alluvium. In: *J. Barry Cooke Volume – Concrete Face Rockfill Dams*, Beijing, 2000.

Noguera, G.; Vidal, L. Puclaro's cut-off wall. In: II Symposium on CFRD-Dams, 1999, Florianópolis. *Proceedings*, Florianópolis: CBDB, 1999.

Oldcop, L.A. *Compresibilidad de escolleras. Influencia de la humedad.* Tesis doctoral, Escuela de Caminos, Canales y Puertos, Universidad Politécnica de Cataluña, 2000.

Olivier, H. Through and overflow rockfill dams – new design techniques. *Proceedings of the Institution of Civil Engineers*, v. 36, paper 7012, pp. 433–471, 1967.

Parkin, A.K.; Adikari, G.S.N. Rockfill deformation from large-scale tests. *Proceedings 10th International Conference on Soil Mechanics and Foundation Engineering,* Estocolmo, v. 4, pp. 727–731, 1981.

Peck, R.B. *Geotechnical Instrumentation News,* USA, Sept. 2001.

Peixoto, M.; Saboya JR., F.; Karan, J. Analysis of differential movements between the face and the embankment body of concrete face rockfill dams. In: II symposium on CFRD-Dams, 1999, Florianópolis. *Proceedings,* Florianópolis: CBDB, 1999.

Penman, A.D.M. *Rockfill.* Building Research Station Current Paper, Department of the Environment, CP 15/71, Apr. 1971.

Penman, A.D.M. Shear characteristics of a saturated silt, measured in triaxial compression. Géotechnique, v. 3, n. 8, pp. 312–328, 1953.

Penman, A.D.M.; Rocha Filho, P.; Toniatti, N.B. Instrumentation of the 145 m Segredo dam – problems and results. In: *Proceedings IV Int. Symposium on Field Measurements in Geomechanics,* Bergamo, Italy, 1995.

Penman, A.D.M.; Rocha Filho, P. Instrumentation for CFRD Dams. In: *J. Barry Cooke Volume – Concrete Face Rockfill Dams,* Beijing, 2000.

Pereira, R.F.; Albertoni, S.C.; Antunes, J. Instrumentation and ascultation of Itapebi CFRD during and after filling the reservior. In: III Symposium on CFRD-Dams Honoring J. Barry Cooke, 2007, Florianópolis. *Proceedings*, Florianópolis, 2007.

Perez, H. J.; Johannesson, P.; Stefansson, B. The Kárahnjúkar CFRD in Iceland instrumentation and first impoundment dam behavior. In: III Symposium on CFRD-Dams Honoring J. Barry Cooke, 2007, Florianópolis. *Proceedings,* Florianópolis, 2007.

Pinkerton, I.L.; Siswowidjono, S.; Matsui, Y. Design of Cirata Concrete Face Rockfill Dam. In: Cooke, J.B.; Sherard, J.L. (Eds.). *Concrete Face Rockfill Dams – Design, Construction and Performance.* Proceedings ASCE Symposium, Detroit, USA, 1985.

Pinto, N.L.S. Very high CFRD dams – Behavior and design features. In: III Symposium on CFRD-Dams Honoring J. Barry Cooke, 2007, Florianópolis. *Proceedings*, Florianópolis, 2007.

Pinto, N.L.S.; Blinder, S.; Toniatti, N.B. Foz do Areia and Segredo CFRD Dams – 12 Years Evolution. *Proceedings International Symposium on High Earth-Rockfill Dams (Especially CFRD),* Beijing, China, October, 1993.

Pinto, N.L.S.; Materón, B.; Marques Filho, P.L. Design and performance of Foz do Areia concrete membrane as related to basalt properties. In: International Congress on Large Dams, 14., 1982, Rio de Janeiro. *Proceedings*, Rio de Janeiro: CBGB-ICOLD, 1982. v. 4. Q. 55, R. 51. pp. 873–905.

Pinto, N.L.S.; Marques Filho, P.L. Estimating the maximum face deflection in CFRDs. *Int. Journal on Hydropower & Dams,* n. 6, pp. 28–31, 1998.

Qian, Chen. Immediate development and future of 300 m high CFRD. CFRD World, *Journal of CFRD International Society,* v. 2, n. 1, 2008.

Ramirez, O.C.A. Mazar Dam: a 166 m high CFRD in an asymmetric. In: III Symposium on CFRD-Dams Honoring J. Barry Cooke, 2007, Florianópolis. *Proceedings*, Florianópolis, 2007.

Ramirez, O.C.A.; Peña, M.E.A. Considerations on the geometric design of plinth. In: II symposium on CFRD-Dams, 1999, Florianópolis. *Proceedings*, Florianópolis: CBDB, 1999. pp. 201–210.

Regalado, G. et al. Alto Anchicayá Concrete Face Rockfill Dam – Behaviour of the concrete face membrane. In: International congress on large dams, 14., 1982, Rio de Janeiro. *Proceedings,* Rio de Janeiro: CBGB-ICOLD, 1982.

Resende, F.; Materón, B. Itá Method – New Construction Technology for the Transition Zone of CFRDs. In: International Symposium on Concrete Faced Rockfill Dams, 2000, Beijing, China. *Proceedings*, Beijing: ICOLD, 2000.

Roberts, C.M. The Quoich Rockfill Dam. In: VI International Congress on Large Dams, 1958, New York. *Proceedings*, New York: ICOLD, 1958. v. 3, pp. 101–121.

Romo, M.P.; Reséndiz, D. Computed and observed deformation of two embankment dams under seismic loading. Conference on the Design of Dams to Resist Earthquake, The Institution of Civil Engineers, London, pp. 219–226, October 1980.

Schewe, L.D.; El Tayeb, A. Merowe dam project CFRD of extended lenght. In: *Proceedings of the Symposium on 20 Years for Chinese CFRD Construction,* Yichang, China, September 2005.

Scuero, A.; Vaschetti, G.; Wilkes, J. Repair of a 10 m High CFRD: Geomembrane installation at Salt Springs, USA. In: III Symposium on CFRD-Dams Honoring J. Barry Cooke, 2007, Florianópolis. *Proceedings*, Florianópolis, 2007.

Seed, H.B.; LEE, K.L. Undrained strength characteristics of cohesionless soils. *Journal of the SMFD,* ASCE, v. 93, n. SM6, pp. 333–360, 1967.

Sierra, J.M.; Ramirez, C.A.; Hacelas, J.E. Design features of Salvajina Dam. In: Cooke, J.B.; Sherard, J.L. (Eds.). *Concrete Face Rockfill Dams – Design, construction and performance* – Proceedings ASCE Symposium, Detroit, USA, 1985.

Sigvaldason, O.T. et al. *Analysis of the Alto Anchicaya dam using the Finite Element Method.* International Symposium Criteria and Assumptions for Numerical Analysis of Dams, Swansea, UK, Sept. 1975.

Silva, S.A.; Casarin, C.; Souza, R.J.B. Xingó Dam – Use of semi-permeable transition under the concrete face. In: II Symposium on CFRD-Dams, 1999, Florianópolis. *Proceedings*, Florianópolis: CBDB, 1999.

Silveira, A. A note on critical hydraulic gradient. Solos e Rochas – Revista Brasileira de Geotecnia, v. 6, n. 2, ago. 1983.

Sowers, G.F.; Williams, R.C.; Wallace, T.S. Compressibility of broken rock and settlement of rockfills. *Proceedings 6th International Conference on Soil Mechanics and Foundation Engineering,* Montreal, v. 3, pp. 561–565, 1965.

Steele, I.C.; Cooke, J.B. Rockfill dams: Salt Springs and Lower Bear River concrete face dams. *ASCE Transactions,* v. 125, pt. 2, pp. 74–116, 1960.

Sun, Yi; Yang, Z. Application of new technologies in Shuibuya CFRD. In: *Proceedings of the Symposium on 20 Years for Chinese CFRD Construction,* Yichang, China, September 2005.

Taylor, D.M. *Fundamentals of soil mechanics.* New York: John Wiley & Sons, 1948. Chap. 6 and 16.

Terzaghi, K. Discussion on settlement of Salt Springs and Lower Bear river concrete face dams. ASCE Transactions, v. 125, n. 2, pp. 139–148, 1960a.

Terzaghi, K. Discussion on "Wishon and Coutright concrete face dams". *Journal of Geotechnical Engineering,* ASCE, v. 125, part II, 1960c.

Terzaghi, K. From theory to practice in soil mechanics. Selections from the writings of Karl Terzaghi. New York: John Wiley & Sons, 1960b.

Thomas, H.H. Flow through and over rockfills. In: Thomas, H.H. *The engineering of large dams.* New York: John Wiley & Sons, 1976. v. 2. Chap. 15.

Troncoso, J. Análisis de estabilidad de la Presa del Embalse de Santa Juana. Chile: Ministerio de Obras Públicas, 1993.

Tschebotarioff, G.P.; Welch, J.D. Lateral earth pressure and friction between soil minerals. *Proceedings 2nd International Conference on Soil Mechanics and Foundation Engineering,* Rotterdam, v. 7, p. 135–138, 1948.

Wilkins, J.K. A theory for the shear strength of rockfill. *Rock Mechanics and Rock Engineering,* v. 2, n. 4, pp. 205–222, 1970.

Wilkins, J.K. Flow of water through rockfill and its application to the design of dams. *Proceedings 2nd Australia-New Zealand Conference on Soil Mechanics and Foundation Engineering,* pp. 141–149, 1956.

Wilkins, J.K. The stability of overtopped rockfill dams. *Proceedings 4th Australia-New Zealand Conference on Soil Mechanics and Foundation Engineering*, 1963.

Wu, G.Y. et al. Face slab construction at Tianshengqiao 1 (China). In: International Symposium on Concrete Faced Rockfill Dams, 2000, Beijing, China. Proceedings, Beijing: ICOLD, 2000a.

Wu, G.Y. et al. Tianshengqiao-1 CFRD – Monitoring & Performance – Lessons & New Trends for Future CFRDs (China). In: International Symposium on Concrete Faced Rockfill Dams, 2000, Beijing, China. Proceedings, Beijing: ICOLD, 2000b.

Xavier, L.V. et al. Campos Novos CFRD – Treatment and behavior of the dam in the second filling of the reservoir. In: III Symposium on CFRD-Dams Honoring J. Barry Cooke, 2007, Florianópolis. Proceedings, Florianópolis, 2007a.

Xavier, L.V. et al. Concrete face rockfill dams – Studies on face stresses through mathematical models. In: III Symposium on CFRD-Dams Honoring J. Barry Cooke, 2007, Florianópolis. Proceedings, Florianópolis, 2007b.

Yuan, H.; Zhang, C. Rehabilitation design of Gouhou CFRD. In: Proceedings of the International Conference of Hydropower, v. 1, Yichang, China, May 2004.

Figure 1.2 Displacement of the access way and the crest of the dam – this measured up to 630 mm.

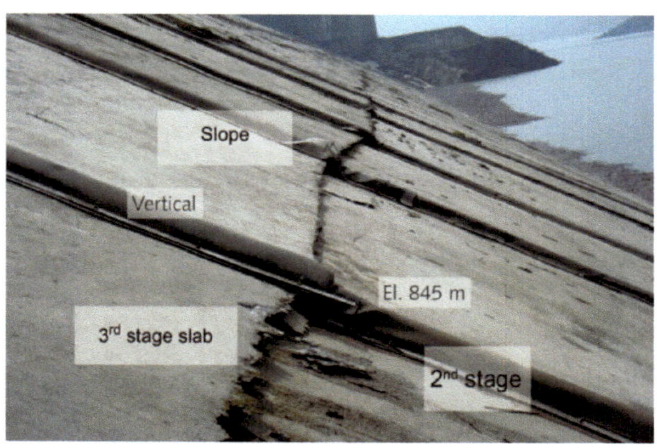

Figure 1.3 Horizontal joint offset at the El. 845 m crossing through 26 slabs.

Figure 1.4 Compressive failure on slabs #23 and #24 at the central part of the dam.

Figure 1.6 Rockfills loosened off on the downstream slope of Zipingpu Dam.

Figure 2.1 Campos Novos Dam.

Figure 3.4 Alto Anchicayá slab construction.

Figure 3.10 Aguamilpa: main slab at third phase construction.

Figure 3.13 Campos Novos Dam: double reinforcement layer close to abutments.

Figure 3.14 Campos Novos Dam: slab failure.

Figure 3.15 Campos Novos Dam: central compressive joints failure.

Figure 3.19 Mohale Dam.

Figure 3.21 Messochora Dam.

Figure 3.23 Equipment used in El Cajón Dam to build 2B transition.

Figure 3.26 Bakún Dam – air view.

Figure 3.28 Golillas Dam.

Figure 3.30 Segredo Dam. Courtesy Copel.

Figure 3.32 Xingó Dam.

Figure 3.36 Itá Dam – aerial view.

Figure 3.37 View of upstream slope protected by the extruded curb, Itá Dam.

Figure 3.39 Machadinho Dam works before diversion.

Figure 3.41 Antamina Dam – detail.

Figure 3.43 Itapebi Dam.

Figure 3.46 Itapebi Dam: slab partial view, built with the platform incorporated to the rockfill.

Figure 3.48 Quebra-Queixo Dam – lateral view.

Figure 3.50 Barra Grande Dam.

Figure 3.52 Hengshan Dam.

Figure 3.60 Slipping forms for plinth construction.

Figure 8.3 Shuibuya Dam: view of the plinth slab on the left steep abutment (upstream and downstream extension).

Figure 8.5 Plinth: details of upstream and downstream and construction joint at the shoulder-deep part (Shuibuya CFRD, China, 2006).

Figure 8.7 BEFC of Kárahnjúkar: geomembrane installation (CARPI) at slab 1st stage.

Figure 8.11 A) Perimeter joint and B) mastic placement in Xingó Dam.

Figure 8.12 TSQ1: A) perimeter joint with fly ash (covered by geotextile and metallic plate on the abutments); B) perimeter joint details.

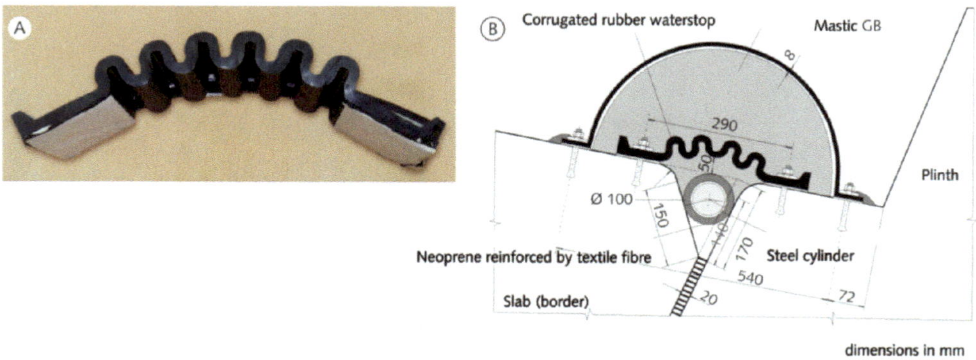

Figure 8.13 A) Corrugated joint adopted in Shuibuya (China) and B) Mazar (Ecuador). Courtesy of Consortium Mazar Management.

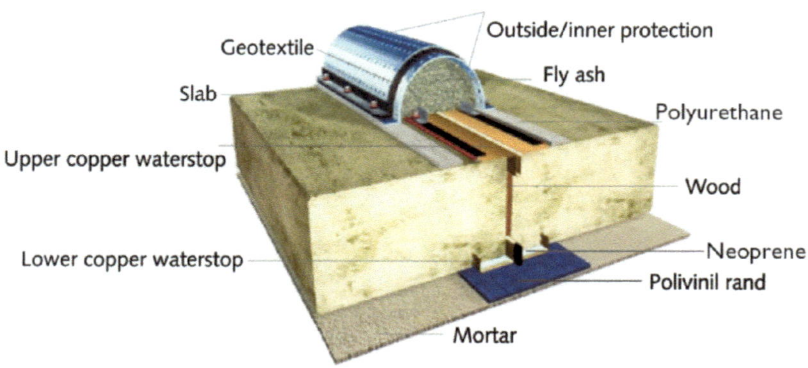

Figure 8.17 Detail of perimeter and vertical joints (zone of tension – abutments) (Mena Sandoval et al., 2007a).

Figure 8.19 Xingó CFRD: parapet wall under construction.

Figure 8.20 Metallic chutes for concrete convey down to the slipping form.

Figure 8.21 Mazar CFRD: detail of electric winches and steel cable traction system.

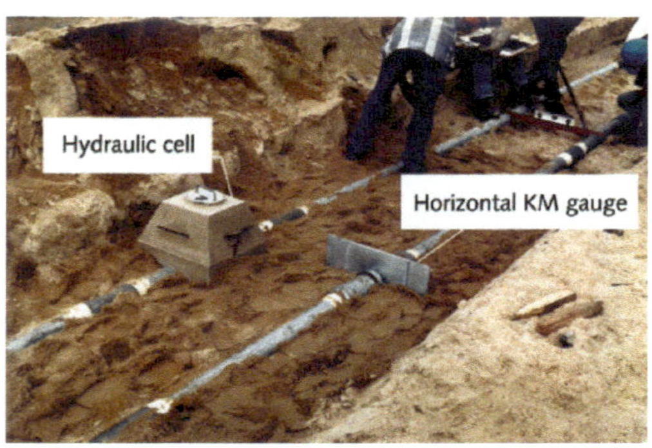

Figure 9.3 Xingó CFRD: settlement (hydraulic) cells and KM (horizontal) displacements gauge placed together (Silveira, 2006).

Figure 9.8 TSQ1: seepage weir before reservoir impounding (Oct., 2000).

Figure 9.9 Campos Novos: seepage weir in operation (Oct., 2007).

Figure 9.10 El Cajón: electrolevel installation on the face slab.

Figure 9.11 El Cajón: inclinometer casing to be cast into slab concrete.

Figure 9.12 El Cajón: inclinometer casing after installation at the dam crest.

Figure 9.13 Three-orthogonal joint meter – front view at Xingó Dam.

Figure 9.14 Three-orthogonal joint meter at El Cajón Dam.

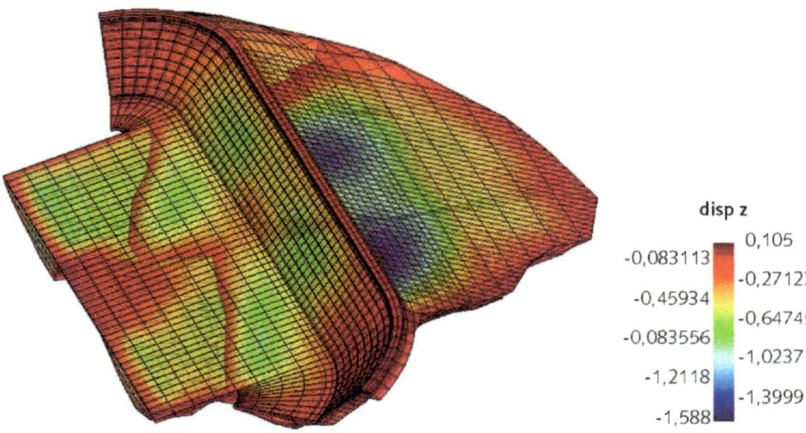

Figure 11.3 Displacements distribution of the dam.

Figure 11.7 Displacement vectors of concrete face slab.

Figure 11.9 River valley at Hongjiadu Dam site.

Figure 11.22 Chahanwusu CFRD: finite element mesh for the analysis.

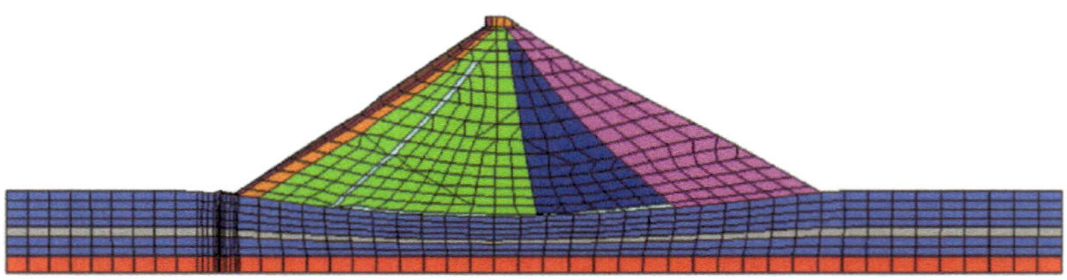

Figure 11.25 Chahanwusu CFRD: deformation of the dam and its foundation.

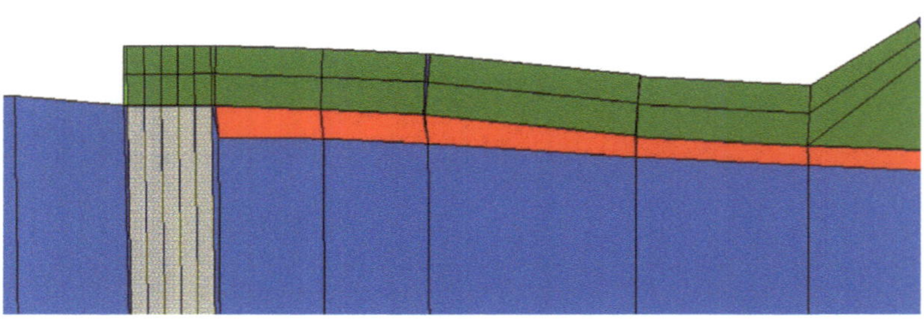

Figure 11.26 Chahanwusu CFRD: deformation of connecting slab (dam construction).

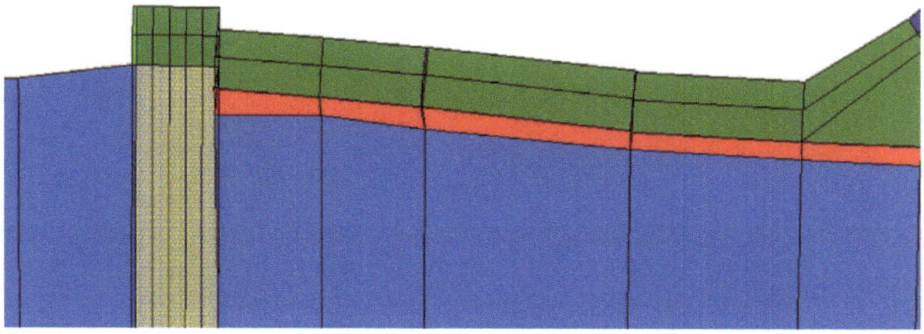

Figure 11.27 Chahanwusu CFRD: deformation of connecting slab (reservoir impounding).

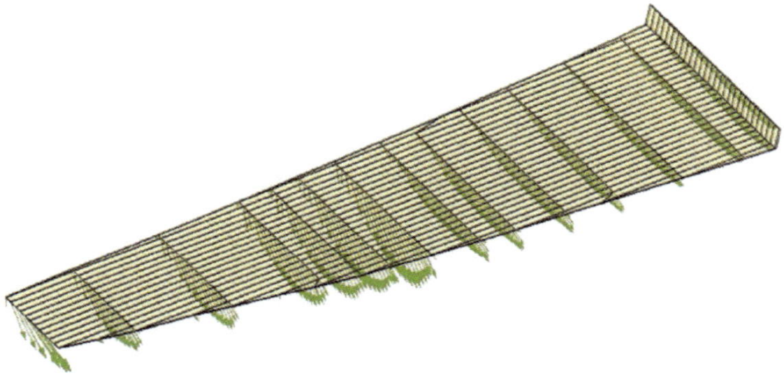

Figure 11.28 Chahanwusu CFRD: deformation of face slab.

Figure 11.29 Chahanwusu CFRD: diaphragm deformation (after construction).

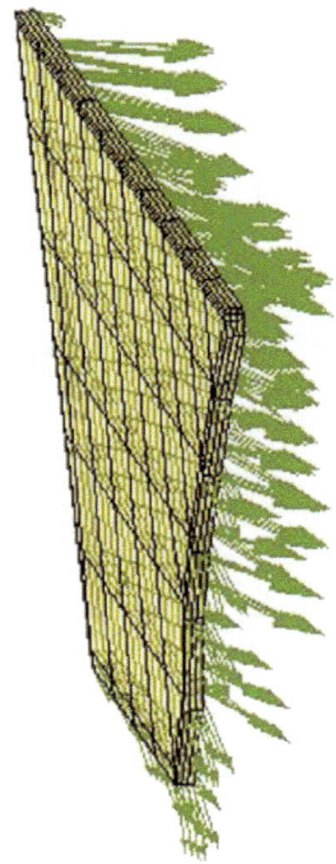

Figure 11.30 Chahanwusu CFRD: diaphragm deformation (reservoir impounding).

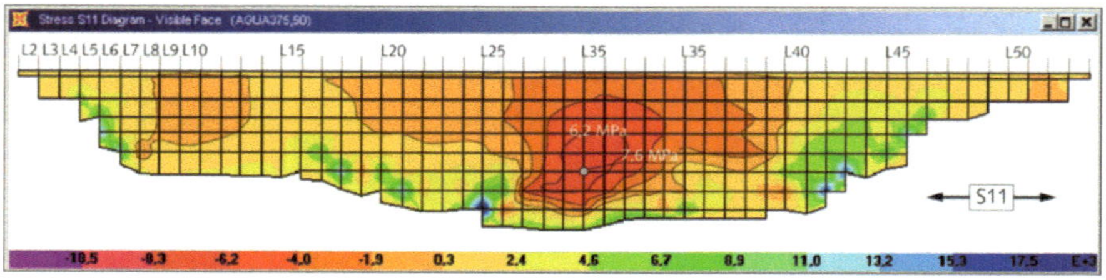

Figure 11.37 Itá CFRD: longitudinal stress diagram in the concrete face (Xavier et al., 2007a).

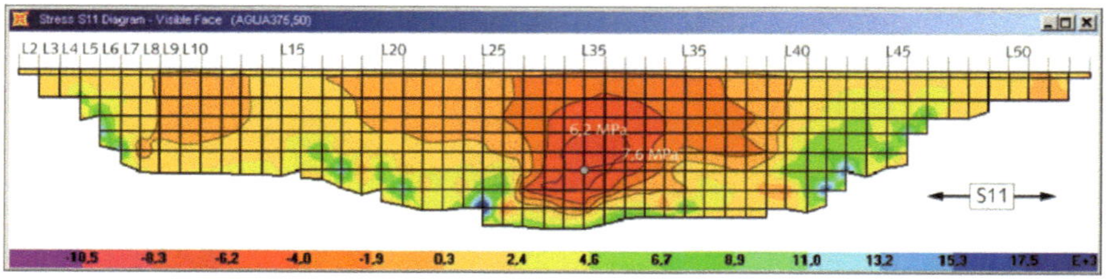

Figure 11.38 Itapebi CFRD: longitudinal stress diagram in the concrete face (Xavier et al., 2007a).

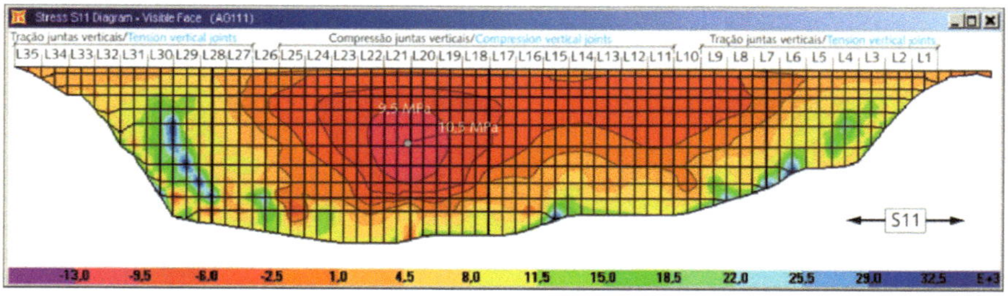

Figure 11.39 Campos Novos CFRD: longitudinal stress diagram in the concrete face (Xavier et al., 2007a).

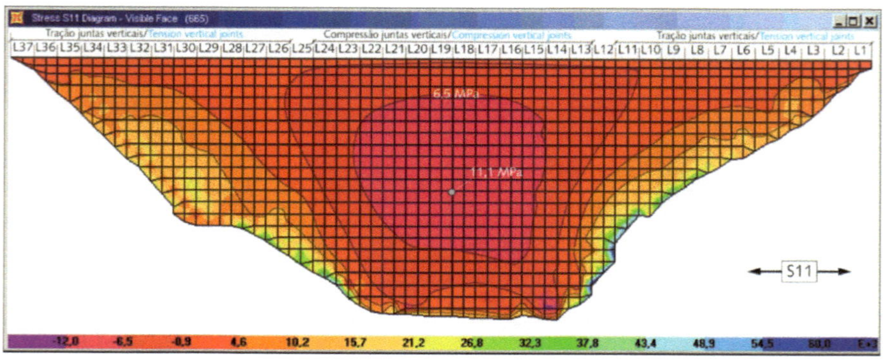

Figure 11.40 Barra Grande CFRD: longitudinal stress diagram in the concrete face (Xavier et al., 2007a).

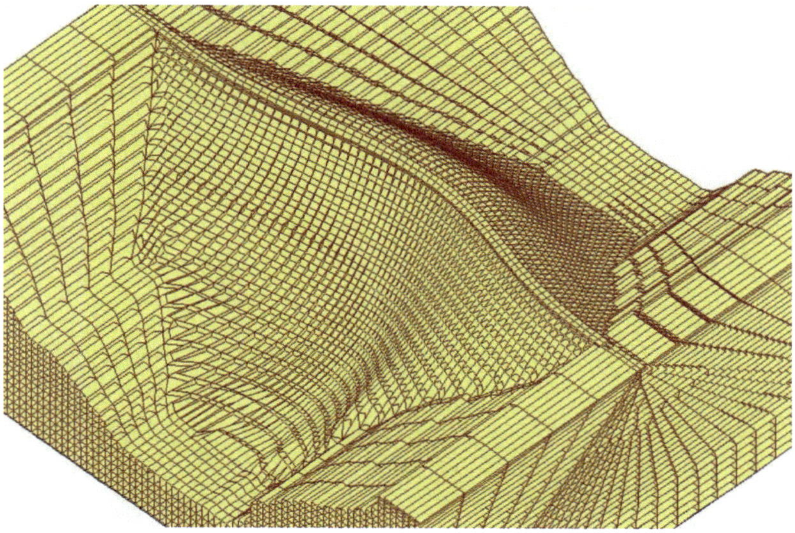

Figure 11.41 Campos Novos Dam: mathematical model 3D (Xavier et al., 2007).

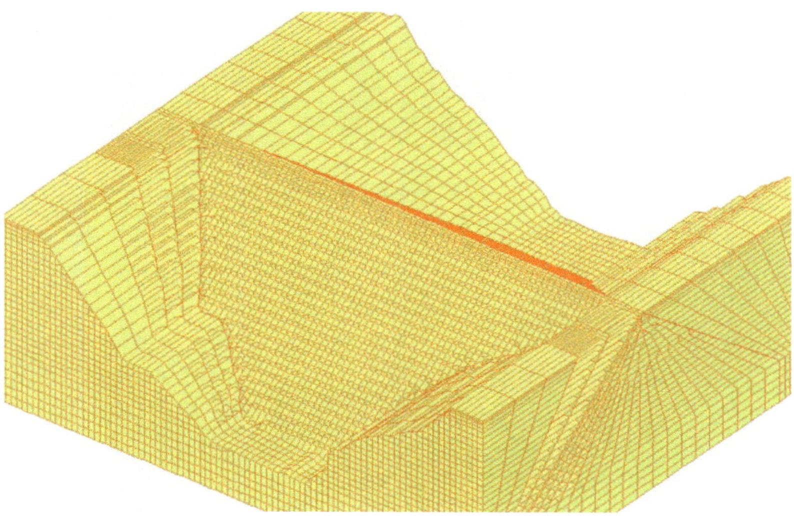

Figure 11.42 Campos Novos Dam: deformed mathematical model (deplacements after reservoir filling (Xavier et al., 2007).

Figure 12.2 Foz do Areia Dam: plinth construction.

Figure 12.3 Itapebi Dam: plinth with internal slab.

Figure 12.4 Plinth slipforming.

Figure 12.5 Santa Juana Dam: articulated plinth on alluvium.

Figure 12.6 Typical guide walls for diaphragm construction.

Figure 12.7 Chisels for rock penetration.

Figure 12.8 Hydrofraise.

Figure 12.9 Merowe Dam: drilling machines for plinth grouting.

Figure 12.10 Messochora Dam: drilling equipment for grouting.

Figure 12.12 TSQ1 Dam: gabion protection for overtopping.

Figure 12.13 TSQ1 Dam: rockfill overtopping.

Figure 12.17 Itapebi Dam: precast parapet.

Figure 12.18 Bakún Dam: downstream slope.

Figure 12.19 Extruded curb construction.

Figure 12.20 Slopes incorporated in the rockfill.

Figure 12.21 Extruded curb machine.

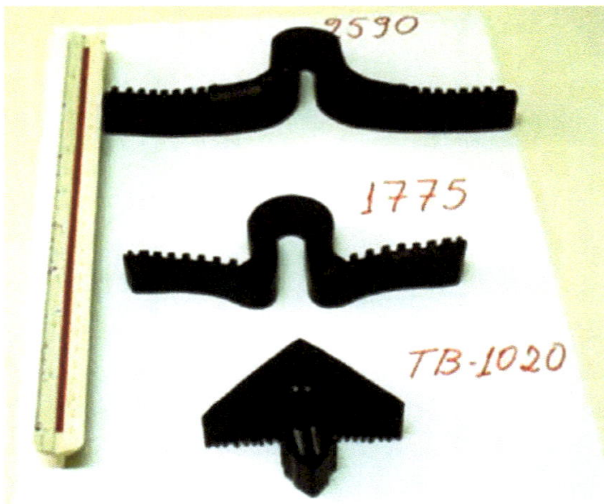

Figure 12.22 External waterstops.

Figure 12.23 Waterstops protection with timber boxes.

Figure 12.24 Corrugated waterstop as used in Merowe CFRD (Sudan).

Figure 12.25 Mastic over the perimetral joint.

Figure 12.26 Structure for slipform transportation from one panel to the next.